# ENVIRONMENTAL CHANGE AND ITS IMPLICATIONS FOR POPULATION MIGRATION

# ADVANCES IN GLOBAL CHANGE RESEARCH

VOLUME 20



*The titles published in this series are listed at the end of this volume.*

# ENVIRONMENTAL CHANGE AND ITS IMPLICATIONS FOR POPULATION MIGRATION

Edited by

Jon D. Unruh

*Indiana University, Bloomington, IN, U.S.A.*

Maarten S. Krol

*University of Twente, Enschede, The Netherlands*

and

Nurit Kliot

*University of Haifa, Israel*

KLUWER ACADEMIC PUBLISHERS
DORDRECHT / BOSTON / LONDON

A C.I.P. Catalogue record for this book is available from the Library of Congress.

ISBN 1-4020-2868-7(HB)
ISBN 1-4020-2877-6 (e-book)
ISSN 1574-0919 Advances in Global Change Research

---

Published by Kluwer Academic Publishers,
P.O. Box 17, 3300 AA Dordrecht, The Netherlands.

Sold and distributed in North, Central and South America
by Kluwer Academic Publishers,
101 Philip Drive, Norwell, MA 02061, U.S.A.

In all other countries, sold and distributed
by Kluwer Academic Publishers,
P.O. Box 322, 3300 AH Dordrecht, The Netherlands.

Cover illustration: based on the adapted version (Beniston, 2004) of a figure from Cavalli-Sforza, Menozzi and Piazza (1994) combined with a diagram from Krol, Fuhr and Döring (2004).

*Printed on acid-free paper*



Printed in China.

# TABLE OF CONTENTS

# PREFACE

## *Synthesis Report of the Wengen-2001 International Workshop on Environmental Change and its Implications for Population Migrations*

On the basis of material supplied by Ellen Wiegandt, Urs Luterbacher, Jon Unruh, John Hay, Thomas Hammer, and Martin Beniston.

Environmental change in general, and climatic change in particular, are likely to impact significantly upon resources such as water and soils, transforming present-day landscapes and their ecological characteristics. As a consequence, disruptions of socio-economic activities in sensitive regions of the globe can be expected in coming decades. Agriculture is at particular risk, especially in areas where prolonged droughts, sea level rise, enhanced natural hazards, or extreme meteorological events such as floods or mudslides threaten marginal existence. Conversely, large-scale movements of people, goods or capital may also disrupt local environments and further contribute to social problems.

One of the direct or indirect effects of global environmental change that is increasingly recognized today is forced migration. One such example includes sea-level rise, whereby populations will be forced to move out of low-lying coastal zones or islands. Migrations can also be triggered when essential resources such as water or food fall below critical thresholds in a given region. In addition, environmental causes can be combined with social causes such as large-scale warfare, civil war, political conflicts, and disputes over resources to produce refugee flows. Social disruption can in itself be at the root of environmental degradation, which then eventually leads to massive out-migration. Various studies in recent years suggest that if environmental change is to be of the projected magnitude and rapidity, there could be as many as 150 million "environmental refugees" by the end of the 21st Century (even if this term is, for the moment, recognized neither legally nor institutionally). In view of the current barriers to migration in most parts of the world, the social, economic, and political consequences of migration at these scales is far from trivial. Most governments are today ill-equipped in legislative terms to deal with this type of situation. The political and economic tensions that will be raised by an increasing number of refugees could lead to conflictual situations in many regions. When investigating the interdisciplinary nature of the problem, it should be stressed that because of the subtle interactions between environmental and economic issues, it becomes difficult to separate different drivers of migration, i.e., political, environmental, economic, ethnic, etc.

It is with these issues in mind that the Wengen-2001 Workshop was held from September 19-21. Over 35 scientists from 16 countries from many different parts of the world attended this meeting, the seventh event in the series since the Wengen Workshops on Global Change were initiated in 1995. Although most of the participants came from academia, there were also a number of representatives from government and international organizations, as well as from industry.

The issues, which were addressed through presentations and discussions, include the possible environmental drivers of population migrations (depletion of resources through environmental changes; sea-level rise; desertification; deforestation, etc.) and socio-economic drivers such as land-use changes, agriculture, mineral and resource exploitation, ethnic issues, etc. Policy and conflictual aspects as a consequence of change also formed an important part of the meeting.

In the complex issues that can lead to migration of populations, it is necessary to distinguish between voluntary migration and forced migration. Voluntary migration can occur for a number of reasons, particularly economic and political or ideological. Forced migration, on the other hand, has a number of root causes, also to be found in political and economic domains (e.g., slavery, war, ethnic strife, etc.). In this context, environmental factors for migration can be considered to be an indirect consequence of decisions taken in the political and/or economic arenas. While sea-level rise is an obvious environmental driver which may significantly impact many low-lying coastal regions and island states around the world, it is necessary to bear in mind that sea-level rise is a consequence of a warming global climate, which is itself in part the consequence of economic and industrial policies that leads to greenhouse-gas emissions. Environmental issues can thus be seen to be an expression of underlying economic and political factors. Along similar lines of thought, population migrations may be triggered by conflicts resulting from resource depletion; in this sense, migration does not occur because of the direct consequence of environmental change but rather as a result of a complex series of interlinked ("snowballing") factors in which single, clear-cut cause-to-effect relations may not be identifiable. Causes of migration are thus seen to be embedded deep within a confluence of processes and patterns.

Migration takes many forms, and the majority occurring as internal migrations, i.e., displacements of populations within national boundaries. Presentations focusing on desertification in China and sub-Saharan Africa, deforestation in the Amazon Basin, and ethnic rivalry in Bangladesh gave accounts of the large numbers of people who left inhospitable regions to seek better conditions within their own countries highlighted these points. International migrations, also take place and conventional wisdom holds that most of these occur from the "South" to the "North", i.e., from the developing countries towards the industrialized countries as a result of perceived economic attractiveness of the industrialized world However, a large proportion of migration takes place within the countries of the "South". A particular example is that of migration from laborers into a region in which mining or forestry is introduced or intensified, as was shown in case studies for Indonesia and Brazil. In these cases, indigenous populations are likely to be forced out of their environment as a result of these commercial activities that transform the traditional resource base. The resulting redistribution of population can in no way be considered to represent an equilibrium, because the out-migration of local inhabitants in the face of new immigration represents a loss of traditional cultures and a profound change in the physical environment. There are therefore "push" and "pull" features of the environment and resource use that can trigger population migrations, "pull" factors representing attraction of migrants into an area, and "push" factors generating out-migration. Push and pull factors can thus be triggered both directly and indirectly by environmental and economic change.

Whatever the direct cause of migration, numerous presentations explicitly or implicitly showed the importance of property structures in shaping migration patterns. Whether indigenous rights were recognized and respected influenced the potential of new outside economic interests (mining, forestry) to move in modify the environment, making it more or less favorable for populations to stay. Pricing structures as a reflection of policy or market forces also influence land use and therefore its preservation, efficient use, or degradation. Population movements themselves have environmental effects and there will thus be a number of economic, political and environmental impacts resulting from the displacement of persons forced from their homelands. Issues such as the sharing of resources between increasing numbers of persons in a region of immigration, land tenure, ethnic rivalry and regional conflicts are likely to emerge as issues that will need urgent attention in the near future, as was shown in several papers presented at the meeting.

The Wengen-2001 Workshop raised the awareness of all participants that the problems under discussion were by no means as "simple" as what the original paragraphs of this synthesis seem to convey. Indeed, this is one of the very first meetings of its kind and, in aggregate, the presentations and discussions which took place in Wengen reveal that the topic of migration and environmental change is both of growing importance and in its initial stages of analysis. The migration and environment literature is far from having a well-developed theoretical or conceptual framework for addressing these issues, even if certain presentations at the Workshop on modeling aspects represent a first step in that direction.

While a number of adaptive measures can be taken to reduce the adverse effects of environmental change and the potential for out-migration that environmental change may induce, there is a need to address some of the root issues in an internationally coordinated manner. In particular, immigration policies in the industrialized world need to be reviewed in order to allow some form of open, well-regulated migrations rather than solutions aimed at keeping migrants out of the prospective host countries. This will require a change of attitude within segments of the population of the host countries in terms of the acceptance of immigrants and their integration within their host society. In the developing world also, policies will need to be altered, in particular in order to remove the "push" factors of migration. This will require a review of current resource-use practices, which are often very poorly managed and which thereby generate considerable environmental degradation. There is also a crucial necessity to improve land policies, valuation and property rights, in order to reduce the wish or the tendency for out-migration. Because the problem is one of balancing peoples' needs and wants with available resources, solutions will likely involve flexible policies about population movement but also about movements of goods and capital to achieve efficient and equitable distributions.

In view of the fact that the topic of environment and human migration is still clearly in an embryonic state, the major question that the Workshop posed relates to the primary avenues of science and policy work that need to be pursued in coming years. As a first response to this question, and as a framework for possible future research, the following points illustrate the common issues that were highlighted at the Wengen-2001 Workshop and which the participants acknowledged required further work. We must address:

- the multidimensional nature of migration, especially that associated with degrading environmental conditions the current lack of agreement in definitions of terms and use of classification procedures,
- the important role of case studies in aiding our understanding of environment-migration relationships,
- data and methodological issues by rapidly increasing our ability to characterize the environmental variables that serve as indicators of pressures which drive migration, for example through remote-sensing techniques, monitoring,
- modeling of environmental change and its impacts,

- the need for research-based studies to provide relevant, timely and useful advice to policy- and decision-makers, which will allow them to respond in a more proactive manner to environment-related immigration issues,
- the important and not always mutually supportive roles of international, national and local governmental agencies,
- the need for those involved in environment-migration studies, policy development, decision making and assistance programs to identify research priorities that will address current gaps in our understanding, reduce uncertainties and increase the usefulness of technical and policy oriented studies related to environment and migration.

The sponsorship of the following organizations is gratefully acknowledged:
University of Fribourg, Switzerland
Swiss Academy of Sciences, Bern, Switzerland
University Institute Kurt Boesch, Sion, Switzerland
Swiss National Science Foundation, Bern, Switzerland
International Human Dimensions Program, Bonn, Germany

# ISSUES RELATING TO ENVIRONMENTAL CHANGE AND POPULATION MIGRATIONS

*A Climatologist's Perspective*

Martin Beniston
*Department of Geosciences, University of Fribourg, Switzerland*

## 1. INTRODUCTION

Human activities in most parts of the world are transforming the global environment. Ecosystems subjected to the stresses of global environmental change become more susceptible to the emergence, invasion, and spread of opportunistic species. When subject to multiple stresses, natural environments may exhibit reductions in resilience and regenerative capabilities. It is sometimes assumed that the time scales associated with environmental change are long and that, in many situations, the environment may find a new equilibrium, if not its original state. Environmental upheavals have occurred in the past, along with species extinctions, and yet the planet has "survived" and evolution has continued. However, it is possible, that anthropogenic pressures are accelerating change and that many systems may not adapt to rapid rates of change, even if they could adapt to the amplitude of change over longer time periods. A number of factors contributing to global environmental change are listed below; many of these factors are essentially human-induced:

- Air pollution and ozone depletion
- Climatic change
- Land use change
- Deforestation
- Desertification
- Loss of biodiversity
- Land degradation
- Fresh water availability
- Hazardous wastes
- War

Many of these elements can act in a synergistic manner, thereby compounding the stress situation for the environment and the adverse effects this may have on human activities and the carrying capacity of a particular region.



*J.D. Unruh, M.S. Krol and N. Kliot (eds.),*
*Environmental Change and Its Implications for Population Migration,* 1–24.

Humans are not only the receptors of environmental change, both directly and indirectly, but they are also in numerous instances the drivers of change. Demography can in addition play a significant role in shaping the course of environmental degradation (Jolly, 1993; Davis and Bernstram, 1993). High population growth in many parts of the world is linked to environmental damage because local inhabitants attempt to maintain or improve their resource base and economic level through the over-exploitation of their environment (Commoner, 1991; Pebley, 1998). This is carried out in general without any long-term environmental management strategy; resources can thus become rapidly depleted or ineffective.

Through technological advances and seemingly adequate resources, the industrialized world in particular lives under the impression that basic life-supporting resources are abundant and quasi-unlimited. However, one is acutely reminded that famine and disease are still widespread in many parts of the world at the beginning of the 21st century, and that over 350 million persons do not have access to potable water (UNCWR, 1998). One tends to forget that even in technologically-advanced societies, water, food and health all constitute basic, interrelated needs for human survival. These elements are all highly dependent on environmental factors such as climate, and are sensitive to even slight shifts in existing environmental conditions. Such changes may upset the delicate balance that, in some countries, favors a high degree of food security, water quality and quantity, and sanitary conditions. Under adverse conditions, it can be surmised that the severe depletion of an essential resource could lead to out-migration, with persons moving from a region affected by resource loss to regions where the resource is still sufficient to sustain both the local and the migrant populations. Because environmental change is recognized to be a possible cause for population migration, the term "environmental refugees" has emerged in recent years (Jacobson, 1988). Although there is no official or legal recognition of this term in the United Nations nomenclature, the High Commission for Refugees (UNHCR, 1993), in the early 1990s, already identified four root causes of refugee flows, which include political instability, economic tensions, ethnic conflict; and environmental degradation. For the first time, a possible link between environmental degradation and refugee flows was acknowledged. Furthermore, the UNHCR recognized that the numbers of displaced persons internationally was much larger than indicated by the statistics on refugee flows; in other words, the accepted causes of migration (political and ethnic strife, economic hardship) could not entirely explain the numbers of refugees identified by the UNHCR. Could these differences thus be explained by environmental refugees?

In this paper it will be shown that, in one form or another, environmental change and its impacts on the basic resources for human sustenance have in the past been a dominant driving force behind population migrations. It will then be suggested that accelerating global change in the 21st century may lead to situations where it once again becomes the major cause for out-migration from particular regions. Migrations recorded during the 19th and 20th centuries, on the other hand, are embedded within complex political, social, and economic contexts and are not solely driven by adverse environmental factors. Some of the 20th century migrations may well have underlying environmental causes as well, but these causes are difficult to disaggregate from the other factors. The subtle links between economic, political, and environmental causes of migration will certainly be present in coming decades, but in some instances the environmental factor may dominate once again in the future as it did in the past.

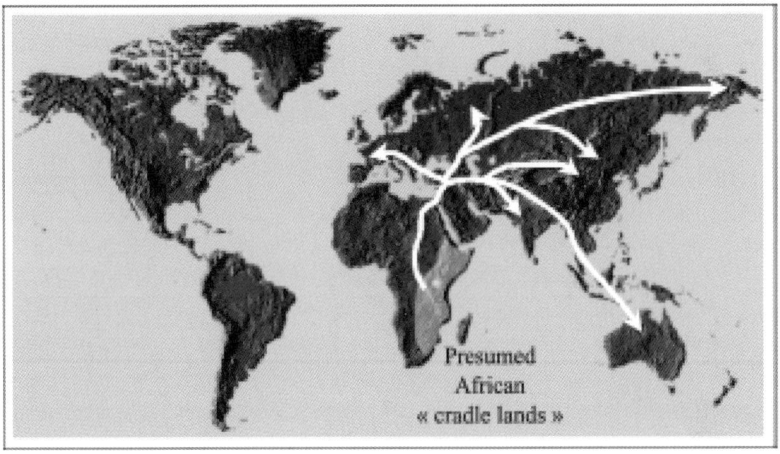


*Figure 1: Possible migration routes of early humankind (adapted from Cavalli-Sforza et al., 1995).*

## 2. POPULATION MIGRATIONS IN THE PAST

It is generally accepted, on the basis of extensive international of archaeological and anthropological research that Homo sapiens emerged from the so-called African "cradle-lands", within the region of the East African Rift Valley. At some stage in a 2-3 million year past, environmental conditions in the Rift Valley favored the evolution of a new species of primate. Migrations of primitive populations began several hundred thousand years ago, perhaps due to epidemics (one of the three basic factors for human sustenance) in a changing climate, or as a result of insufficient food for the increasing populations (Gibson, 1996). The routes followed by early man over several centuries reveal trails that lead to Europe, Central Asia, and perhaps also Australia, as shown in Figure 1. Genetic evidence provides evidence that any given population retains clues to its ancient roots, and common ancestries can be confirmed and human migrations traced by comparing the DNA frequencies of present-day populations (Cavalli-Sforza et al., 1995). Early migrations of modern humans out of Africa have been traced by analysis of DNA sequences; more recent human migrations have been followed through genetic trails as well.

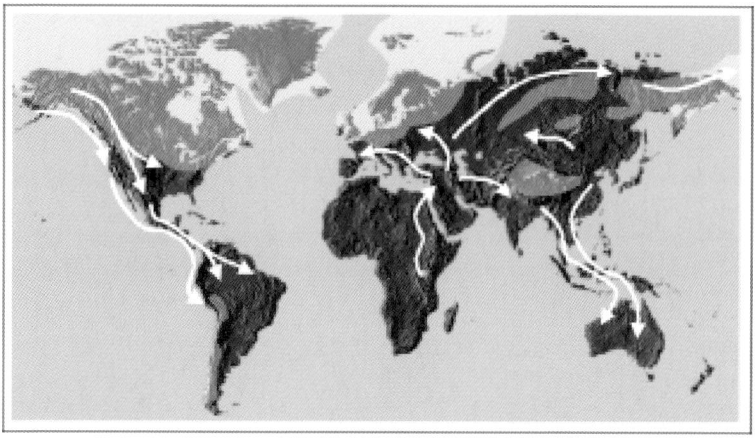

*Figure 2: Migration routes during the last glacial maximum (adapted from Elias, 1997).*

About 120 thousand years ago, a more advanced species of primitive man (the Neanderthal) appeared during the last interglacial period, and spread throughout Europe, Africa and Asia, probably in search of sufficient resources to sustain the growing members of these tribes (Harris, 1991). These were replaced progressively by modern people originating in Africa (Pavolov et al., 2001), about 40 thousand years BP (before present). At the height of the last major ice age, about 25 thousand years BP, a new wave of migrations began, as populations searched for regions with less extreme climates than those affecting many parts of the Northern Hemisphere. Archaeological evidence indicates that land routes across Central Asia, and coastal routes into Australia and along the shores of the Americas circumvented the extensive ice-caps during the last glacial maximum, as illustrated schematically in Figure 2 (Elias, 1997). These population movements led to the settlement of populations with enhanced technological capabilities (more efficient tools; clothing to resist the cold; more efficient hunting weapons; boats capable of navigating in coastal waters) in different parts of the world (Monastersky, 2000).

During this same period, certain areas of the Northern Hemisphere mid-latitudes were less affected by the extreme conditions of the last glacial maximum than others. One such region is the Dordogne and Aquitaine of south-western France, which has undergone continuous human colonization for over 400 thousand years – a record for Europe. Climatic conditions in the Dordogne Valley appear to have been reasonably clement, thereby allowing settlement rather than out-migration, and the progressive development of primitive skills and hunting techniques. Archaeological findings suggest that around 50 thousand years BP (before present), the first ritual burials of the dead took place in the caves near Padirac in the Dordogne Valley. Such rites mark the cultural transition to modern man, and this took place in a region where environmental conditions remained favorable to human development over a sufficient time span to allow cultural development to take place above and beyond day-to-day survival (Bordes, 1973).

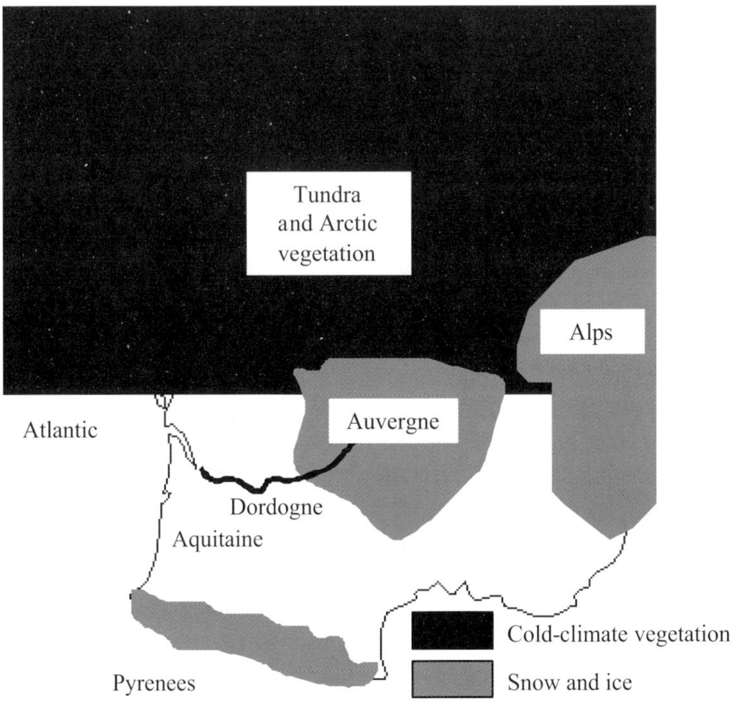


*Figure 3: The Dordogne and Aquitaine regions of southwest France at the height of the last ice age (adapted from Clottes, 2001).*

During the last glacial maximum, the Dordogne and Aquitaine regions of France were probably isolated from the rest of the world. To the south, the snow-covered Pyrenees were certainly a daunting physical obstacle to migration; to the east, the Massif Central was probably an inhospitable highland area swept by high winds and extreme cold for most of the year; to the north, extending towards the ice fields which reached as far south as London, the vegetation cover was essentially reminiscent of today's northern tundra, i.e., yielding little in terms of food; and to the west, the Atlantic was a hostile and stormy ocean (Fig. 3, adapted from Clottes, 2001). With little options open to them, the inhabitants of the Dordogne made the best of their environment; indeed, all the "ingredients" for the survival of early civilizations were to be found here: abundant water and food, a relatively mild climate in the middle of very cold conditions, sufficient protection from enemies in the cliff-and-cave landscape of the Dordogne, and yet sufficient open spaces to enable hunting and thereby to ensure sustenance (Bordes, 1973).

*Figure 4: 25,000-year old cave paintings in Pech Merle, in the Lot Valley of southwestern France (Courtesy: Centre de Préhistoire de Pech Merle, F-46330 Cabrerets, France).*

It is during this period, from about 30 thousand years BP onwards, that pictoral evidence of a radical change in human culture emerges through the use of art to record and, perhaps, to convey to later generations elements of the living environment to which the Dordogne inhabitants were accustomed.

The techniques, style and colors used by these painters in caves such as Lascaux or Pech Merle (Fig. 4) bear witness to the fact that the "climatic oasis" of the Dordogne during the last ice age was a key region for human cultural and technological development. However, it is also a sign that if the carrying capacity of the environment is sufficiently great, then major migrations do not necessarily occur.

About 11 thousand years BP, the present interglacial period, also known as the Holocene, began. Many of the large continental glaciers melted, and as a consequence, the ocean levels rose by 100 m, and new species of animals and plants replaced others which were becoming rapidly extinct, such as the mammoth and the saber-toothed tiger. In parts of the Middle East, in what is today Syria and Iraq, people took up two different ways of life, namely nomadic and sedentary (Teitelbaum and Winter, 1993).

Nomads herded sheep, cattle and other animals. They also gathered plants, but meat and milk were their staple diet. Many of the nomads lived on the steppes stretching from Europe across Asia to northern China. As climates changed, nomads were sufficiently mobile to move on to more hospitable and productive regions. Simultaneously, other populations began sedentary agriculture by planting seeds such as barley, wheat, rice and maize, which not only allowed an improvement in their diets, but could also be stored and eaten throughout the year (McCorriston and Hole, 1991). As a result, migrating in search of food became less of a necessity, and sedentary agriculture led to the building of the first permanent settlements. In some areas, however, soils would tend to lose their fertility after a few years and the farmers moved on, burning down forests and shrub-lands to make new fertile ground for a few more years (Armelagos, 1991). This was the start of the destruction of the world's forest which began in the Mediterranean area (Greece, for example) and which, for the tropical rain forest, is still continuing today. The early destructions of certain biomes are also the first sign that human activities are not only influenced by the environment, but that the environment is capable of being significantly disturbed by mankind (Cohen, 1977).

During the Holocene optimum, which lasted from about 6,000 to 5,000 years BP, temperatures were between 2 and 4°C warmer than currently; precipitation patterns were also different from those of today, with geographical zones such as the Sahara experiencing far more rainfall than today (Crowley and North, 1991). Indeed, rock paintings in the Tassili N'Ajjer of southern Algeria depict hunting scenes, bearing witness to the fact that grasslands and game such as deer were present in what is currently one of the most arid regions of the planet. In Iraq, Egypt, India and China, river valleys flooded every spring and replaced the minerals of the soil, thereby allowing soils to remain fertile for many years, allowing villages to grow into cities. The settled way of life led to an increase in food production and population. Technologies developed rapidly, from farming tools, clay pottery and bricks, wheels, cast bronze tools and weapons, to reading and writing. As in the earlier times of the Dordogne, the carrying capacity of the environment in the Middle East and southern Asia allowed populations to evolve culturally and technologically without moving on to other regions by necessity (Clottes, 2001).

The relative stability of environmental conditions, allowing rapid progress in human technology and culture, was not the only influencing factor for reducing or limiting out-migration. As food supplies became sufficient for larger numbers of people due to farming practices and the storage of food to avert famine during more difficult times, and because technology and shelter were available to a much larger extent, the size of populations grew far more rapidly than in the past (Cohen, 1977). As a result, in some communities, the carrying capacity of the environment became insufficient once a critical threshold of population was exceeded. If demographic growth began to outpace food supply, then a time inevitably came when there was insufficient food supply to sustain an entire, growing population, and out-migration would become once again an option. Strife or epidemics, which could sometimes curtail increasing populations and allow the survivors to remain in their usual environment, could in some instances avert the necessity for out-migration (Armelagos, 1991).

*Figure 5: Ruins of Anasasi cave-village in Mesa Verde National Park, Colorado (USA). (Courtesy: National Park Service, Washington, D.C., USA).*

It is believed that the sudden disappearance of the Anasasi Indians, in the American Southwest, may be an early example of the combined environmental and social upheavals leading to population out-migration. The Anasasi were known for their advanced community structures (Jones and

Cordell, 1985), exemplified by the superb architectural remains in Mesa Verde (Colorado; Fig. 5) or Bandelier (Arizona), for example. They were essentially farmers, as the large threshing surfaces and communal ovens found in Mesa Verde indicate. Over the several hundred years of Anasasi presence in the Southwest United States, paleo-climate reconstructions suggest that a number of severe droughts intervened, each one capable of severely depleting food stocks. However, until the sudden disappearance of the Anasasi, food stockpiled to survive periods of drought were generally sufficient to sustain the indigenous populations. According to one hypothesis (e.g., Cordell, 1984), it is only when the Anasasi population exceeded a certain critical level that food stocks were no longer capable of nourishing the entire population during the droughts that ensued. This could explain their rapid disappearance, perhaps through migrations to various destinations which were perceived as less inhospitable.

The examples that have been succinctly presented here show that populations have responded to environmental stresses, particularly climatic change and its impacts on food and water resources, and health, by migrating to areas with more abundant food and water, and improved health conditions. Following the last major ice age, a more clement climate and the discovery of agriculture, reduced the need to migrate, until the capacity of sedentary agricultural practices to sustain rapidly-growing populations was no longer possible.

## 3. POPULATION MIGRATIONS IN THE 20TH CENTURY

The 20th century has seen unprecedented social and economic changes, many of which have been the primary drivers of migration. Two world wars and many regional conflicts, highlighting opposing ideologies and shifting alliances, have also been major actors in population displacements. Throughout the 20th century, the interactions between economics, politics and environmental stress have become increasingly complex, thereby rendering more difficult than previously any single cause-to-effect relationship. Earlier deterministic cause and effect models where a set of environmental stresses resulted in a particular response, namely migration, from individuals and communities are no longer the norm where out-migration is the result of numerous concurrent causes. Some 20th century population displacements may have occurred with certain forms of environmental catastrophe, where there is no option but to move, but the most visible migrations have definitely been linked to conflict and economic hardship. In addition, simple cause-to-effect models break down in more complex social structures since, in general, levels of sensitivity and perception by communities to a major stress are typically high, and thus there exist different levels of ability to cope with environmental stresses. The tolerance thresholds of populations are highly variable, being surpassed very readily in some, and being almost insurmountable in others. A proper appreciation and understanding of the complexity and diversity of human responses to environmental degradation is essential in order identify the complexities of migration in past decades.

Figure 6 shows the trends in the numbers of persons currently under the protection of the UN High Commission for Refugees (UNHCR, 2000), and in particular the spectacular rise from less than 5 million in the late 1970s to over 25 million in 2000. This is only part of the overall picture, however, because the mandate of the UNHCR is basically to offer aid and protection to victims of political persecution and conflicts. Furthermore, it is necessary to differentiate between internally-displaced persons (IDPs, who migrate within national boundaries in order to escape the consequences of civil war, for example), and forced migration of persons who flee across international boundaries. In terms of the actual numbers of migrants worldwide, in addition to those officially registered by the UNHCR, figures vary widely and range in the early 1990s from about 80 million according to the International Office of Migrations to over 100 million in 1993 (Castles and Miller, 1993).

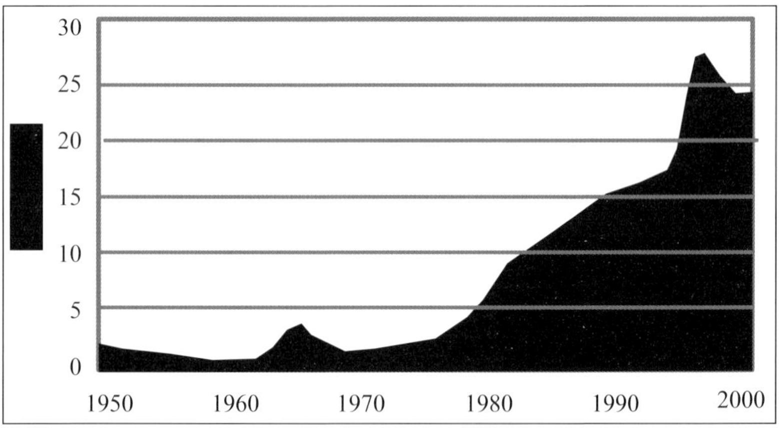


*Figure 6: Refugees under the protection of the United Nations High Commission for Refugees (UNHCR, 2000).*

The UNHCR (1995) acknowledges that collecting accurate statistical data on refugees and asylum-seekers is "one of the most problematic issues" confronting the agency. Using 1990 data from UN sources, Westing (1992) for example notes that in addition to officially recognized refugees (roughly 40% of total displaced persons), there were cross-border refugees (10%), and internal refugees (51%). In total, the 1990 figures amount to 41.5 million persons, which represents a 60% increase on the numbers of displaced persons estimated in 1986. On the basis of these figures, Westing (1992) further speculates that the growth is due to the addition of "environmental refugees." Jacobson (1988) noted already that "environmental refugees have become the single largest class of displaced persons in the world", a statement that is partially upheld by the UNHCR (1993) which admits that there are "clear links between environmental degradation and refugee flows", as we have already seen. Other drivers for migration in the 20th century are linked to demography, economic restructuring and increasing economic disparities. In the latter part of the 20th century, these disparities have been the root cause of many displacements of people, from the developing countries to the industrialized world because of perceived economic attactivity of North America and Europe, for example, and "South-to-South" migration as persons move within countries or groups of countries within the developing world.

It is perhaps symptomatic of the large number of social and economic upheavals witnessed in the 20th century that the focus of migration studies has been these more obvious driving forces. This should not overshadow the fact that, however, there have been environmental factors that have led to forced migrations in the 20th century also. These include natural disasters such as earthquakes, hurricanes, droughts and floods; industrial accidents, leading, for example, to the evacuation of a 30-km radius around the Tchernobyl nuclear power station in the Ukraine, following the 1996 disaster; development projects such as dams and irrigation projects which have displaced over the last 25 years more than 20 million persons in India alone (Fornos, 1991), etc.

# 4. POSSIBLE ENVIRONMENTAL CHANGE AND POPULATION MIGRATIONS IN THE 21ST CENTURY

## 4.1 Introduction

In coming decades, population growth, increasing pressures on resources, and persistent inequalities in resource access imply that scarcities will affect many environmentally sensitive regions with a severity, speed, and scale unprecedented in history. Many countries lack the social institutions that are essential prerequisites to ensure social and technical solutions to face up to problems of scarcity. In the face of significant external stress, population displacement often indicates the breakdown of social resilience. For example, in the context of food security, displacement and coping strategies reflect an extreme manifestation of vulnerability. Coping strategies generally represent short-term adaptations to extreme events; they are usually involuntary, and rarely prepare the way to reducing a population's vulnerability to future famine situations.

Dwindling resources in an uncertain political, economic and social context are capable of generating conflict and instability, but the causal mechanisms are often indirect. Scarcities of cropland, fresh water, and forests constrain agricultural and economic productivity. Such situations are capable of generating population movements (Myers, 1993). In the extreme, such a situation can contribute to local or regional conflicts, which may increase over time as environmental scarcities worsen; while such conflicts may not be as conspicuous as wars at an international level, there is nevertheless a potential for significant repercussions for the security interests in both the developing and the industrialized countries. Such internal, resource-based conflicts can affect international trade relations, produce humanitarian disasters, and lead to growing numbers of refugee flows (Parnwell, 1993).

In the complex issues that can lead to migration of populations in the 21st century, it will be increasingly necessary to distinguish between voluntary migration and forced migration. Voluntary migration can occur for a number of reasons, particularly economic and political or ideological. Forced migration, on the other hand, has a number of root causes, also to be found in political and economic domains, in particular war and ethnic strife, etc. In this context, environmental factors for migration can be considered to be an indirect consequence of decisions taken in the political and/or economic arenas. While sea-level rise is an obvious environmental driver that may significantly impact many low-lying coastal regions around the world, sea-level rise is a consequence of a warming global climate, itself in part the result of economic and industrial policies leading to enhanced greenhouse-gas emissions.

Environmental issues are thus seen to be an expression of underlying economic and political factors. Similarly, population migrations may be triggered by conflicts resulting from resource depletion. In this sense, migration does not occur because of the direct consequence of environmental change but rather as a result of a complex series of inter-linked factors in which single, clear-cut cause-to-effect relations may not be identifiable. Causes of migration are thus seen to be embedded deep within a confluence of processes and patterns.

## 4.2 Water resources

If present consumption patterns are sustained, the global annual water demand is likely to increase from 4,130 km3 in 1990 to 5,190 km3 in 2020 (Shiklomanov, 1993). Because of increasing population, the additional demand will be accompanied by a sharp decline in water availability per capita. While a consumption of 1,000 m3 of water per year and per capita is considered a standard for "well-being", currently less than 50% of the world's population reaches this level (UNCWR, 1998).

Against this backdrop of social and ecological problems, water resources will most likely come under increasing pressure. Significant changes in climatic conditions will affect demand, supply and water quality. In regions which are currently sensitive to water stress (arid and semi-arid mountain regions), any shortfalls in water supply will enhance competition for water use for a wide range of economic, social, and environmental applications. In the future, such competition will be sharpened as a result of larger populations, which will lead to heightened demand for irrigation and perhaps also industrialization, at the expense of drinking water (Noble and Gitay, 1998).

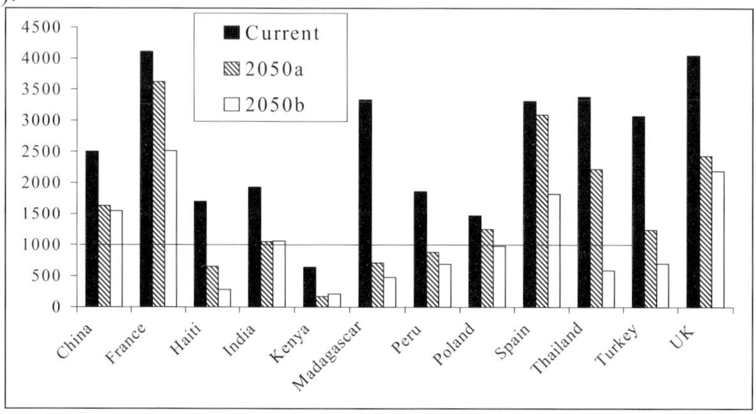


*Figure 7: Water availability per capita by 2050 for selected countries in the industrialized and the developing world, for scenarios with constant climate (2050a) and with changing climate (2050b), following IPCC (1998).*

Projections of annual water availability per capita by the 2020s point to a declining trend in all parts of the world, including those that are considered to have ample water resources (IPCC, 2001). Figure 7 illustrates the possible changes in water availability per capita for a number of countries around the world. It is seen that in most cases, water amount will dwindle in large part due to population growth; the superimposed reductions resulting from climatic change will exacerbate the situation stressed by demography but will certainly not be the dominant factor leading to reduced water availability. The World Bank (1995) estimates that about 600 million people in Africa will suffer from water shortage and dwindling water quality. Shifting precipitation belts account for only a fraction of this reduction; rapid population growth, urbanization and economic expansion place additional burdens on water supply.

In Latin America, which accounts for 35% of global freshwater, the impacts of climate change are expected to occur in the more arid regions of the continent, which are often associated with the rain-shadow influences of the Andes ranges. Shifts in water demands will depend on population growth, industrial expansion, and agricultural potential. In many countries of the region, water availability is expected to decline, which is likely to generate potential for international conflicts. The IPCC (1998) estimates that water availability per capita and per annum will decrease from 4,750 m3 in 1990 to 2,100 m3 in Mexico by 2025, without any change in climate, i.e., due to population and economic growth. Based on several GCM simulations, projected shifts in precipitation in a warmer climate yield a range of 1,740 - 2,010 m3. For Peru, the respective set of figures are 1,860 m3, 880 m3 resulting from demography alone, and 690 - 1,020 m3 with climatic change, i.e., close to or below the minimum requirements for « well-being ».

Water resources in tropical Asia are very sensitive to tropical cyclones and fluctuations in their trajectories and intensity. The dominant effect of the Monsoon may be perturbed in a changing climate. Runoff in the Ganges, for example, is more than 6 times that of the dry season. As elsewhere in the world, water resources will become increasingly vulnerable to increasing population growth, urbanization, industrial development and agriculture, as shown by Schreir and Shah (1996). An impacts assessment study by Mirza (1997) for a number of Himalayan basins contributing to the Ganges has shown that changes in the mean runoff in different sub-basins ranged from 27 to 116% in a climate forced by a doubling of CO2 concentrations relative to their pre-industrial levels. The sensitivity of basins to climate change was seen to be greater in the drier catchments than in the wetter ones. However, water demand is greatest in the dry season in India, and demand cannot be met by supply in this season. Shifts in the timing and intensity of the Monsoon, and the manner in which the Himalayan range intercepts the available precipitable water content of the atmosphere, will have major impacts on the timing and amount of runoff in river basins such as the Ganges, the Brahmaputra or the Irrawaddy. Amplification or weakening of the Monsoon circulations would indirectly impact upon agriculture and fisheries, freshwater supply, storage capacity, and salinity control.

In Africa, where over 60% of the continent's rural population and 25% of its urban population does not have access to safe drinking water (IPCC, 1998), climatic change will exacerbate the current situation. The expected degradation will in part be due to increasing population pressures and insufficient financial resources to ameliorate the existing problems of water supply and quality. Furthermore, Africa has a large number of rivers that cross or form international boundaries, so that the sharing of a dwindling resource could ultimately lead to regional conflicts. Armed disputes over water resources may well be a significant social consequence in an environment degraded by pollution and stressed by climatic change, as in the Middle East (Lonergan and Brooks, 1994). The potential for forced migration, either directly because of insufficient water to sustain growing populations, or indirectly because of armed conflicts, is clearly present.

A more rational approach to the conservation and use of what is probably the most vital single resource for humankind is thus crucial. New initiatives are aimed at conservation and distribution of water within the developing world so that their inhabitants, in particular women, can avoid spending a large part of their working lives merely carrying drinking water for their families (Engelman and Leroy, 1993).

## 4.3 Food security

Food security is also threatened by climatic change, both directly by changing temperature and precipitation patterns, and indirectly through the losses of agricultural land by sea-level rise, enhanced eolian and fluvial erosion. In addition, human-induced land-use change, linked to deforestation and desertification, have already reduced the agricultural potential of many regions in the world (IPCC, 2001).

The world food system involves a complex dynamic interaction of producers and consumers, inter-linked through global markets. World agriculture generates over US$1,300 billion in annual revenue. Related activities include input production and acquisition, transportation, storage and processing, and these generate an additional revenue estimated at US$ 4,000 billion on an annual basis (IPCC, 1998). Although agricultural productivity has increased to keep pace with growing world population over the last century, there are still close to 1 billion people who are undernourished. Furthermore, agriculture is probably the most vulnerable of all human activities to weather and climate variability; the chief controls on agricultural yields include temperature, precipitation, soil moisture, carbon dioxide levels, and disease and pests (themselves largely climate-dependent). Any changes in one or more of these controlling factors may have profound, non-linear effects on productivity. The Food and Agricultural Organization (FAO, 2000) has warned that by 2020, agricultural yields should almost double in order to keep up with demographic trends and the diversification of consumption patterns. It is unlikely that the "green revolution" of the 20th century will be reiterated, even when taking into account new technologies such as genetic engineering, because competition for land and climatic change may negate all or part of the progress made in agricultural productivity.

The impacts of climatic change on agriculture can be assessed at different scales including crop yield, farm or sector profitability, regional economic activity or hunger vulnerability. Impacts depend on biophysical and socioeconomic responses. The effects of climate on agriculture in individual countries cannot be considered in isolation, because in addition to the close links between agriculture and climate, it is necessary to take into account the international nature of food trade and food security, and the need to consider the impacts of climate change in a global context.

Agricultural production will be affected by the severity and pace of climate change. If change is gradual, there will be time for political and social institutions to adjust. Slow change may also enable natural biota to adapt. Many untested assumptions lie behind efforts to project the potential influence of climatic change on crops. In addition to the magnitude and rate of change, the stage of growth during which a crop is exposed to drought or heat is important. Moreover, temperature and seasonal rainfall patterns vary from year to year and region to region, regardless of long-term trends in climate. Temperature and rainfall changes induced by climate change will likely interact with carbon dioxide levels, fertilizers, insects, plant pathogens, weeds, and the soil's organic matter to produce unanticipated responses. In many parts of the world, generally warmer temperatures and longer hot periods will put additional stress on certain crops. Corn, for example, has a stress limit at about 35°C which, when exceeded for any length of time, can exert irreversible physiological damage to the plant. The United States Midwest, one of the world's principal cereal-producing regions, could be particularly vulnerable to prolonged heat, heightening the potential for crisis in the global food supply. The 1988 drought in the Midwest resulted in severe shortfalls in corn yields, and for the first time since the Second World War, the United States was a net importer of cereals rather than an exporter. A warmer and drier climate at critical times of the year could increase the frequency of crop failures.

Rainfall, however, remains the major limiting factor in the growth and production of crops worldwide. Adequate moisture is critical for plants, especially during germination and fruit development. Any changes to rainfall patterns will also reduce soil water contents. In certain semi-arid and arid zones, the soil moisture often allows plants to survive a short drought period; a warmer climate, accompanied by more evaporation, less precipitation and associated reductions in soil moisture recharge, would spell disaster for regions where agriculture is only just viable today.

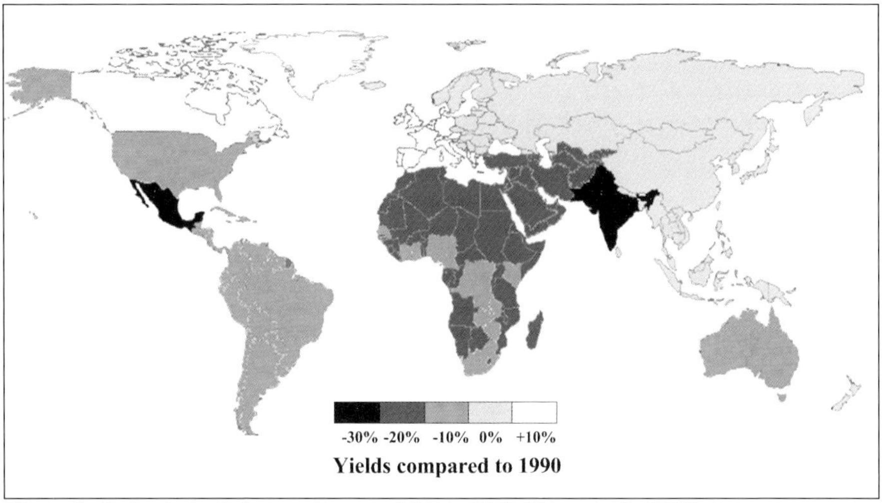


*Figure 8: Projections of changes in corn yields in 2050 (Parry et al., 1999).*

Based on projections of economic, technological, and climatic trends, maps of yield changes for corn have been compiled by Parry et al. (1999) for various periods in the future. Figure 8 illustrates the changes in wheat yields in 2050 compared to the 1990 baseline; with the exception of western Europe, Canada, and the Russian Federation, almost all countries are projected to suffer yield reductions. Shortfalls range from 10% in Argentina or Australia, to over 25% in India and Mexico (Parry et al., 1999). The collateral effects of severe shortfalls include financial fragilization, as the less wealthy countries import food at the expense of vital hard-currency reserves.

A large range of extreme weather events, which may increase in frequency and severity in certain parts of the world, may compound the stress effects of a warming average climate. Drought, fire and heat waves are one category of extremes which need to be considered, while heavy precipitation and hail are another category that can severely affect agricultural production. These events may be offset to some extent in colder regions by a lower frequency of spring frosts, which are often damaging to plants at the beginning of their growth cycle.

While these issues related to changes in agricultural yields do not necessarily imply that shortfalls in basic foodstuffs would lead to out-migrations, they do convey the dramatic situation in which countries with growing populations are likely to find themselves in the future. Internal migrations are likely, as populations move out of the agricultural zones which are affected the most into other rural areas where conditions are less acute or, possibly, into urban areas in search of work, shelter, food and water.

Vulnerability is low in regions where agriculture is well adapted to current climate variability, or where market and institutional factors allow a redistribution of agricultural surpluses to make up for shortfalls. In order to plan ahead and reduce the impacts of climatic change on agriculture, long-term agricultural policy options should be carried out in parallel to other concerns, such as erosion, loss of topsoil, salinization, soil and water pollution. Furthermore, improved water management and irrigation practices should be implemented to help reduce the adverse effects of droughts and heat-waves that are likely to be on the increase in a warmer global climate.

## 4.4 Health

In terms of human health, there are many determinants that can all interact with one another (McMichaels et al., 1996). Human biological and psychological factors are primary determinants, but ecological and global systems are also involved, as are economics and access to health care, which shape vulnerability of societies to disease. Shifts in environmental conditions, interacting with the biology of disease agents, can exert profound effects. Changes in how land is used affect the distribution of disease carriers, such as rodents or insects, while climate influences their range, and affects the timing and intensity of outbreaks. Changing social conditions, such as the growth of multimillion-inhabitant cities in the developing world and widespread ecological change, are today contributing to the spread of infectious diseases. This will be exemplified here by looking into one of the best-known vector-borne diseases, malaria.

The occurrence of vector-borne diseases is widespread, ranging from the tropics and subtropics to the temperate climate zones. With few exceptions, they do not occur in the cold climates of the world, and are absent above certain altitudes in mountain regions of the tropical and equatorial belt. Currently in most parts of Africa, at elevations above roughly 1,300 m, the Anopheles mosquito cannot breed nor survive; as a result, malaria is almost totally absent in many highland regions of the continent. The occurrence of vector-borne diseases is largely determined by the abundance of vectors and intermediate and reservoir hosts, the prevalence of disease-causing parasites and pathogens suitably adapted to the vectors, the human or animal host, and local environmental conditions, especially temperature and humidity; and the resilience and behavior of the human population.

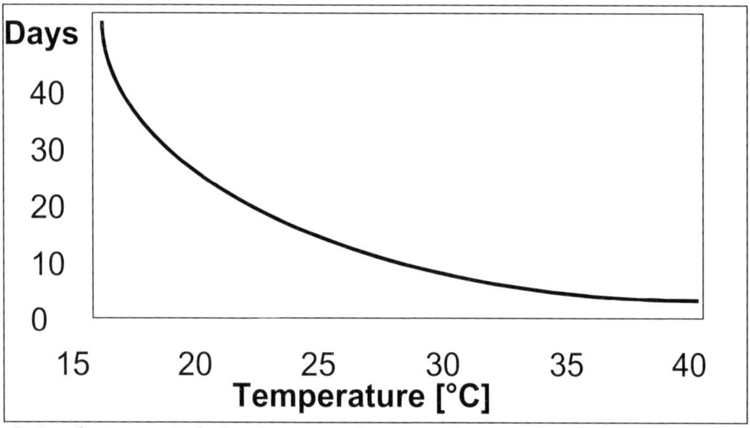


*Figure 9: Incubation period of malaria as a function of temperature (adapted from Martens et al., 1995).*

Vectors require specific ecosystems for survival and reproduction. These ecosystems are influenced by numerous factors, many of which are climatically-controlled. Changes in any of these factors will affect the survival and hence the distribution of vectors (Kay et al., 1989). Global climatic change projected by the IPCC (2001) may have a considerable impact on the distribution of vector-borne diseases. A permanent change in one of the abiotic factors may lead to an alteration in the equilibrium of the ecosystem, resulting in the creation of either more or less favorable vector habitats. At the present limits of vector distribution, the projected increase in average temperature is likely to create more favorable conditions, both in terms of latitude and altitude for the vectors, which may then breed in larger numbers and invade formerly inhospitable areas.

Malaria is currently one of the largest causes of mortality in many parts of the developing world; the increase in the incidence of the disease, resulting from both population growth and environmental changes, is, and will increasingly be, a cause for concern. Figure 9 shows that the relation between the incubation period for malaria and temperature is exponential (WHO, 1990); small increases in temperature can lead to a sharp reduction in the number of days of incubation. Regions at higher altitudes or latitudes may thus become hospitable to the vectors; disease-free highlands, such as parts of Ethiopia or Kenya, may be invaded by vectors as a result of an increase in the annual temperature. If this were to occur, then the number of persons infected by malaria would increase sharply, since the infection rate is a function of temperature; higher temperatures augment quasi-exponentially the rate at which persons exposed to the vector may contract the disease (WHO, 1990).

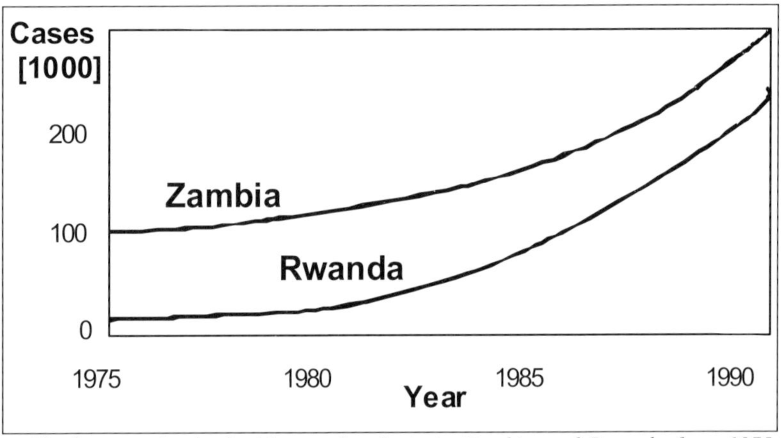


*Figure 10: Increase in the incidence of malaria in Zambia and Rwanda from 1975-2000 (Loevinsohn, 1994).*

The World Health Organization (WHO, 1990) has investigated the possible changes in the distribution of the principal vector-borne diseases. Increases in temperature and rainfall would most probably allow malaria vectors to survive in areas immediately surrounding their current distribution limits. How far these areas will extend both in terms of altitude and latitude depends upon the extent of warming. The IPCC (1998) has published maps of increases in the incidence of malaria in Africa, and concludes that the regions with the sharpest rise in the rate of infection are those which lie above 1,000 m, i.e., those that are still relatively free of malaria because climatic conditions are currently unfavorable to the development and propagation of the disease. In these highland regions, even a modest rise in temperature will lead to a spread of the disease into hitherto disease-free regions. Figure 10 illustrates that the trend is already discernible in a number of highland regions of Africa, such as Zambia and Rwanda (Loevinsohn, 1994). It is seen here that there has been a sharp increase in the incidence of malaria, which is, at least in part, attributable to changing climatic conditions since the mid-1970s.

Africa is not the only continent to be affected by the increase in vector-borne diseases; in certain countries where the disease has been eradicated in the second half of the 20th Century, particular strains of malaria are resurging. There are reports from various low to medium elevation upland sites in Turkey, Tajikistan, Uzbekistan, Turkmenistan and the Urals that malaria is being transmitted in rural populations, with close to epidemic proportions in south-eastern Anatolia (Wilson et al., 2001).

### 4.5 Summary

In conclusion, there will be numerous interacting root causes, from politics and economics, to profound changes in the environment (sea-level rise, deforestation, soil degradation, and climatic change) that are likely to impact heavily upon water resources and food security, as well as on human health. The extent to which the reductions in water supply and shortfalls in agricultural yields, or changing patterns of disease, may actually force extensive out-migration, is a matter of debate. The IPCC (2001) notes that one of the greatest effects of climate change may be on human migration, as millions of people will be displaced due to shoreline erosion, coastal flooding, agricultural disruption, and dwindling water resources. Following the first assessment report of the IPCC (1990) which had already raised this issue, Myers (1993) projected that about 150 million "environmental refugees" may represent one of the direct consequences in the "greenhouse world" of the 21st century. There is no certainty associated with this particularly figure, and it may be an overstatement. But it helps to raise the awareness to these issues and to stimulate thought and action in order to institutionally and legally prepare for refugees in larger numbers than hitherto experienced.

## 5. CONCLUSIONS

As societies have become more complex in cultural and technological terms in recent decades, the interactions between environmental and economic issues have become more subtle, and as a consequence, it has become increasingly difficult to separate different drivers of migration, i.e., political, environmental, economic, ethnic, from one another. Population movements themselves have environmental effects and there will thus be a number of economic, political and environmental impacts resulting from the displacement of persons forced from their homelands (Westing, 1992; Tickell, 1990). Issues such as the sharing of resources between increasing numbers of persons in a region of immigration, land tenure, ethnic rivalry and regional conflicts are likely to emerge as issues that will need urgent attention in the near future.

Based on the conclusions of an international workshop held in Switzerland in September 2001 (Wengen Workshop, 2001), a number of issues were highlighted and possible avenues leading to solutions to alleviate certain problems were discussed. In particular, while a number of adaptive measures can be taken to reduce the adverse effects of environmental change and the potential for out-migration that such change may induce, there is a need to address some of the root issues in an internationally-coordinated manner. Immigration policies in the industrialized world need to be reviewed in order to allow some form of open, well-regulated migrations rather than solutions aimed at keeping migrants out of the prospective host countries. These will require a change of attitude within segments of the population of the host countries in terms of the acceptance of immigrants and their integration within their host society. In the developing world also, policies will need to be altered, in particular in order to remove the "push" factors of migration. A review of current resource-use practices, that are often very poorly managed and thereby generate considerable environmental degradation, will be needed. There is also a crucial necessity to improve land policies, valuation and property rights, in order to reduce the wish or the tendency for out-migration. Because the problem is one of balancing peoples' needs and wants with available resources, solutions will likely involve flexible policies not only related to population movement but also to movements of goods and capital to achieve efficient and equitable distributions.

In view of the current barriers to migration in most parts of the world, the social, economic and political consequences of migration at these scales is far from trivial. Most governments are today ill-equipped in legislative terms to deal with this type of situation, and it is at this level that urgent attention is already needed today to prepare for the future.

## REFERENCES

Armelagos, G.J. 1991. „The origins of agriculture: population growth during a period of declining health." *Popul. Environm.* 13:9-22.

Bordes, F. 1973. "On the chronology and contemporaneity of different Paleolithic cultures in France." Pp. 217-226 in *The explanation of culture change: Models in Prehistory*, editor C. Renfrew. London: Duckworth.

Castles, S. and M.J. Miller. 1993. *The Age of Migration: International Population Movements in the Modern World*. New York: The Guildford Press.

Cavalli-Sforza, L.L., P. Menozzi and A. Piazza. 1994. *The history and geography of human genes*. Princeton, USA: Princeton University Press.

Clottes, J. (ed.) 2001. *La Grotte Chauvet, l'Art des Origines*. Paris: Editions du Seuil.

Cohen, M. 1977. *The food crisis in prehistory*. New Haven: Yale Univ. Press.

Commoner, B. 1991. "Rapid Population Growth and Environmental Stress." *International Journal of Health Services* 21(2):199-227.

Cordell, L.S. 1984. *Prehistory of the Southwest*. New York: Academic Press.

Crowley, T.J. and G.R. North. 1991. *Paleoclimatology*. New York: Oxford University Press.

Davis, K. and M.S. Bernstam (eds.) 1991. *Resources, Environment, and Population: Present Knowledge and Future Options, Population and Development Review: a Supplement to Volume 16*. The Population Council, New York: Oxford University Press.

Elias, S. 1997. *The Ice-Age history of Southwest National Parks*. Washington DC: Smithsonian Institution Press.

Engelman, R. and P. LeRoy. 1993. *Sustaining Water Population and the Future of Renewable Water Supplies*.

Parry, M., C. Rosenzweig, A. Iglesias., G. Fischer and M. Livermore. 1999. "Climate change and world food security: A new assessment." *Global Environ. Change* 9:51-67.

FAO. 2000. *The State of Food and Agriculture*. Rome: FAO Agriculture Series.

Fornos, W. 1992. *Desperate Departures: The Flight of Environmental Refugees*. Washington DC: The Population Institute.

Gibson, K.R. 1996. The biocultural brain, seasonal migrations, and the emergence of the upper Paleolithic." Pp. 33-46 in *Modelling the early human mind*, editors P. Mellars and K.R. Gibson. Cambridge University Press.

Gleick, P. 1990. "Environment, resources and international security and politics." Pp. 501-523 in *Science and International Security: Responding to a Changing World*, editor E.H. Arnett. Washington DC: American Association for the Advancement of Science.

Harris, M. 1991. *Cannibals and kings, the origins of cultures*. New York: Vintage Books.

Homer-Dixon, T. 1991. "On the threshold: environmental changes as causes of acute conflict." *International Security* 16(2):76-116.

IPCC. 1996. *Climate Change. The IPCC Second Assessment Report*. Cambridge University Press, Cambridge and New York. Volumes I (Science), II (Impacts) and III (Socio-economic implications).

IPCC. 1998. *The regional impacts of climate change*. Cambridge and New York: Cambridge University Press.

IPCC. 2001. *Climate Change. The IPCC Third Assessment Report*. Cambridge and New York: Cambridge University Press. Volumes I (The Scientific Basis), II (Impacts, Adaptation, and Vulnerability) and III (Mitigation).

Jacobson, J. 1988. *Environmental Refugees: A Yardstick of Habitability*. Worldwatch Paper No. 86, Washington DC.

Jolly, C.L. May. 1993. "Four Theories of Population Change and the Environment." *Population and Environment* 16(1).

Jones, D. and L.S. Cordell. 1985. *The Anasazi world*. Portland, USA: Graphic Arts Center Publishing Company.

Kay, B.H. et al. 1989. „Rearing temperature influences flavivirus vector competence of mosquitoes." *Med. Vet. Entomol*. 3:415-422.

Loevinsohn, M. 1994. "Climatic warming and increased malaria incidence in Rwanda." *The Lancet* 343:714-718.

Lonergan, S.C. and D. Brooks. 1994. *Watershed: The Role of Freshwater in the Israeli-Palestinian Conflict*. Ottawa: IDRC Press.

Lonergan, S. and M. Parnwell. 1999. *Environmental degradation and population displacement*. Environment and Security.

Martens, P., L.W. Niessen, J. Rotmans, T.H. Jetten and A.J. McMichael. 1995. "Potential impact of global climate change on malaria risk." *Environmental Health Perspectives* 103:458-464.

McCorriston, J. and F. Hole. 1991. "Ecology of seasonal stress and the origins of agriculture in the Near East." *Amer. Anthropology* 93:46-69.

McMichael, A.J., A. Haines, R. Slooff and S. Kovats. (eds.) 1996. *Climate Change and Human Health*. Geneva, Switzerland: WHO/WMO/UNEP.

Mellars, P. 1991. "Cognitive changes and the emergence of modern humans in Europe." *Cambridge Archaeological Journal* 1:63-76.
Mirza, M.Q. 1997. "The runoff sensitivity of the Ganges river basin to climate change and its implications." *Journal of Environmental Hydrology* 5:1-13.
Monastersky, S. 2000. "Drowned lands hold clues to first Americans." *Science News* 157:85-90.
Myers, N. 1993. "Environmental refugees in a globally warmed world." *Bioscience* 43:752-761.
Noble, I. and H. Gitay. 1998. "Climate change in desert regions." Pp. 191-217 in *The Regional Impacts of Climate Change*, IPCC, editors R.T. Watson, M. Zinyowera and R. Moss. Cambridge: Cambridge University Press.
Parnwell, M.J.G. 1993. *Population Movements and the Third World*. London: Routledge.
Parry, M.L. (ed.) 2000. *Assessment of potential effects and adaptations for climatic change in Europe*. The ACACIA Report. Brussels: EU Publications.
Pavolov, P., J.I. Svendsen and S. Indrelid. 2001. "Human presence in the European Arctic nearly 40,000 years ago." *Nature* 413:64-67.
Pebley, A.R. 1998. "Demography and the Environment." *Demography* 35:377-389.
Schreir, H. and P.B. Shah. 1996. "Water dynamics and population pressure in the Nepal Himalayas." *Geojournal* 40(1-2):45-51.
Shiklomanov, I. 1993. "World freshwater resource." Pp. 13-24 in *Water in crisis: A guide to the World's freshwater resources*, editor P. Gleick. Oxford, UK: Oxford University Press.
Teitelbaum, M. and J. Winter. 1993. "The Missing Links: The Population-Environment Dynamics in Historical Perspective." Pp. 17-31 in *Population Environment Dynamics: Ideas and Observations*, editors G.D. Ness, W.D. Drake and S.R. Brechin. Ann Arbor, Michigan: The University of Michigan Press.
Tickell, C. 1990. "Environmental Refugees: The Human Impact of Global Environmental Change." In *Greenhouse Glasnost: The Crisis of Global Warming*, editor T. Minger. New York: The Ecco Press.
UNCWR. 1998. *The United Nations Conference on Water Resources*. Paris
UNHCR. 1993. *The State of the World's Refugees: In Search of Solutions*. United Nations High Commissioner for Refugees. Geneva, Switzerland.
UNHCR. 1995. *The State of the World's Refugees: In Search of Solutions*. United Nations High Commissioner for Refugees. Geneva, Switzerland.
UNHCR. 2000. *The State of the World's Refugees: In Search of Solutions*. United Nations High Commissioner for Refugees. Geneva, Switzerland.
Westing, Arthur H. 1992. "Environmental refugees: a growing category of displaced persons." *Environmental Conservation* 19(3):201-207.
WHO. 1990. *Potential health effects of climatic change*. Report of a WHO Task Group. World Health Organization. Geneva.
Wilson, M.L., B. Mahanty, A. Wannebo, P. MacDonald, A. Gleason, R. Smith and S. Aksoy. 2001. *Vector-borne Disease Associated with Irrigation, Agriculture, and Environmental Change in Southeastern Turkey: Application of Satellite Image Analysis*. Yale-New Haven Medical Center Report.
World Bank. 1995. *Towards environmentally sustainable development in Sub-Saharan Africa*. Washington DC, USA: The World Bank.

# INNOVATION AS AN ALTERNATIVE TO MIGRATION?

## *Exemplary Results from a Multiple-Agent Programming Model applied to Chile*

Thomas Berger
*University of Bonn, Center for Development Research*

## 1. INTRODUCTION

Several authors have questioned a direct relationship between environmental degradation and migration. In her literature review, Kliot (2001) argues that immediate causation is usually taken for granted but not accompanied by documented evidence. What is typically interpreted as forced environmental migration in developing countries is often an institutionalized mechanism to cope with resource scarcity. Households locate their members in different areas and labor markets so as to diversify risks that cannot be privately insured (Stark, 1991). Remittances may furthermore serve to finance investments in new technologies aimed at diminishing the household's dependency on fragile natural resources. Against this background, combining migration and innovation is a highly suitable household strategy, especially when households lack access to capital markets. Facilitating credit as a policy measure might therefore provide strong incentives for households to invest in innovations such as improved land use practices and to substitute for migration.

Usually, migration and innovation are associated with substantial structural changes in the rural economy and especially in agriculture. The number of farm households declines considerably, resulting in modified labor/land ratios. Land markets, as well as resource use efficiency, and finally local income levels are directly affected. The outcome of this complex adjustment process is not predictable in a straightforward way. Cumulative causation (or positive feedback) play important roles both for migration and innovation and impinge on the direction of the adjustment path. A conceivable outcome is a new equilibrium in the sending region, both in terms of environmental status and agricultural incomes. Another possible equilibrium outcome is unmitigated environmental depletion and total population displacement.



*J.D. Unruh, M.S. Krol and N. Kliot (eds.),*
*Environmental Change and Its Implications for Population Migration,* 25–46.

In the last decade, economists have intensively discussed such dynamic phenomena, usually referred to as multiple equilibria or path-dependency (Cowan and Gunby, 1996; Brandes et al., 1997). One implication is that policy measures can involuntarily lead to inferior development paths that are afterward difficult to abandon. Adjustment costs are then too high and cause a lock-in situation. Hazell and Fan (2001) also point to the policy-relevance of this issue in the context of environmental stress, innovation and migration. Though innovation in agriculture is a key strategy for many environmentally fragile areas with growing populations, policy-makers should avoid inadvertently locking too many people into these marginal areas when long-term prospects are limited.

Building quantitative models to forecast the households' responses to environmental changes and to identify policy interventions leading to lock-in still remains a challenge. An ideal model would incorporate biophysical as well as socio-economic processes and capture the dynamic effects of complementary migration and innovation decisions. It should make allowance for the potentially path-dependent feature of adjustment; be capable of exploring the likely impacts of different technology and policy options; and thus generate useful information for policy formulation and analysis. Migration and several crucial environmental processes in agriculture – such as soil erosion, nutrient leaching and flows of irrigation water – are spatial phenomena. An ideal model would also be capable of considering the related biophysical changes in a spatially explicit manner. Integrated simulation models based on the multiple-agent systems approach are promising candidates to meet these criteria. The next section discusses in more detail the basic theoretical concepts of positive feedback in migration and innovation processes. Section 3 briefly introduces multiple-agent systems and explains how this modeling technique has been used in agricultural economics. Section 4 describes the implementation of a multiple-agent model for a rural area in Chile characterized by water scarcity and high potentials for migration and innovation. Section 5 discusses some simulation results that show the model's applicability to policy-related research questions. The last section concludes with final remarks.

## 2. ENVIRONMENT, MIGRATION, AND INNOVATION

Isolating the influence of environmental stress on migration in developing countries poses conceptual difficulties, as a great deal of complexity characterizes the underlying decision-making processes. A multitude of factors and economic motives affect the households' choices among different alternatives of action. Typically, decisions have to be made in an information-poor setting under a considerable degree of uncertainty. Since information search costs are usually high, households have to rely and build their perceptions on rather vague judgments and the subjective experiences of others. Some authors, cited in Kliot (2001), therefore object to any attempt to draw a linear deterministic relationship between environmental degradation and population migration. Rather, a disaggregate micro-level approach is called for to disentangle the complex decision-making process.

Fischer et al. (1997) provide a comprehensive account of micro-economic research on migration decisions. To understand how households manage the information problem and how they arrive at their decisions, a behavioral model based on the theory of investment under uncertainty has been broadly applied (Dixit and Pindyck, 1994). As with long-term invest-ments in capital goods, migration and innovation typically imply high initial costs and uncertain later returns. Households compare the perceived costs and benefits of all possible investments according to their own decision rules. Following Stark (2001), three fundamental motives underlie the household decision rules in this context: (1) potential increase of income; (2) relative deprivation, i.e., the household's economic status compared to other households in the reference community; (3) personal desire or preference for migration or innovation independent of direct economic considerations. Though household decision-making is evidently a dynamic process, most studies on migration take a comparative-static view so as to reduce the complexity of analysis. For future research, Fischer et al. (1997) suggest putting more emphasis on the dynamic repercussions that earlier household decisions have on the determinants influencing the decision-making process of others.[1] The remainder of this section explains that one possible way of taking into account these dynamic effects builds on a model of information exchange, which has been, until now, applied separately to migration and innovation.

Faist (1997) describes the sociological concept of chain migration and illustrates the underlying mechanism by means of an S-shaped migration curve. Pioneer migrants, who maintain the social ties to their home community, are crucial for communicating the working and living conditions abroad. If the information they communicate is positive, this information contagion can start off a kind of self-sustained process leading to more and accelerated out-migration. The pioneers encourage their relatives and friends to migrate; these encourage other relatives and friends to migrate, and so on. The terms "pioneers" and "information contagion" are also well-known to agricultural economists, though in a slightly different context. Cochrane

(1979) developed a model of technical and structural change in agriculture, where farmers learn from the pioneers' experiences in new technologies. Since the "innovators" enjoy some additional profits through technology adoption, their competitive advantages on the land market improve over the "laggard" farmers who will then gradually be driven out of business.

Both approaches base themselves on the uncertainty-reducing effect of new information when a few pioneering households start engaging in a novel activity.[2] The location of particular households within the chain of information contagion can be estimated empirically with the so-called network threshold approach (Rogers, 1995). Making this communication effect endogenous to a micro-economic model of migration and innovation decisions would then enable forecasting of the likely impacts of conceivable environmental and policy changes. Accordingly, the idea that will be developed below in more detail is to extend Cochrane's approach of agricultural change by explicitly including migration, capturing the decision-making processes of all households involved, and applying this empirically-parameterized model to a potential sending region. As stated in the introduction, migration and several crucial environmental processes in agriculture are spatial phenomena that impinge on the choice of land and water use options. The biophysical environment in which potential migrant households interact should therefore be captured in a spatially explicit manner. A very effective way of encoding this integrated spatial model in computational form is employing a multi-agent system. The next section briefly introduces this novel modeling technique and describes how it has been used for the simulation of land-use/land-cover change.

## 3. AGENT-BASED MODELING OF LAND-USE / LAND-COVER CHANGE

Multi-Agent Systems (MAS) is a quite recent concept, originated in the computer sciences, that has rapidly diffused to other disciplines and is now more broadly applied to the analysis of complex systems. In the social sciences, MAS gave rise to a completely new field of research, namely, computer simulations of the social mechanisms that supposedly underlie human societies (Gilbert and Troitzsch, 1999). A prominent example is Sugarscape, an artificial society of simple computational agents that harvest and consume sugar, trade it with a second resource, migrate, reproduce and may even engage in tribal wars (Epstein and Axtell, 1996).

MAS is also of great interest for the empirical study of human-environment interactions, because it is highly suitable for representing interlinked socioeconomic and biophysical processes. Parker et al. (2002) review current applications of MAS to the modeling of human resource use decisions and subsequent land-cover changes. They define agent-based models of land-use/land-cover change as consisting of two key components. The first component is a cellular model that represents the landscape under study. This cellular model may draw on a number of specific modeling techniques such as cellular automata, spatial diffusion models, and Markov models. The second component is an agent-based model that represents human decision making and interactions. It consists of autonomous decision making units (computational agents), an environment through which the agents interact, rules that define the relationship between agents and their environment, and rules that determine the sequencing of actions. An agent typically represents a land manager who combines individual knowledge and values, information on soil quality and topography (the bio-physical landscape environment), and an assessment of the land-management choices of neighbors (the spatial social environment) to calculate land-use decisions. The model agents may also represent higher-level entities or social organizations such as village assemblies, local governments, or a neighboring country. In the place of differential equations at an aggregate level, agent-based models of land-use/land-cover change represent the decision rules, such as income maximization or minimum subsistence levels, of each human actor, their environmental feedbacks, and carryover of spatially distributed resources.

The particular strength of agent-based models is their suitability to account for heterogeneity and interdependencies among agents and their environment. The cellular model component provides a spatial framework to link biophysical simulation models with socio-economic decision models. Agent-based models are generally implemented via object-oriented programming languages, which provide an efficient and transparent way of organizing large amounts of data to handle complex model dynamics. Furthermore, their high degree of flexibility makes it possible to incorporate a wide range of agent decision rules. Especially in agricultural and resource economics, mathematical programming has been widely used to represent the microeconomic decision rules of real-world land managers (Hazell and Norton, 1986). Mathematical programming, in a nutshell, is a constrained optimization technique. It maximizes a function of independent variables (here: hectares of land put under certain land use types) subject to a priori limitations on the allowed values of independent variables (for example: hectares of land must be a nonnegative number and all farming activities together must not exceed the total area of arable land). The function to be optimized (the "objective" function) can reflect different decision rules. If for example profit-maximization is sought for, it corresponds to the sum of profits of all available land use activities. The solution to the programming problem is then the profit-maximizing ("optimal") choice of land use activities. In principle, a wide range of decision rules can be formalized in mathematical programming so as to reflect other and more complex goals and decision rules of real-world land managers.

The combination of mathematical programming – to mimic farmer's decision making – and agent-based modeling has been pioneered by Balmann (1997). He developed a hypothetical farm sector model and showed the theoretical effects that the spatial distribution of farms has on land rents and the speed of structural change in agriculture. As the next section shows, such integrated agent-based land-use models allow to parameterize a migration-innovation model with empirical data and to address the policy research questions outlined above.

# 4. MODEL IMPLEMENTATION IN CHILEAN CASE STUDY

This section briefly sketches a prototype MAS application to a rural area in Chile with high potential for both migration and innovation. For a listing of model parameters and equations consult Berger (2000).

## 4.1 Problem and Research Questions

As noted in the first section, agricultural intensification and, in particular, higher levels of efficiency in water and land use are key elements for improving the livelihood of rural households in potential sending regions. Both generally require some form of innovation, such as farm investments in superior land-use practices and irrigation methods, agricultural extension, and institutional changes. Several authors have argued that viewing land and water use improvements as exogenous technical change can result in misleading policy recommendations and certainly to an under emphasis on farm investment as a policy issue. In line with this argument, the model focuses on the diffusion of water-saving irrigation methods in a watershed; the effects of innovation and migration on the farm structure; and the impacts of possible government interventions aimed at supporting farm-households to improve their resource use efficiency.

## 4.2 Methodological Pre-Considerations

Though currently only a prototype, the model is in principle designed to provide policy-relevant information, especially regarding the impacts of policy on different farm and resource user groups. By means of computer simulations, it should facilitate the exploration of suitable policy options and forecasting out-migration and natural resource use changes. This explorative and predictive purpose has clearly impacted the level of abstraction and complexity in the representational model. It works at a highly disaggregated level, since the phenomena under study – diffusion of innovations, change in farm sizes and migration – require the modeling of heterogeneous farm-households and inter-household linkages. The spatial context figures prominently – e.g. upstream-downstream water uses, local water and land markets – and therefore spatial relationships must also be included.

Accordingly, several socioeconomic and biophysical processes such as decision making and interactions of individual agents, land markets, migration as well as irrigation water flows and agronomic relationships are endogenous to the model. However, socio-political phenomena such as rule formation, group decision making, and institutional change are treated exogenously.

## 4.3 System under Study

To test its applicability, the model was first applied to the Melado River Catchment in Chile with a size of about 670 $km^2$ and 5,400 farm holdings. Irrigation water is scarce and only sufficient for extensive cropping and livestock farming. An overall switch of production toward higher-value irrigation systems would require first the introduction of water-saving irrigation techniques and second the reallocation of water rights among farmers. Currently, many farmers grow traditional crops such as cereals with relatively inefficient irrigation techniques and accordingly make only limited use of their water rights. The situation might, however, change rapidly in the next years. In 1996 Chile signed an agreement with the South-American trade union "Mercosur" that will result in reductions of tariffs by 30%, on average, over a period of 17 years. As a consequence, relative prices in agriculture will change and considerably affect the profitability of different farming practices. The new market environment implies both strong incentives for shifting production systems toward high-value crops irrigated with modern water-saving technologies and disincentives for growing traditional crops with rather inefficient irrigation techniques. But it is questionable whether these incentives will also induce the traditional farm households to innovate and to modernize their farming systems. Instead, empirical studies based on household interviews have predicted a high potential for out-migration in this household group.

In order to capture the complete process of market integration in agriculture, the temporal period being modeled is 19 years – starting in 1997.[3] To model the adjustment at the farm-level, disaggregated land use types in agriculture and forestry are being included: 5 soil types, 3 technological levels, and 160 cropping, forestry, and livestock systems. Since the catchment's farmers only employ surface water for irrigation, and other water uses do not figure importantly, the model concentrates on surface water flows in agriculture. The model limits itself only to the farm-households and non-farm landowners who engage in land and water markets and whose plots belong to different irrigation sections within the Melado water user association. Each resource user – or household to be more precise – is represented individually – i.e. the model is disaggregated to the farm-household level. Other real world agents – such as farm workers and *minifundistas* with farmland of less than 2.5 ha – are not included since they do not contribute significantly to the resource use decisions and market dynamics.

## 4.4 Model Implementation

The spatial resolution at which the model operates is 158 * 158 m – i.e. the size of one grid cell is 2.5 ha –, and the time step is one month. This rather fine spatio-temporal resolution had to be chosen because rented farm plots are typically of this size, and crop water requirements are usually modeled based on a monthly time interval.

The model contains three basic functional types of agents that stand for campesino family farms, commercial farm holdings and non-agricultural landowners. Empirical analysis in Chile revealed that both holding types represent two distinct communication networks. In each network, five subgroups were identified corresponding to different positions within the chain of information contagion. The position of a particular household is measured by its network threshold, defined as the percentage of all other households within its reference group that must previously engage in a novel activity before the household eventually adopts this behavior (Valente, 1995). Innovators, for example, adopt a new technology when the percentage of adoption in their communication network is still low; they have low thresholds. Laggards with a relatively high demand for interpersonal information adopt the same technology only when the percentage of adoption is close to hundred percent; they have high thresholds. These empirically estimated thresholds can be used to predict the communication of information in similar decision problems (Berger, 2001). The same applies, in principle, to migration decisions.[4]

The agent's decision rules, chosen to represent the behavior of real farm-households, are based on the general framework of maximization of expected utility. The household agents allocate their labor and land resources so as to maximize the expected household income over time. The decision model abstracts from intra-household bargaining and assumes a household head that behaves altruistically on behalf of all household members. In comparison with other decision models in the migration literature, this model is dynamic. The household agent decides on whether to move or to stay at the end of each year, after the harvesting is done. An entire agent household sells out all its assets and moves out of the region either because of earning differentials (voluntary migration) or insolvency (forced migration). Here, insolvency means the inability to attain a minimum subsistence income. Since environmental changes directly translate into production costs, the household agent attempts to anticipate them and forms expectations. The decision model captures some of the uncertainty of migration and innovation decisions by relating them to the experiences of network partners.

Household members or entire agent families move to other locations or adopt new technologies when the expected increase in family income is positive and a sufficient proportion of peers in their network has previously moved or adopted. The dynamics of the agent's decision making is captured through comparisons of adoption levels for migration or new technologies with the agent's network threshold. In the Chilean model the decision rule at the agent level is as follows: (1) monitor the present adoption level and compare it with the individual threshold; (2) if the network threshold is reached, calculate the household's net benefits from adoption; (3) if the expected net benefits are positive, then adopt. By varying parameters such as network thresholds, expectation coefficients, and household-specific opportunity costs, different motives for migration and innovation can be implemented. If the model households, for example, compute the expected incremental income in relation to the average income in their reference group, the effect of relative deprivation is captured. As a special case, the model households may also behave according to the standard economic theory, which implies they have complete information and perfect foresight with respect to farm prices; maximize expected income; and migrate whenever their market opportunity costs are higher.[5] As will be shown in the next section, comparing the standard economic scenario with frequency-dependent scenarios reveals multiple equilibria and potential lock-in and thus provides useful policy information.

Besides innovation and migration, the decision-making process of real world farmers also contains rather simple problems that are frequently repeated and involve much lower degrees of uncertainty. Such problems are, for example, choice of crops, distribution of water for irrigation, and renting of plots. Exchange of peer-to-peer information is usually not a pre-requisite for decision making. Hence, representing these decision problems does not require network thresholds; maximization of expected income as in the standard economic approach will in most cases be a good approximation of real farm household behavior.

The representational model of decision making is encoded in computational form by means of a recursive whole-farm mathematical programming routine (Hazell and Norton, 1986). Each farm-household agent has its own objective function, resource constraints and updates its expectations for prices and water availability. A mixed-integer linear programming solver is used for the farm investment and land rental decisions. In this respect, the model has characteristics similar to the independent representative farm models described by Hanf (1989). However, there are two important features that distinguish the present model from the conventional independent farm approach: (1) there is a one-to-one correspondence between real-world and modeled agents, that is, each single real world farm-household is represented by a single model agent; (2) several types of interactions among agents are endogenous to the model, such as contagion of information, exchange of land and water resources, and return-flows of irrigation water.

This one-to-one MAS representation facilitates the consideration of agricultural production at a very fine spatial resolution, as well as the representation of bilateral and direct interactions between agents. Including these direct interactions among agents broadens the scope of resource use modeling significantly because – apart from the frequency-dependent effects in migration and innovation – other economic phenomena that conventional models cannot easily address are now explicitly modeled.

First, as has been maintained above, migration and innovation affect the local land markets; in some areas the level of rent tends to rise, in others, to decrease. Around each plot being offered on the land market, internal transport costs from the plot to a farmstead shape a kind of "von Thünen ring." As a consequence, only a small number of neighboring farmers compete for a plot, which may lead to excessive land prices when several farmers with high land productivity attempt to simultaneously expand the size of their farm. Internal transport costs thus impinge on the level of rent by limiting competition on the land market. Here, the model captures the agents' location and internal transport costs through a raster-based geographical information system. Each grid cell corresponds to one farm plot held by one single landowner. This direct ownership representation was chosen to implement land and water markets in a spatially explicit way. Due to internal transport costs, only neighboring model agents compete for each offered plot. Finally the agent with the highest bid receives a particular plot, providing his bid is higher than the asking price.[6]

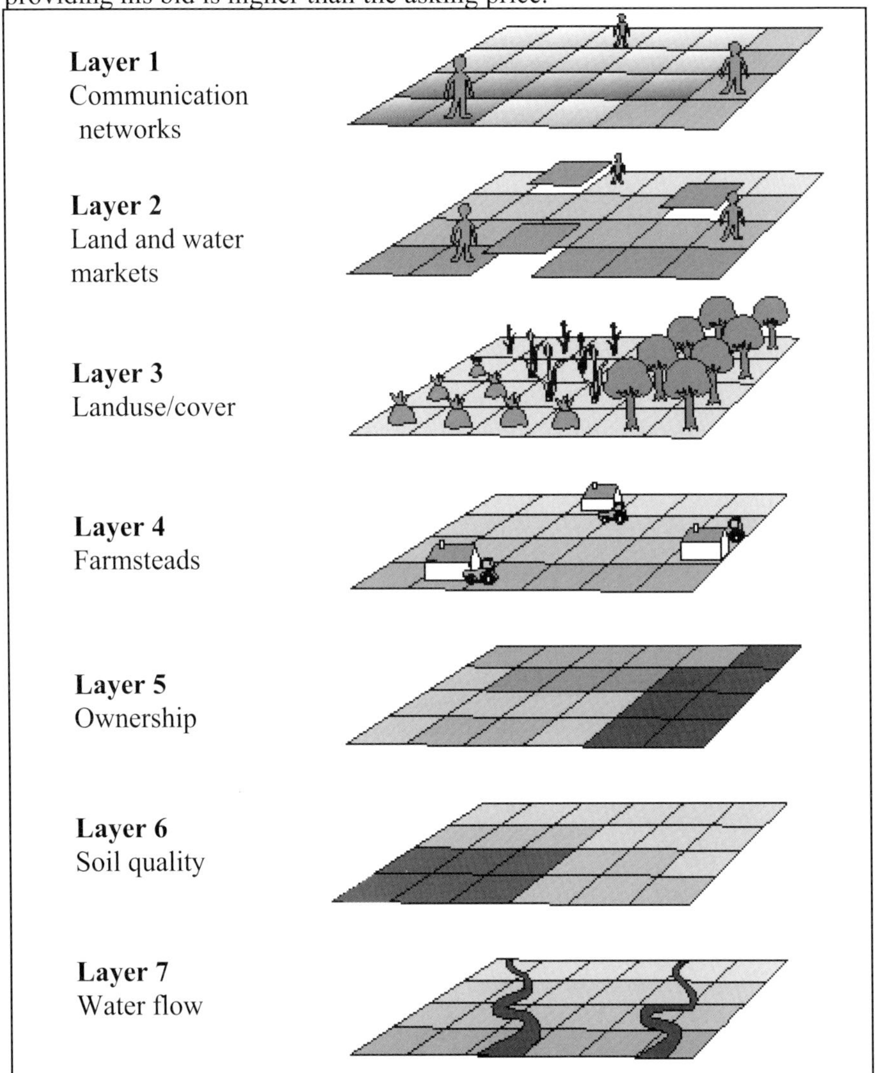


*Figure 1: Spatial data representation and interdependencies.*

Second, feedbacks that stem from the spatial distribution of irrigation water flows are included in the model. Monthly return flows affect downstream water availability, and may force model farmers to temporarily undersupply their crops or even to abandon them completely. The model farmers then have an incentive to employ more efficient, water-saving irrigation technologies. In reality, water shortages usually hit downstream farmers harder than upstream farmers, because upstream farmers often take more water than their irrigation quotas allow. The model reflects either a perfect water allocation – the farm agents receive their quota of irrigation water exactly – or more realistically, at least in the Chilean context, deficiently enforced water rights where parts of the return flows are uncontrolled. The spatial interactions of the water resources system are represented at a much coarser scale than the ownership of parcels, because grid cells are grouped to hydrologic units of an average size of about 32 $km^2$.

Figure 1 summarizes the spatial data representation together with the heterogeneity, interdependencies and hierarchies of the model. There is spatial heterogeneity (soil quality, irrigation water supplies, ownership of land parcels and water user rights), technological heterogeneity (farming equipment of different technological levels), and social heterogeneity (different managerial capacity, several social networks). Interdependencies are spatial (return flows, land and water markets) and social (communication networks). The land cover/use and water supply of a particular grid cell are results of the decision-making process at the farm level, which reflects technical, financial and higher-level social constraints.

## 4.5 Verification and validation

Having calibrated the model to a base year, validation experiments were performed to test the model's validity. As the model operates on various scales simultaneously, a previous aggregation of input data to one common level of aggregation was not necessary. This implies the thorough testing of its ability to approximate real-world observations on the micro-level (farm-households), meso-level (hydrologic units) and regional-level (river catchment). Since reliable remote sensing data were not available at that time, only aspatial statistical analyses were conducted that revealed a sufficient "goodness of fit". Since the model has many degrees of freedom and contains highly recursive dynamics, extensive robustness experiments and statistical tests were also conducted. Finally, comparisons of performance with other models and expert validation helped to create trust in the model's behavior and results.[7]

The often quoted advantage of the mathematical programming approach in merging different data sources was fully exploited (Hazell and Norton, 1986: 3). An extensive farm-household survey, in-depth interviews, social network analyses and results from farm trials were used to derive a consistent farm data set. Based on a water engineering study for the Chilean Ministry of Public Works, the hydrologic units, equations and model parameter were defined. Spatial data at the hydrologic unit level had to be disaggregated to plot level using a random data generator constrained by *a priori* information. The registry of the local water user association was consulted to assign water rights to model agents.

## 4.6 Technical aspects

The author developed the multiple-agent programming software for this study, drawing on Balmann's (1997) source code. The new source code is written in the C++ object-oriented programming language and has MS-Windows 32 bit and UNIX portability. Input and output files are in ASCII-text format and can be processed with common spreadsheet and graphics programs (http://www.zef.de/mas.htm).

Usually, encoding heterogeneity in the carryover of farm resources, different technical coefficients, interest rates, objective functions as well as storing spatial data poses difficulties in farm programming models. The object-oriented programming language, in contrast, allows a much more transparent organization of data and model dynamics. By implementing agents as objects, the computational model can be encoded in a clear modular form. Using an object-oriented programming language typically reduces model development costs and numerical difficulties. As Harrington (1995) shows with the instructive example of a simple program for calculating debt servicing, the object-oriented implementation considerably increases the extendibility and portability of previous verified source code. The code of Berger's (2000) multi-agent model can therefore be extended relatively comfortably by ecological constraints or interfaces with GIS-applications, for example.

# 5. DISCUSSION OF SIMULATION RESULTS

By representing the resource users' own decision making in a spatially explicit way, the multi-agent model forecasts competing land and water uses over time and investigates in particular the role of innovation and migration in agriculture. Some simulation results that demonstrate the type of information the model generates regarding the following topics will be briefly discussed:

- What is the range of possible water and land use changes? How fast will water-saving irrigation methods diffuse?

- Will these innovations reach the traditional farmers? Will they generate additional incomes and induce them to stay?
- Will the situation on marginal lands deteriorate? Will the farm income on these lands change in comparison to the regional average?
- Under what conditions will migration increase or decrease? Can innovation be an alternative to migration?

Figure 2 shows the "possibility space" of water/land-use changes by comparing the frequency of several irrigation methods with different on-field efficiencies. The right-hand graph reflects a hypothetical situation without technical change, where farmers are reluctant to innovate and refuse any technology adoption.[8] The left-hand graph reflects the boundary scenario of ideal technical change where the model's farmers adjust smoothly as predicted by the standard economic models.[9] Ideal technical change leads to a sizeable expansion of modern water-saving irrigation within ten years.

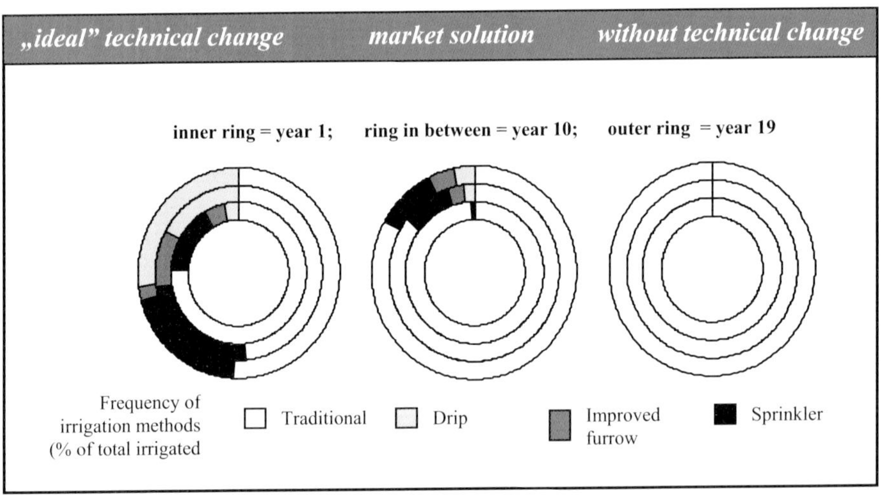


*Figure 2: Frequency of water-saving irrigation techniques under different technological scenarios.*

Almost half of the irrigated area would then be efficiently irrigated; the rest are soils of poor quality where only extensive rain-fed land uses, such as grasslands and forest, are profitable. In contrast, the middle graph shows the expansion of modern irrigation techniques under "bandwagon" conditions, that is, when the model farmers rely on interpersonal communication and learn from their peers' experiences.[10] The diffusion of water-saving innovations is then significantly slowed down and reaches only a sixth of the irrigated area over twenty years. Many model farmers with high network thresholds do not adopt these irrigation technologies, though they would under ideal conditions. One might therefore conclude that agricultural water use is locked into low efficiencies and could call for policy interventions to speed up the diffusion of water-saving innovations. In other scenarios, not shown here, the effects of different policy programs, such as the one demanded by the Chilean farmers association, were analyzed. The program includes special credit schemes to facilitate the adoption of water-saving innovations, public investments in irrigation facilities as well as fertilizer subsidies. A comparison of this program with other programs showed that temporary path-dependencies can eventually be broken down but may demand considerable financial resources.

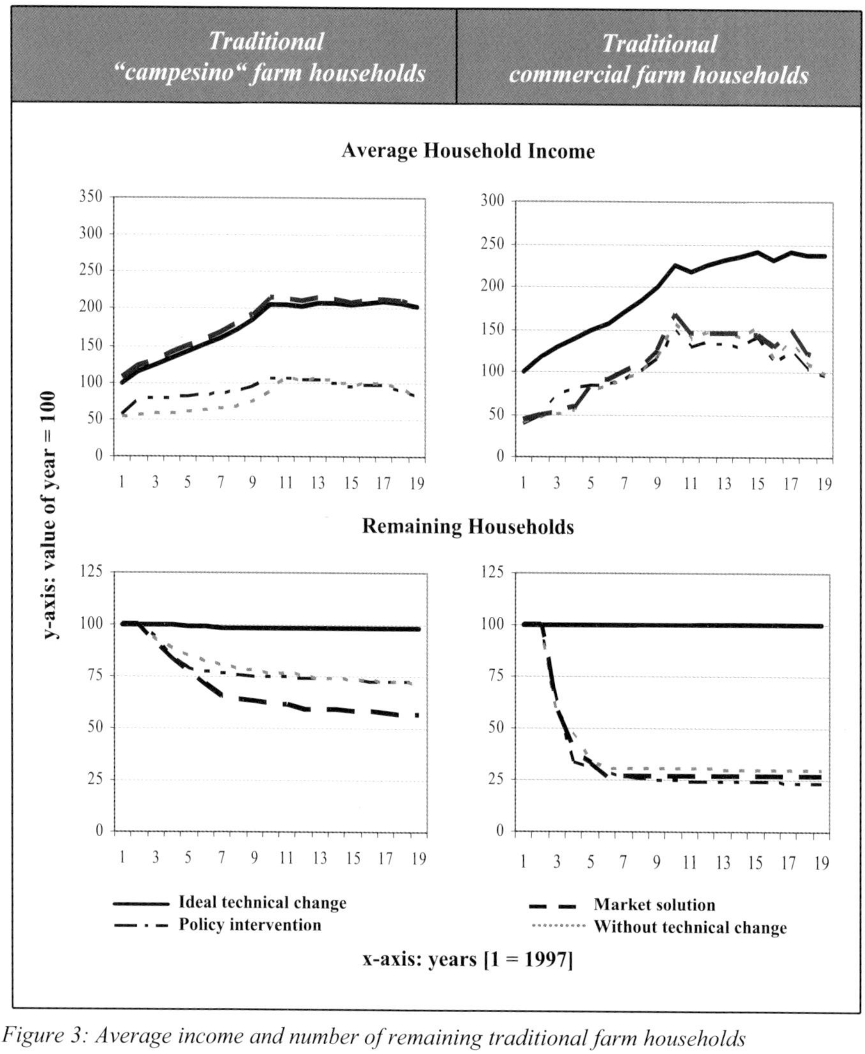


*Figure 3: Average income and number of remaining traditional farm households*

Figure 3 shows the effects of technical change on the traditional farm households in both communication networks. Under ideal technical conditions, traditional households adopt water-saving innovations and almost double average family income as compared to the scenarios without technical change. As the lower left-hand graph indicates, they have no incentive to move and remain in the farming business. Under market conditions, however, the picture changes completely. In contrast to other household groups, technical change does not reach traditional households and many model households decide to migrate permanently. Approximately 25% of the traditional *campesino* and 75% of the traditional commercial households move out. Although traditional *campesino* and commercial households behave similar under bandwagon market conditions, they respond differently to policy interventions.

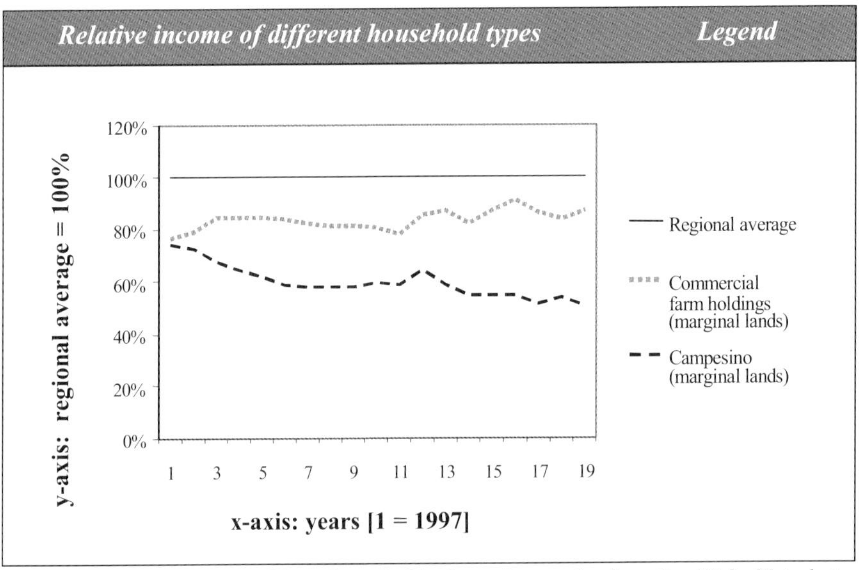


*Figure 4: Relative income of farm-households depending on land quality ("ideal" technical change).*

For example, the program of the Chilean farmers association only slightly affects the average household income and the rate of out-migration of traditional commercial households. But the same policy intervention induces a notable response of traditional *campesino* households. The average household income rises to levels as high as under ideal technical conditions and out-migration increases significantly. The explanation is that traditional *campesino* households are relatively deprived through policy intervention because the average farm income of the more innovative households in their network groups increases more than proportionally. As a consequence, the traditional *campesino* households with the lowest farm incomes move out and thereby raise the average income of the remaining households in this group.

Figure 4 indicates that the relative income of households operating on marginal lands will likely deteriorate over time. In the ideal technical scenario, the relative income of the mean *campesino* farm-household declines progressively and finally reaches only 50% percent of the regional average in this holding type. In the group of commercial holdings, differences in relative income are less pronounced but still amount to 15% percent. The relative deterioration of household incomes is also accompanied by significant changes in the number of farms. For example, among the commercial holdings on marginal lands, almost 4.6% of the traditional farms go out of business per annum and migrate to other destinations. In the bandwagon scenarios, income differences increase even further. These rather discouraging forecasts suggest policy interventions are needed to prevent an increasing gap between the incomes of households farming marginal and average lands. As already mentioned in the first section, policy interventions often imply the risk of locking too many farm-households into these fragile areas. When long-term prospects are limited, switching to alternative paths might become more difficult than without previous policy interventions. Again, the model helps to explore the dynamic effects of alternative policies and can thus inform policy formulation. In the Chilean case, potential path-dependencies could be found in the diffusion of innovations, but social hardships seem not to arise. For more details on the policy analysis and especially the role of land/water markets refer to Berger (2000).

Finally, Figure 5 illustrates the interplay of migration and innovation measured by the labor employed in agriculture. This regional indicator reflects all forms of migration – in and out, temporary and permanent – and is here compared with the labor capacity of all farm-households in 1997, the simulations' starting period. In the boundary scenario without technical change, approximately one third of the initial labor force is used off-farm and migrates to the non-agricultural sector permanently or temporarily. Ideal technical change, in contrast, attracts additional farm labor and converges to a higher equilibrium level of labor allocation in agriculture. Innovation is therefore an alternative to migration and has the potential of turning a sending region into a receiving region. But if the farm-households continue to depend on their communication networks in decision-making – the market solution with bandwagon –, then off-farm labor allocation only diminishes slightly over time, i.e. the effects of innovation and migration almost compensate. In this respect, the simulation results underline the importance of interpersonal communication and of cumulative causation in the process of innovation and migration.[11] Farm labor as an indicator for migration, however, does not capture the underlying structural change in the farm sector. As indicated in Figure 3, many traditional farms move out of the region, especially those in the group of commercial holdings. Growing farm holdings, especially in the middle size classes, absorb these lands, adopt new technologies and employ temporary farm workers. See Berger (2000) for a more detailed discussion of these and other indicators of structural change especially those related to land and water markets.[12]

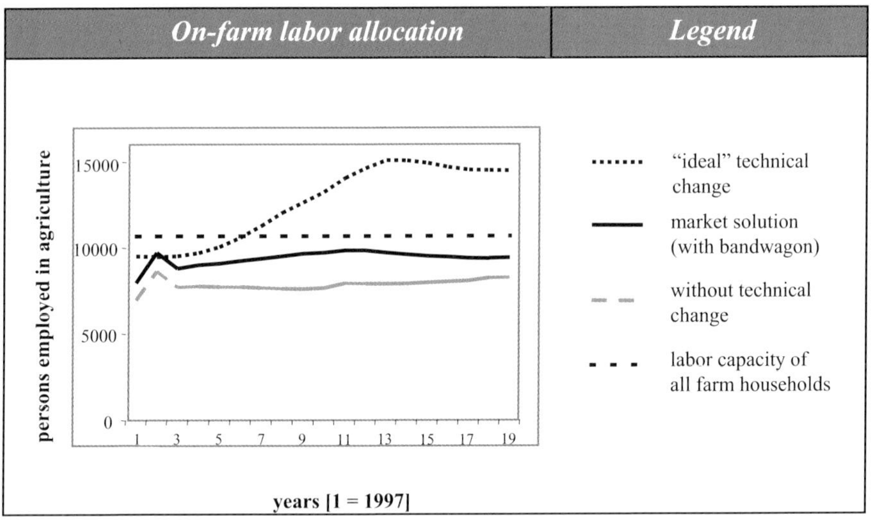


*Figure 5: Farm Labor employed in the study region*

## 6. CONCLUSIONS

This paper presents an integrated simulation model that addresses the complex relationship between environmental stresses, migration and innovation. In line with the literature of new economics of migration, the model focuses on the decision making of rural households in developing countries who regularly make use of migration and innovation to cope with environmental hazards. It also considers the uncertainty-reducing effect of information becoming available when a few pioneering households start engaging in novel activities. Since migration and several crucial environmental processes in agriculture – such as evapotranspiration and flows of irrigation water – are spatial phenomena, the model integrates these biophysical processes in a spatially explicit way. By representing the farm-households' decision making and their local environment over time, the model forecasts land and water use changes that might emerge under different technological, political and environmental scenarios.

The model is encoded as a multiple-agent system, a relatively new concept of implementing complex computer models with the help of object-oriented programming languages. There is a one-to-one correspondence between real-world and modeled agents, that is, each single real world farm-household is represented by a single model agent. Several types of interactions among agents are endogenous to the model such as the contagion of information, exchange of land and water resources, and return-flows of irrigation water. A recursive whole-farm mathematical programming routine is used to mimic the decision making of farm-households. This one-to-one multi-agent representation facilitates examining the spatial context of agricultural production at a very fine resolution, as well as bilateral and direct interactions between agents. Including these direct interactions among agents broadens the scope of resource use modeling significantly because – apart from the frequency-dependent effects in migration and innovation – other economic phenomena such as the limited competition on land markets and upstream-downstream trade-offs in irrigation are now explicitly modeled.

To test its applicability, a model prototype is applied to the Melado River Catchment in Chile. Irrigation water is scarce and only sufficient for extensive cropping and livestock farming. An overall switch of production toward higher-value irrigation systems would first require the introduction of water-saving irrigation techniques and second the reallocation of water rights among farmers. Currently, many farm-households grow traditional crops with relatively inefficient irrigation techniques and locate some of their members in different areas and labor markets. Empirical studies have also found a high potential of out-migration among the traditional farm households.

By including networks of communication in the model, the simulation experiments indicate that interpersonal communication slows down the diffusion of innovations significantly and thereby accelerates out-migration. Far fewer farm-households remain in the farming business and adopt water-saving technologies than predicted by the standard modeling approach. One might therefore conclude that agricultural water use is locked into low efficiencies and that policy interventions to speed up the diffusion of water-saving innovations are called for. The paper reports some results of the policy analyses and illustrates the income effects on marginal lands as compared to the regional average. Under prevailing environmental conditions, innovation and migration seem to act as antagonists. Favorable conditions for technology adoption lead to increased employment in agriculture and might even turn a potential sending region into a receiving region.

The pilot study demonstrates the usefulness of the multi-agent programming approach in incorporating the complexity of humans' responses to environmental changes. The approach integrates biophysical as well as socio-economic processes and captures the dynamic effects of complementary migration and innovation decisions. It makes allowance for potentially path-dependent adjustments; is capable of exploring the likely impacts of different technology and policy options; and generates useful information for policy analysis. Further testing of this model class and coupling with models of regional environmental change is called for.

## NOTES

[1] Fischer et al. (1997) name two related fields of research that are also to a great extent unexplored: the behavioral dynamics of the migration decision itself – when, why and how often do individuals ask themselves whether to migrate –, and the formation of individual expectations about disadvantages and advantages of migration – how is information gathered and up-dated. This paper focuses only on inter-household linkages because this approach can be grounded on empirical parameters and implemented in a straightforward way in a computer simulation model.

[2] This uncertainty-reducing effect figures importantly in different strands of literature such as technology diffusion (Metcalfe, 1988), path-dependence (Cowan and Gunby, 1996), and social networks (Valente, 1995).

[3] Choosing the starting point in potentially path-dependent systems may predetermine the development path and thereby affect the model forecasts. Numerical robustness tests with random parameters showed, indeed, the model's sensitivity to starting conditions on the individual household level. On the level of household types, however, where policy impacts are usually evaluated, the absolute forecasted values differ, but the relative trends are uniform. As a consequence, the model seems robust enough to compare different scenarios under identical starting conditions. More details on the robustness experiments and supportive statistical tests can be found in Berger (2000).

[4] To the author's knowledge, there is yet no empirical study available supporting the threshold model of adoption for migration. The Chilean model parameters for migration are therefore based on ad hoc values and tested with sensitivity analyses. The simulation results that will be presented below in the next section, are derived from a model specification with two sub-networks for innovation decisions but only one overall network for migration decisions.

[5] To be precise, they almost behave like rational decision makers with perfect foresight. Since the model contains non-convexities, the employed decision making routines may not under all circumstances converge toward the global optimum. 'Market opportunity costs' refer here to the household income that could be obtained if the household closed down its farm and migrated to a new destination.

[6] The agents base their bid on economic calculus and compute the parcel's "shadow price" (Berger, 2000).

[7] More details on verification and validation are given in Berger (2000).

[8] This scenario is implemented by setting all network thresholds to hundred percent.
[9] Here, smoothly means that interpersonal communication is ignored by setting all network thresholds to zero. Hence, this scenario corresponds to the so-called equilibrium diffusion concept that postulates a priori complete information sets (Metcalfe, 1988). Differences in adoption behavior are explained by indivisibilities and minimum farm sizes.
[10] The bandwagon scenario reflects the disequilibrium diffusion concept (Metcalfe, 1988) and employs network thresholds as estimated in the empirical network study. See Berger (2001) for more details.
[11] Policy interventions that facilitate favorable conditions for technology adoption, especially investment and extension programs, and thereby encourage the employment of additional labor in agriculture are discussed in Berger (2000).
[12] Since empirical data were not available on how farm-households in the study region respond to rising opportunity costs, several model parameters are still based on ad hoc assumptions. A follow-up study will aim at clarifying these simulation results.

# REFERENCES


Balmann, A. 1997. "Farm-based modelling of regional structural change: A cellular automata approach." *European Review of Agricultural Economics* 24: 85-108.

Berger, T. 2000. Agentenbasierte räumliche Simulationsmodelle in der Landwirtschaft. Anwendungsmöglichkeiten zur Bewertung von Diffusionsprozessen, Ressourcennutzung und Politikoptionen. *Agrarwirtschaft Special Issue* 186, Agrimedia.

Berger, T. 2001. "Agent-based spatial models applied to agriculture: a simulation tool for technology diffusion, resource use changes and policy analysis" *Agricultural Economics* 25 (2/3), 245-260.

Brandes, W., G. Recke and T. Berger 1997. *Produktions- und Umweltökonomik – traditionelle und moderne Konzepte*. Stuttgart.

Cochrane, W.W. 1979. The *Development of American Agriculture*. Minneapolis: University of Minnisota Press

Cowan, R. and P. Gunby 1996. "Sprayed to Death: Path dependence, Lock-In and Pest Control Strategies." *Economic Journal*, 106: 521-542.

Dixit, A.V. and R.S. Pindyck 1994. *Investment Under Uncertainty*. Princeton, New Jersey: Princeton University Press.

Epstein, J.M. and R. Axtell 1996. Growing Artificial Societies. Social Science from the Bottom Up. Washington, DC: Brookings Inst. Press.

Faist, T. 1997. The Crucial Meso-Level. In: Hammar, T. et al. (eds. *International Migration, Immobility and Development. Multidisciplinary Perspectives*. Berg, Oxford, 187-218.

Fischer, P.A., R. Martin and T. Straubhaar 1997. Should I Stay or Should I Go? In: Hammar, T. et al. (eds.) *International Migration, Immobility and Development. Multidisciplinary Perspectives*. Berg, Oxford, 49-90.

Gilbert, G.N. and K.G. Troitzsch 1999. *Simulation for the Social Scientist*. Buckingham: Open University Press.

Hanf, C.-H. 1989. *Agricultural Sector Analysis by Linear Programming Models*. Kiel: Wissenschaftsverlag Vauk.

Harrington, J.L. 1995. *C++ and the Object-Oriented Paradigm. An IS Perspective*. New York: Wiley.

Hazell, , P.B.R. and S. Fan 2001. Balancing Regional Development Priorities to Achieve Sustainable and Equitable Agricultural Growth. In: Lee, D.R. and C.B. Barett (eds.). *Tradeoffs or Synergies? Agricultural Intensification, Economic Development and the Environment*. Wallingford, New York: CABI Publishing.

Hazell, P.B.R. and R.D. Norton 1986. *Mathematical Programming for Economic Analysis in Agriculture*. New York: Macmillan.
Kliot, N. 2001. Environment, Migration and Conflict: A Critical Review. Paper presented at the Wengen 2001 Workshop on Migration and Environmental Change, Wengen Sept. 19–21
Metcalfe, J. S. 1988. The Diffusion of Innovations: An Interpretative Survey. In: Dosi, G., et al. (Eds.). *Technical Change and Economic Theory*, 560-589.
Parker, D., T. Berger and S. Manson (in press.) *Agent-Based Models of Land Use / Land Cover Change*. LUCC Report Series No. 6. LUCC International Project Office, Louvain-la-Neuve.
Rogers, E. 1995. *Diffusion of Innovations*. 4. Ed., New York: The Free Press
Stark, O. 1991. *The Migration of Labor*. Cambridge, Mass: Basil Backwell.
Stark, O. 2001. *On the Economics of Migration: Beyond Wage Differentials*. ZEF Public Lecture on the occasion of receipt of the Humboldt Prize, Center for Development Research, Bonn, March 20, 2001.
Valente, T.W. 1995. Network Models of the Diffusion of Innovations. New Jersey: Hampton Press.

# HUMAN CAPITAL AND INDUCED AND FORCED MIGRATION

Raymond Dacey
*University of Idaho, Moscow, Idaho 83844-3161*

## 1. INTRODUCTION

Many kinds of individuals migrate for many reasons. This paper is focused on a particular group of individuals and a set of fairly specific reasons. The analysis presented here is focused on individuals who possess human capital and are induced to migrate. An individual possesses human capital if he or she is educated or trained to the level of professional or skilled performance. For example, such people include doctors, lawyers, engineers, scientists, scholars, electricians, graphic and industrial designers, and computer technicians.

Induced migration occurs in one of two basic ways. An individual is induced to migrate if he or she is "pushed" out of his or her home country and/or is "pulled" toward a target country for any combination of environmental, political, economic, religious, or social reasons. The extreme form of induced migration is forced migration. An individual is forced to migrate if remaining in the home country is not an option. Forced migration due to environmental change was the primary concern of the Wengen-2001 Workshop on Environmental Change and Population Migrations. Environmentally forced migration is a particular form of forced migration wherein the option of remaining in the home country is not available due to environmental change. This paper presents models of induced and forced migration, and derives results of relevance to both individuals facing the migration decision and the governments of the countries involved in those decisions.

Most analyses of immigration view the immigrant as an invader and the government of the target nation as a gatekeeper (Borjas 1999, 2000; Faist 2000; Thomas 1973). These analyses view the immigrant not as an individual decision maker, but a member of a flood, and the national government not as a recruiter, but as the defender of a land under invasion. Luterbacher and Theler (1992) provide a game-theoretic account of the invasion model. International trade theory and growth theory show that the wealth of a nation depends upon the human capital it can produce or attract and retain (Grossman and Helpman 1991; Helpman 1981; Helpman and



*J.D. Unruh, M.S. Krol and N. Kliot (eds.),*
*Environmental Change and Its Implications for Population Migration,* 47- 68.

Krugman 1985, 1989; Krugman 1990, 1991, 1992; Kwok and Leland 1982; Romer 1986, 1987, 1990). From this perspective, the immigrant possessing human capital is a prize to be won and the national government is a recruiter. This view supports one simple claim - a key component of a developed nation's policy regarding international trade and domestic economic growth is that nation's policy toward the development of its human resources through education (i.e., production) and immigration (i.e., acquisition). Put differently, the fundamental immigration policy examined here is one that is designed to improve (or at least maintain) the well-being of the government in office. (See, e.g., Borjas 1999, pp. 174f for a discussion of who benefits from an immigration policy.)

In the standard view, the developed countries provide the "pull" via higher wages and better living conditions and the home countries provide the "push" through environmental, economic, political, and/or social degradation. However, the home country can provide both a push and a pull. Skilled workers are particularly valuable to the home country, and their loss via migration tends to add to the degradation that provided the initial push. Thus, a home country is naturally interested in retaining or recapturing citizens who possess human capital, and can exert a domestic "pull" of its own, through government action (Rogers 1981).

The view taken here on the attraction and retention of migrants is not new. Almost thirty years ago Brinley Thomas put this point as follows: "Advanced countries are keen to attract qualified manpower and they erect barriers against the entry of the unskilled; the actions of governments suggest that immigration policy has come to resemble tariff policy as an instrument for the pursuit of national gain ..." (1973, p. 306). The logic behind this view rests on the simple observation that there are relatively few individuals in the world who possess human capital and that these individuals, when successful, increase the welfare of the country in which they work. Therefore, the number of such immigrants will be rather small, and the number of citizens threatened by the immigration of such individuals is too small to have negative political consequences for a government. However, an immigrant who possesses human capital but employs it unsuccessfully is likely to displace a native worker who possesses only labor. Since there are many such individuals, the bloc of citizens threatened by the immigration of these workers is large enough to have serious negative political consequences for a government. Therefore, the government of a target country finds it beneficial to attract immigrants who possess human capital, to retain those who are successful, but to repel those immigrants who are unsuccessful.

Migration is generally viewed as being in one of two distinct categories - forced and voluntary migration. The distinction usually rests on the presence or absence of physical coercion "because nation-states use distinct criteria to separate forced migrants as refugees from voluntary migrants as labour migrants" (Fiast 2000, p. 23). This distinction is too severe for our purposes. For example, an ecological or economic disaster causes people to migrate but does not involve physical coercion. Our concern here is with the distinction between induced migration, wherein "the migrants retain some decision making power to decide whether or not to leave" and forced migration, wherein the migrants do not have this power (Petersen 1958, p. 261). The distinction between induced and forced migration can be modeled via decision problems wherein the potential migrant does and does not have the option of remaining in, or returning to, the individual's home country.

This paper presents a basic model of the economic migration decision problem. The model permits the analysis of the kinds of policies that generate the push and pull forces of relevance to the resolution of the migration decision. The paper is structured as follows. Section 2 presents the sequential migration decision problem with the possibility of returning to the home country. This model is sufficiently rich to permit an analysis of the push and pull forces that operate on an individual who possesses human capital. The model is also sufficiently flexible to support an analysis of forced migration. Section 3 presents an analysis of forced migration. Section 4 presents some concluding remarks. Section 5 discusses some extensions and limitations of the model.

## 2. THE MIGRATION DECISION PROBLEM

An individual can migrate, and, having migrated, can re-migrate either by moving to another country or by returning to the home country (Thomas 1973, pp. 308-309). The decision tree for the general migration problem is very large, and is not particularly informative.

To simplify the problem I will presume that the individual migrates at most twice, that the first decision involves a choice between migrating to another country and remaining in the home country, and that the second decision involves a choice among migrating to a second country, remaining in the new country, and returning to the home country. The second decision is made on the basis of the individual's experience after migrating. I will label the countries as 'home', 'country #1', 'country #2, and 'country #3', respectively. The decision tree for this problem is presented in Figure 1.

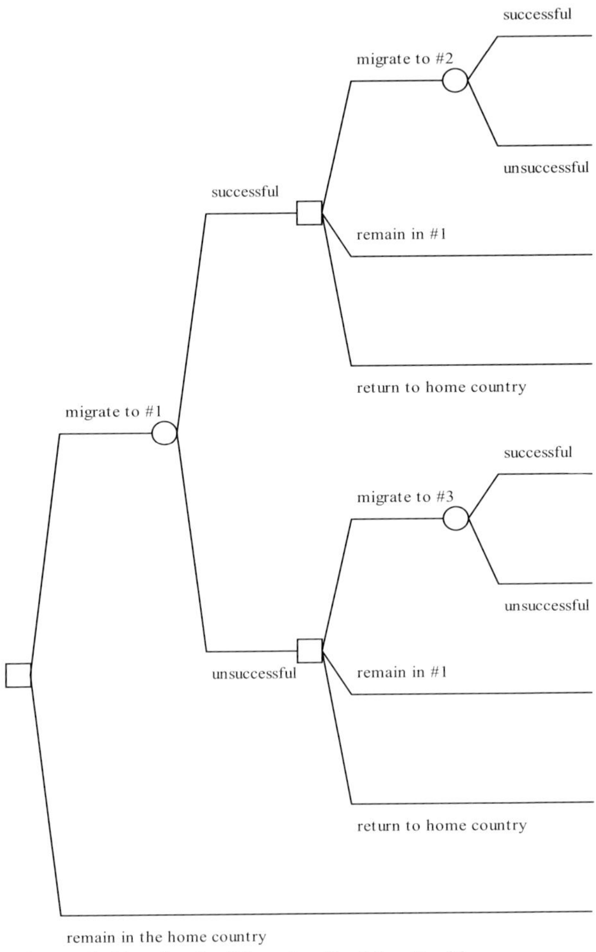


*Figure 1: Tree for the Migration Decision Problem*

The payoffs received by the potential migrant depend upon both the individual's choice of country and whether the individual is successful or unsuccessful in that country. For notational convenience I will use upper case R's to denote the present value of the lifetime increment to income if the individual is successful, and I will use lower case r's to denote the present value of the lifetime increment to income if the individual is unsuccessful. Then $R_1$, $R_2$, and $R_3$ are the increments to income if the individual is successful in countries #1, #2, and #3, respectively. By presumption, the individual is successful in the home country. (Later I will comment on modeling non-success in the home country due to the stripping of credentials.) I will denote the increment to income in the home country by $R_h$. Similarly, I will use lower case r's to denote the increment to income if the individual is unsuccessful. Thus, $r_1$, $r_2$, and $r_3$ are the increments to income if the individual is unsuccessful in countries #1, #2, and #3, respectively. I will also denote the cost of migrating from one country to another by an upper case C followed by two lower case indicators of the countries. Thus, $C_{h1}$ is the cost of migrating from the home country to country #1, and $C_{12}$ and $C_{13}$ are the costs of migrating from country #1 to countries #2 and #3, respectively, and $C_{1h}$ is the cost of re-migrating from country #1 to the home country.

The first migration decision (i.e., the decision at node 1) is made when the individual has T years of working time remaining in his or her life. The second migration decision (i.e., the decisions at nodes 2 and 3) are made when the individual has already spent $T_1$ of those T years and has $T-T_1$ years remaining. Let $t_1 = \frac{T_1}{T}$ and let $t_2 = \frac{T - T_1}{T}$. Then $t_1$ and $t_2$ are percentages, and $t_1 + t_2 = 1$. Finally, I will denote the probability of success in each country by an upper case P followed by a country designator. Thus, $P_1$, $P_2$, and $P_3$ represent the probabilities of success in countries #1, #2, and #3, respectively.

For convenience, the foregoing notation is gathered together in the Table 1.

*Table 1: Notation for the Migration Decision Problem*

| | |
|---|---|
| the incremental income in the home country | $R_h$ |
| the incremental income in country #1 if successful | $R_1$ |
| the incremental income in country #1 if unsuccessful | $r_1$ |
| the incremental income in country #2 if successful | $R_2$ |
| the incremental income in country #2 if unsuccessful | $r_2$ |
| the incremental income in country #3 if successful | $R_3$ |
| the incremental income in country #3 if unsuccessful | $r_3$ |
| the cost of migrating from the home country to country #1 | $C_{h1}$ |
| the cost of re-migrating from country #1 to the home country | $C_{1h}$ |
| the cost of re-migrating from country #1 to the country #2 | $C_{12}$ |
| the cost of re-migrating from country #1 to the country #3 | $C_{13}$ |
| the probability of success in country #1 | $P_1$ |
| the probability of success in country #2 given success in country #1 | $P_2$ |
| the probability of success in country #3 given failure in country #1 | $P_3$ |
| the remaining professional lifetime of the individual | T |
| the time spent in country #1 before making a remigration decision | $T_1$ |
| the percent of professional lifetime spent in country #1 | $t_1$ |

It remains to specify the relationships between and among the parameters. The basic idea behind the model is quite simple. If the individual migrates to country #1 and is successful, then the individual can reasonably consider migrating to another country, here called country #2, where the rewards to successful professionals are greater than in country #1. Thus, country #2 is such that $R_2 > R_1$ but $P_2 < P_1$. Contrariwise, if the individual migrates to country #1 and is unsuccessful, then the individual can reasonably consider migrating to another country, here called country #3, where he or she is more likely to be employed as a professional, albeit at a lower level than the reward to successful professional employment in country #1. Thus, country #3 is such that $P_3$, the probability of professional employment, is reasonably high, but the reward to professional employment, $R_3$, is lower than the reward to professional employment in country #1, $R_1$. The individual considers an intermediary country, i.e., country #1, because the probabilities of success in countries #2 and #3 depend upon the realization of success or failure in country #1. In technical terms, $P_2$ and $P_3$ are posterior probabilities, and are thereby based upon $P_1$. For convenience, the basic presumptions are gathered together in Table 2.

*Table 2: Presumptions for the Migration Decision Problem*

| | |
|---|---|
| country #1 and the home country | $R_1 > R_h$ |
| country #1 and country #2 | $R_2 > R_1$, $P_2 < P_1$ |
| country #1 and country #3 | $R_3 < R_1$, $P_3 > P_1$ |
| time | $T_1 + T_2 = T$, $t_1 + t_2 = 1$ |

The overall decision problem is resolved backwards with respect to time. Thus, we begin by resolving the sub-problems at nodes 2 and 3. These decision problems concern remigration from country #1 to countries #2 and #3, respectively. Having resolved these sub-problems, we then move to node 1 and resolve the overall decision problem.

Consider the decision made at node 2. This decision involves an individual who has migrated to country #1 and has been successfully employed as a professional. The individual is considering three options -- re-migrating to country #2, remaining in country #1, and returning to the home country. Re-migrating to country #2 is a gamble where the individual receives the payoff $R_2$ with probability $P_2$ and the payoff $r_2$ with probability $1 - P_2$. Since $R_2 > R_1$, re-migrating to country #2 is, at least, an attractive gamble. Remaining in country #1 and returning to the home country are both sure-thing actions where the individual is certain to be employed as a professional. The decision at node 2 is made after the individual has been in country #1 for $T_1$ years, and the choice made at node 2 depends only on the payoffs for the remaining time.

Note that an individual who is successful in country #1 is exactly the kind of individual that country #1 seeks to retain. The government of country #1 retains the individual if the payoff to remaining in country #1 is greater than both the expected value of the gamble resulting from re-migrating to country #2 and the sure-thing payoff to returning to the home country. The payoff to remaining in country #1 is $t_2 R_1$; the expected payoff to re-migrating to country #2 is $P_2 (t_2 R_2 - C_{12}) + (1 - P_2) (t_2 r_2 - C_{12})$; and the sure-thing payoff to returning to the home country is $t_2 R_h - C_{1h}$. Thus, the individual remains in country #1 if $t_2 R_1 > P_2 (t_2 R_2 - C_{12}) + (1 - P_2) (t_2 r_2 - C_{12})$ and $t_2 R_1 > t_2 R_h - C_{1h}$. This amounts to the following: the individual chooses to remain in country #1 if

$$t_2 R_1 > P_2 (t_2 R_2 - C_{12}) + (1 - P_2) (t_2 r_2 - C_{12})$$

and

$$t_2 R_1 > t_2 R_h - C_{1h}.$$

There are two interesting results here. First, country #1 has an unconditional advantage in retaining a successful immigrant vis-à-vis the individual returning to the home country. Since $R_1 > R_h$, clearly $t_2 R_1 > t_2 R_h$ and thereby $t_2 R_1 > t_2 R_h - C_{1h}$. Thus, an individual who migrates to country #1 and is successful will not return to the home country for economic reasons.

Second, country #1 has a conditional advantage in retaining a successful individual vis-à-vis country #2. Country #1 retains the successful immigrant vis-à-vis country #2 if

$$t_2 R_1 > P_2 (t_2 R_2 - C_{12}) + (1 - P_2) (t_2 r_2 - C_{12}).$$

This rearranges to

$$\frac{1}{t_2} C_{12} > [P_2 R_2 + (1 - P_2) r_2] - R_1,$$

which is the condition under which country #1 retains a successful individual.

The foregoing equation states that country #1 will retain a successful immigrant only if the gap between the expected gain in country #2, $P_2 R_2 + (1 - P_2) r_2$, and the gain in country #1, $R_1$, is not sufficient to cover the amortized cost of migrating from country #1 to country #2.

By assumption, country #2 provides higher income to a successful individual than does country #1 (i.e., $R_2 > R_1$). If country #2 is also provides higher income to unskilled workers than country #1 provides to skilled workers (i.e., if $r2 > R_1$), then $P_2 R_2 + (1 - P_2) r_2$ is greater than $R_1$, so that the immigrant is retained only if $\frac{1}{t_2} C_{12}$ is large. Contrariwise, if country #1 provides higher income to skilled workers than country #2 provides to unskilled workers (i.e., if $R_1 > r_2$), then $R_1$ is greater than or less than $P_2 R_2 + (1 - P_2) r_2$ depending upon the particular values of $R_1$, $P_2$, $R_2$, and $r_2$. Thus, given that $R_1 > r_2$, country #1 retains a successful migrant vis-à-vis country #2 under the condition that

$$[P_2 R_2 + (1 - P_2) r_2] - \frac{1}{t_2} C_{12} < R_1.$$

I will presume that success in country #1 is a signal that the individual is likely to be successful in country #2. Thus, in the case at hand, we can assume that $P_2$ is reasonably high. If $P_2$ is close to 1, then $P_2 R_2 + (1 - P_2) r_2$ is close to $R_2$, and thereby $P_2 R_2 + (1 - P_2) r_2$ is likely to be greater than $R_1$. Nonetheless, if $C_{12}$ is large or $t_2$ is small, so that $\frac{1}{t_2} C_{12}$ is large, then country #1 can retain the individual.

In summary, the primary result in the case of remigration from country #1 to country #2 is that country #1 will retain the successful immigrant only if the amortized cost of migrating to country #2 is large. The amortized cost of migrating, $\frac{1}{t_2} C_{12}$, is large if either $\frac{1}{t_2}$ or $C_{12}$ is large. Since country #1 does not control either $\frac{1}{t_2}$ or $C_{12}$, and since only country #2 can make $C_{12}$ small, country #1 can retain a successful immigrant, but not entirely by its own actions.

Let me make a side comment on the costs of migrating and re-migrating. There is an aspect of the costs of migration and remigration that is particularly relevant here. The cost of migrating has two components -- the cost of transportation and the cost of meeting requirements imposed in the target country. The latter include governmental and non-governmental barriers to entering the country and to entering his or her profession or skilled craft. These barriers include statutes imposed by governments and certification requirements and education and experience requirements imposed by professional and labor organizations (Chang 2002). The former, called governmental barriers, involve requirements imposed upon all potential immigrants. For example, the US immigration system favors an individual who has relatives already living in the US. The latter, called non-governmental barriers, involve requirements imposed on individuals on the basis of their profession or craft. For example, an individual who is a recognized medical doctor in one country may be required to complete additional years of professional education and/or specialized training before being licensed to practice in another country. (The requirements imposed on foreign trained medical doctors who migrate to the US are presented by the American Medical Association at http://www.ama-assn.org/ama/pub/category/1544.html.) The costs of overcoming these barriers can be formidable. In particular, these barrier costs can make the cost of migrating (e.g., $C_{h1}$) or re-migrating (e.g., $C_{12}$) remarkably high.

One reviewer raised the interesting case of an individual being stripped of his or her professional credentials in the home country as a form of forced migration, and asked how the model presented here can be adapted to account for this case. If the individual has already been stripped of his or her credentials, then we replace the assumption of success in the home country with the presumption of failure in the home country. This makes migrating to country #1 an easy decision, and the analysis is quite simple. However, if the individual anticipates that he or she may be stripped of the credentials, then we must adapt the model used here and incorporate the probability of having the credentials removed. This makes the analysis more complex, but not radically so. Since there are no migration costs attached to staying in the home country, this case can be modeled by viewing the payoff to remaining in the home country as a convex combination of professional and non-professional wages determined by the probability of having his or her credentials stripped. As such, the payoff to remaining in the home country will be lower than the rewards to success in the home country. This will make migrating to country #1 an easy decision, and the analysis is again quite simple.

Now, consider the decision at node 3. This decision involves an individual who has migrated to country #1 and has been unsuccessfully employed as human capital. The individual is considering re-migrating to country #3, remaining in country #1, and returning to the home country. Re-migrating to country #3 is a gamble, where the individual receives $R_3$ with probability $P_3$ and $r_3$ with probability $1 - P_3$. Recall that, by assumption, $R_3 < R_1$ but $P_3 > P_1$ and $P_3$ is reasonably high. Thus, re-migrating from country #1 to country #3 can be an attractive gamble vis-à-vis remaining in country #1. Similarly, if $R_h > r_1$, then the sure-thing payoff from returning to the home country can also be attractive vis-à-vis remaining in country #1. The decision at node 3, like that at node 2, is made after the individual has been in country #1 for $T_1$ years.

Note that an individual who is unsuccessful in country #1 is exactly the kind of individual that country #1 seeks to repel. The government of country #1 successfully repels the individual if the payoff to remaining in country #1 is less than both the expected value of the gamble resulting from re-migrating to country #3 and the sure-thing payoff resulting from returning to the home country. The payoff to remaining in country #1 is $t_2 r_1$; the expected payoff to re-migrating to country #3 is $P_3 (t_2 R_3 - C_{13}) + (1 - P_3) (t_2 r_3 - C_{13})$; and the sure-thing payoff to returning to the home country is $t_2 R_h - C_{1h}$. Thus, the individual will r-migrate from country #1 if $t_2 r_1 < P_3 (t_2 R_3 - C_{13}) + (1 - P_3) (t_2 r_3 - C_{13})$ and $t_2 r_1 < t_2 R_h - C_{1h}$. This amounts to the following: the individual chooses to leave country #1 if

$$t_2 r_1 < P_3 (t_2 R_3 - C_{13}) + (1 - P_3) (t_2 r_3 - C_{13})$$

or

$$t_2 r_1 < t_2 R_h - C_{1h}.$$

The interesting results here are as follows: country #1 does not have an unconditional advantage, and may have an unconditional disadvantage, in repelling an unsuccessful individual.

Consider now the decision involving remaining in country #1 and re-migrating to the home country. (Note that country #2 is not considered here because failure in country #1 makes failure in country #2 virtually certain. Thus, it is assumed that the expected value of migrating to country #2 is too low to make the choice of migrating to country #2 viable.) Country #1 repels the unsuccessful immigrant back to the home country if

$$t_2 r_1 < t_2 R_h - C_{1h},$$

which rearranges to

$$\frac{1}{t_2} C_{1h} < R_h - r_1.$$

Since $R_h > r_1$, the foregoing condition is met as long as the amortized cost of returning home, $\frac{1}{t_2} C_{1h}$, is not large. Country #1 can keep the amortized cost from becoming large by making the cost of returning to the home country, $C_{1h}$, small. By presumption, the individual already meets the requirements for successful employment as human capital in the home country. Thus, by presumption, the professional barrier component of $C_{1h}$ is small. Thus, country #1 has a conditional advantage in repelling an unsuccessful immigrant vis-à-vis the home country. The condition is that $\frac{1}{t_2} C_{1h}$ is not large.

Consider, now, the decision involving remaining in country #1 or re-migrating to country #3. By assumption, country #3 is such that $R_3 < R_1$ but $P_3$ is reasonably high. The individual migrates from country #1 to country #3 if

$$t_2\, r_1 < P_3\, (t_2\, R_3 - C_{13}) + (1 - P_3)\, (t_2\, r_3 - C_{13}).$$

This rearranges to

$$r_1 < P_3\, R_3 + (1 - P_3)\, r_3 - \frac{1}{t_2} C_{13}.$$

If country #3 is such that $r_3 < R_3 < r_1 < R_1$, then $r_1 > P_3\, R_3 + (1 - P_3)\, r_3$ and thereby $r_1 > P_3\, R_3 + (1 - P_3)\, r_3 - \frac{1}{t_2} C_{13}$. Therefore, if $r_3 < R_3 < r_1 < R_1$, then the foregoing condition is not met and the individual remains in country #1. Hence, under the condition that the individual is better off employed as labor in country #1 than as a professional in country #3 (i.e., the condition that $r_1 > R_3$), then country #1 has an unconditional disadvantage in repelling an unsuccessful immigrant vis-à-vis country #3.

Contrariwise, if country #3 is such that the individual is better off employed as a professional in country #3 than as labor in country #1, i.e., if $r_1 < R_3$, then the individual will migrate to country #3 if $P_3$ is reasonably high and the amortized cost of migrating is reasonably low. The former condition is met by assumption, and the latter can be met if country #1 subsidizes the cost of migrating to country #3. Therefore, under the condition that $r_1 < R_3$, then country #1 can repel an unsuccessful immigrant. Put differently, country #1 can repel an unsuccessful immigrant vis-à-vis country #3 given that country #3 provides the individual with a reasonable economic alternative. In summary, country #1 does not have an unconditional advantage, and may have an unconditional disadvantage, in repelling an unsuccessful individual.

Consider, now, the choice between returning to the home country and re-migrating to country #3. The individual will choose to re-migrate to the home country over re-migrating to country #3 if

$$t_2\, R_h - C_{1h} > P_3\, (t_2\, R_3 - C_{13}) + (1 - P_3)\, (t_2\, r_3 - C_{13}),$$

which rearranges to

$$R_h - \frac{1}{t_2} C_{1h} > P_3\, R_3 + (1 - P_3)\, r_3 - \frac{1}{t_2} C_{13}.$$

Since we have made no assumptions about the relationship of the home country and country #3, we must examine various cases. We will treat two basic cases.

First, suppose that country #3 is such that the income to a successful individual in the home country is greater than the income to a successful individual in country #3 (i.e., suppose that $R_h > R_3 > r_3$). Then $R_h > P_3 R_3 + (1 - P_3) r_3$, and the individual will choose to return to the home country given that the costs $C_{1h}$ and $C_{13}$ are similar.

Second, if country #3 is such that the incomes paid to successful and unsuccessful individuals straddle the income to a successful individual in the home country (i.e., if $R_3 > R_h > r_3$), then the individual's decision depends upon the values of the parameters. If we assume, as before, that $P_3$ is reasonably high, then $P_3 R_3 + (1 - P_3) r_3$ is reasonably close to $R_3$. If the costs $C_{1h}$ and $C_{13}$ are similar, then it is likely that the individual will prefer to re-migrate to country #3 in preference to re-migrating to the home country.

In summary, if the individual migrates to country #1 and is unsuccessful, then the individual will choose between returning home and migrating to country #3 on the basis of the parameter values. Neither country #3 nor the home country has an unconditional advantage or disadvantage in attracting or repelling the individual, respectively.

Finally, consider the overall decision, i.e., the decision at node 1. Since the decisions at nodes 2 and 3 depend upon the values of the parameters, we must examine various cases. We will examine the two cases wherein country #1 achieves its goal of (i) attracting and retaining individuals who successfully employ human capital, and (ii) repelling those who are unsuccessful. The first case is where the individual chooses to migrate to country #1, and if successful, chooses to remain in country #1, and if unsuccessful, chooses to return home. The second case is where the individual, if unsuccessful, chooses to migrate to country #3. These cases are of immediate interest because they allow us to determine the best policy for the government of country #1. Since this is the policy the government of country #1 would institute, it is the policy that a potential migrant would face.

The condition for the first case is

$$P_1 [R_1 - C_{h1}] + (1 - P_1) [t_1 r_1 + t_2 R_h - C_{h1} - C_{1h}] > R_h,$$

which rearranges to

$$P_1 R_1 + (1 - P_1) [t_1 r_1 + t_2 R_h - C_{1h}] - R_h > C_{h1}.$$

Let $C_{h1}^{*} = P_1 R_1 + (1 - P_1) [t_1 r_1 + t_2 R_h - C_{1h}] - R_h$. From the perspective of country #1, $C_{h1}$ is a random variable. (Country #1 can influence, but cannot determine, the value of $C_{h1}$.) Then the probability of attracting and retaining a successful individual and repelling an unsuccessful individual to the home country, is

$$\Pr(\text{immigration}) = \Pr(C_{h1} < C_{h1}^{*}) = \int_0^{C_{h1}^{*}} f(C_{h1}) dC_{h1} .$$

The probability of immigration is a function of $P_1$, $R_1$, $r_1$, $R_h$, and $C_{h1}$. Therefore, the government is interested in how the probability of migration responds to changes in P, $R_1$, $r_1$, $R_h$, and $C_{1h}$.

The influence of each of these parameters upon the probability of migration can be found via Liebinitz's rule. The rule states that the partial derivative of $\Pr(C_{h1} < C_{h1}^{*})$ with respect to a parameter x is

$$\frac{\partial \Pr(C_{h1} < C_{h1}^{*})}{\partial x} = \int_0^{C_{h1}^{*}} \frac{\partial f(C_{h1})}{\partial x} dC_{h1} + f(C_{h1}) \frac{\partial C_{h1}^{*}}{\partial x} - f(0) \frac{\partial 0}{\partial x} .$$

Since $\frac{\partial f(C_{h1})}{\partial x} = 0$ and $\frac{\partial 0}{\partial x} = 0$ for x = $P_1$, $R_1$, $r_1$, $R_h$, and $C_{1h}$, we have

$$\frac{\partial \Pr(C_{h1} < C_{h1}^{*})}{\partial x} = f(C_{h1}^{*}) \frac{\partial C_{h1}^{*}}{\partial x}$$

for each parameter. Since $f(C_{h1}^{*}) \geq 0$ for all values of $C_{h1}^{*}$, the sign of $\frac{\partial \Pr(C_{h1} < C_{h1}^{*})}{\partial x}$ is equal to the sign of $\frac{\partial C_{h1}^{*}}{\partial x}$ for each parameter. Finally, since

$$C_{h1}^{*} = P_1 R_1 + (1 - P_1) (t_1 r_1 + t_2 R_h - C_{1h}) - R_h > 0,$$

we have the following:

$$\text{sign}[\frac{\partial \Pr(C_{h1} < C_{h1}^{*})}{\partial P1}] = \text{sign}[R_1 - (t_1 r_1 + t_2 R_h - C_{1h})] > 0,$$

$$\text{sign}[\frac{\partial \Pr(C_{h1} < C_{h1}^{*})}{\partial R_1}] = \text{sign}[P_1] > 0,$$

$$\text{sign}[\frac{\partial \Pr(C_{h1} < C_{h1}^{*})}{\partial r_1}] = \text{sign}[(1 - P_1)\, t_1] > 0,$$

$$\text{sign}[\frac{\partial \Pr(C_{h1} < C_{h1}^{*})}{\partial R_h}] = \text{sign}[(1 - P_1)\, t_2 - 1] < 0,$$

$$\text{sign}[\frac{\partial \Pr(C_{h1} < C_{h1}^{*})}{\partial C_{1h}}] = \text{sign}[- (1 - P_1)] < 0.$$

Thus, the government of country #1 seeks to increase $P_1$, $R_1$, and $r_1$, and to decrease $R_h$ and $C_{1h}$.

The second case is similar. The condition for the second case is

$$P_1\, [R_1 - C1] + (1 - P_1)\, [P_3\, (t_1\, r_1 + t_2\, R_3 - C_{h1} - C_{13}) + (1 - P_3)\, (t_1\, r_1 + t_2\, r_3 - C_{h1} - C_{13})] > R_h,$$

which rearranges to

$$P_1\, R_1 + (1 - P_1)\, [P_3\, (t_1\, r_1 + t_2\, R_3) + (1 - P_3)\, (t_1\, r_1 + t_2\, r_3) - C_{13}] - R_h > C_{h1}.$$

Let $C_{h1}^{**} = P_1\, R_1 + (1 - P_1)\, [P_3\, (t_1\, r_1 + t_2\, R_3) + (1 - P_3)\, (t_1\, r_1 + t_2\, r_3) - C_{13}] - R_h$. Then the probability of attracting the individual, and retaining a successful individual and repelling an unsuccessful individual to country #3, is

$$\Pr(\text{immigration}) = \Pr(C_{h1} < C_{h1}^{**}) = \int_0^{C_{h1}^{*}} f(C_{h1})dC_{h1}\,.$$

As before, the partial derivatives can be found by Liebniz's rule,

$$\frac{\partial \Pr(C_{h1} < C_{h1}^{**})}{\partial x} = \int_0^{C_{h1}^{**}} \frac{\partial f(C_{h1})}{\partial x} dC_{h1} + f(C_{h1}) \frac{\partial C_{h1}^{**}}{\partial x} - f(0)\frac{\partial 0}{\partial x}\,,$$

and, as before, $\frac{\partial f(C_{h1})}{\partial x} = 0$ and $\frac{\partial 0}{\partial x} = 0$ for x = $P_1$, $R_1$, $r_1$, $R_3$, $r_3$, and $C_{13}$.
The relevant partial derivatives are new, but the signs are the same as before. Since

$$C_{h1}^{**} = P_1 R_1 + (1 - P_1) [P_3 (t_1 r_1 + t_2 R_3) + (1 - P_3) (t_1 r_1 + t_2 r_3) - C_{13}] - R_h > 0,$$

we have the following:

$$\text{sign}[\frac{\partial \Pr(C_{h1} < C_{h1}^{**})}{\partial P_1}] = \text{sign}[R_1 - P_3 (t_1 r_1 + t_2 R_3) - (1 - P_3) (t_1 r_1 + t_2 r_3) - C_{13}] > 0,$$

$$\text{sign}[\frac{\partial \Pr(C_{h1} < C_{h1}^{**})}{\partial R_1}] = \text{sign}[P_1] > 0,$$

$$\text{sign}[\frac{\partial \Pr(C_{h1} < C_{h1}^{**})}{\partial r_1}] = \text{sign}[(1 - P_1) t_1] > 0,$$

$$\text{sign}[\frac{\partial \Pr(C_{h1} < C_{h1}^{**})}{\partial R_3}] = \text{sign}[(1 - P_1) P_3 t_2] > 0,$$

$$\text{sign}[\frac{\partial \Pr(C_{h1} < C_{h1}^{**})}{\partial r_3}] = \text{sign}[(1 - P_1) (1 - P_3) t_2] > 0,$$

$$\text{sign}[\frac{\partial \Pr(C_{h1} < C_{h1}^{**})}{\partial C_{13}}] = \text{sign}[- (1 - P_1)] < 0.$$

Thus, the government of country #1 seeks to increase $P_1$, $R_1$, $r_1$, $R_3$, and $r_3$, and to decrease $C_{13}$.

## 3. POLICY IMPLICATIONS

The policy implications in each of the foregoing cases are similar. In both cases, the government of country #1 seeks to increase $P_1$, $R_1$ and $r_1$, and to reduce the costs of remigration (i.e., $C_{1h}$ and $C_{13}$). However, in the former case, the government of country #1 seeks to decrease $R_h$, whereas in the second case, the government seeks to increase $R_3$ and $r_3$. The ability of the government of country #1 to influence $R_h$, $R_3$, and $r_3$ may be very limited. For example, if wages are market determined in country #1, then the government of country #1 cannot influence $R_1$ and $r_1$. On the other hand, if $R_1$ and/or $r_1$ are set by the government, then the government of country #1 can determine $R_1$ and $r_1$. Clearly, the government of country #1 can, at best, influence only indirectly the wages in the home country (i.e., $R_h$) and those in countries #2 and #3 (i.e., $R_2$, $r_2$, $R_3$, and $r_3$).

The government of country #1 can influence, but not fully determine, the costs of migration and remigration. The government of country #1 can reduce the transportation component and the barrier component of $C_{h1}$ and $C_{13}$. The government of country #1 can achieve the former via subsidies and the latter via the reduction of governmental barriers. In most cases, the government of country #1 will not be able to influence the non-governmental barriers. The government of country #1 seeks to increase the cost of re-migrating from country #1 to country #2. However, the government of country #1 typically cannot increase this cost.

## 4. FORCED MIGRATION

The model provides a structure to make the meanings of induced and forced migration reasonably precise. Recall that our concern here is with the distinction between induced migration as migration wherein "the migrants retain some decision making power to decide whether or not to leave" and forced migration as migration where the migrants do not have this power (Petersen 1958, p. 261). The distinction between induced and forced migration, so specified, can be modeled via decision problems wherein the potential migrant has and does not have the option of remaining in, or returning to, the individual's home country, respectively. The foregoing model represents induced migration. Forced migration is then formalized within the foregoing model by removing the branches in the decision tree related to the home country. The decision tree for the forced migration problem is then the foregoing decision tree with the branches labeled "Return to the home country" and "Remain in the home country" pruned away.

The consequence of removing the home country option is different at each of the three nodes. The removal of the home country option at nodes 1 and 3 has major consequences, whereas it is of no consequence at node 2. Let me consider these in turn.

At node 1, given no option of staying in the home country, the individual has no choice but to migrate. Note that the removal of the home country option amounts to the presumption that the individual who is forced to migrate from the home country can migrate to the country that would have been the best alternative if the individual were not forced to migrate. This is not a particularly powerful assumption, and I will make it here. Note that country #1 is simply better than the other countries; it may not be better than remaining in the home country were migration not forced. Thus, country #1 may be a step down from what the individual realized in the home country before the events that forced his or her migration. The removal of the option to remain in the home country at node 1 can thereby have major consequences. Australia, Canada, the United States, and various other countries were the beneficiaries of forced migration of this kind during the 1930's and 1940's. Thus, at node 1, both the migrant and the government of country #1 face uncertain futures. However, the situation of the migrant is not nearly as rosy as that of the government of country #1.

At node 2, the removal of the home country option plays no role. Here, the individual would not choose to return to the home country even if the option were available. The foregoing analysis shows that at node 2 the individual will remain in country #1 or re-migrate to country #2. Therefore, the removal of the home country option is of no consequence at node 2.

At node 3, the removal of the home country option plays a major role. Here, the individual would choose between moving to country #3 (where $R_3 < R_1$) or returning to the home country (where $R_h < R_1$). The removal of the home country option forces the individual to choose between remaining in country #1 unsuccessfully employed as human capital or migrating to country #3 to gamble on successful employment as human capital. The individual is in the uncomfortable position of being the kind of person country #1 seeks to repel. Recall that if country #3 is such that $r_3 < R_3 < r_1$, then country #1 has an unconditional disadvantage in repelling an unsuccessful immigrant and the individual would remain in country #1. This is the worst case for both the migrant and the government of country #1; the individual, who possesses human capital, is employed as an unskilled worker, and having no better place to which to migrate, cannot be repelled by country #1.

Also recall that if country #3 is such that $r_1 < R_3 < R_1$, then the individual will migrate to country #3 if the probability of success in country #3 (i.e., $P_3$) is reasonably high and the amortized cost of migrating is reasonably low. The condition on $P_3$ is met by assumption and plays no role here. However, the latter condition, i.e., that $\frac{1}{t_2}C_{13}$ is reasonably low, can be met if country

#1 subsidizes the cost of migrating to country #3 or if $t_2$ is small. This is the next worst case for both the migrant and the government of country #1. The individual may be "induced" into gambling on successful employment as human capital in country #3 if country #1 can make the cost of migrating low enough. Thus, the migrant may be induced to select a gamble and country #1 may face the costs of inducing the migrant to take the gamble.

# 5. EXTENSIONS AND LIMITATIONS OF THE MODEL

The foregoing model can be extended to treat various issues not treated here. These issues include the migration of unskilled workers, internal migration, non-economic rewards, and attitude toward risk.

## 5.1 Unskilled Workers

The model does not directly account for the migration decisions of unskilled workers. By presumption, a country seeks to attract skilled workers and simultaneously to repel unskilled workers. However, the economic conditions that attract professionals are likely to result in conditions that attract unskilled workers. Thus, a target country must maintain a balancing act whereby the unskilled workers are repelled but skilled workers are attracted. We can model this via either of two approaches. In the first approach, we include the unskilled workers with the skilled workers. In the second case, we consider the unskilled workers alone.

The first case is fairly straightforward. In order to add the analysis of unskilled workers to the foregoing analysis, we must account for the case wherein the unskilled worker is repelled. This is the situation where the individual is better off staying at home than migrating to country #1. This case is modeled by the equation

$$r_1 - C_{h1} < r_h,$$

which rearranges to

$$r_1 - r_h < C_{h1}.$$

Simply put, the unskilled individual remains at home if the cost of migrating is greater than the incremental income. Unskilled workers can then be included in the analysis by including the foregoing equation within the model.

The second case is very straight forward. We simply view the migration decision from the perspective of an unskilled individual, and view the rewards denoted by upper and lower case letters as the rewards paid to unskilled workers when successful (i.e., employed at the minimum wage) and unsuccessful (i.e., unemployed and on welfare). We then seek the condition under which the potential immigrant is deterred from migrating. This amounts to reversing the inequality between the expected rewards to migrating and the rewards to remaining in the home country.

The model does not account for the use of non-economic barriers erected to stop the migration of unskilled workers. For example, Canada employs a point system to "filter out" unskilled workers (Borjas 1999, 58-61). The model employed here cannot be used, without serious alteration, to account for non-economic filters. However, the model can be altered to account for this case. We must add a third time period wherein the individual, while in the home country, invests in his or her human capital, so as to overcome the non-economic barrier. As noted, this is a serious alteration in the model.

## 5.2 Internal Migration

The model does not deal directly with internal migration. The model, however, is easily adapted to model this common and important form of migration. The traditional pattern of rural-to-urban-to-foreign migration is modeled by viewing the home country as the rural area, country #1 as the urban area in the home country, and countries # 2 and #3 as foreign countries. This renaming does not change the analysis, and the results presented above hold. Indeed, the model makes clear why an individual would migrate to the urban area and then re-migrate to another country. The simple reason is cost -- typically, one can migrate to a city within the home country at low cost, and, in many countries, success in the city is a strong indicator of success in another country. The former holds for obvious reasons, and the latter holds because cities provide employment opportunities for learning-by-doing that are unavailable in rural areas. This not to suggest that the analysis of internal migration is a simple case of migration.

## 5.3 Non-Economic Rewards

The foregoing model does not directly account for non-economic rewards. For example, the United States gives preference to family members of individuals who have already migrated. As such, the model does not account for all potential migrants; it accounts only for those induced by economic rewards. However, as the schematic presented by Byerlee (1974,

p. 553) shows, these non-economic influences can be captured in the expected present value of migration.

## 5.4 Attitude towards Risk

Migrants are typically viewed as being very "gutsy," i.e., being risk takers. The foregoing analysis presumes, for simplicity, that the potential migrant is risk neutral. If we presume that the potential migrant is either risk averse or risk taking over all payoffs, then the foregoing analyses hold because the values treated there can be viewed simply as the arguments of the individual's utility function. However, if we presume that the potential migrant has the kind of hybrid risk attitude reflected in prospect theory (Kahneman and Tversky, 2000), then the foregoing model, but not the foregoing analysis, can account for the risk-related behavior of migrants. The foregoing model employs the same account of risk as that employed in Todaro (2000a, chapter 8; 2000b). A variant of the relevant analysis of the hybrid risk attitude of prospect theory is presented in Dacey (2003) and is not repeated here.

# 6. CONCLUDING REMARKS

The foregoing analysis shows that even a simple model of the migration decision problem can be informative. Most striking is the result that the country to which the individual first migrates retains an advantage over other countries and the home country with respect to retaining successful individuals. This result is of particular relevance to primary target countries such as Canada, Australia, the United States, the United Kingdom, Norway, The Netherlands, and Switzerland (Thomas, 1973, 308-309). These countries have been major beneficiaries of the attraction and retention of remarkably talented individuals, even though they were not the countries providing the highest wages.

The foregoing analysis is also capable of generating relevant results for those countries seeking to induce the return of former residents. This case is modeled as remigration to the home country. Of late, Ireland has been very successful in attracting high technology workers of Irish birth to return home and help fuel a rapidly growing economy. Ireland provides an interesting case in so far as, historically, Ireland represents the paradigmatic case of economically induced out-migration.

Despite its shortcomings, the model and analysis of the migration decision problem presented here provides interesting and useful results. In particular, the model provides a coherent account of the basic distinction between induced and forced migration.

The results of the foregoing model regarding induced migration can be summarized as follows. First, country #1 has an unconditional advantage in retaining a successful immigrant vis-à-vis the home country and a conditional advantage in retaining a successful individual vis-à-vis a country with higher wages (i.e., country #2 where $R_2 > R_1$). The analysis of node 2 shows that an individual who migrates to country #1 and is successful will not re-migrate to the home country, and will not re-migrate to country #2 if the net gain from re-migrating is not sufficient to cover the amortized cost of re-migrating. As noted, these costs can be high, particularly in countries like country #2. Thus, country #1 has at least a conditional advantage in retaining a successful immigrant.

Second, country #1 has a conditional advantage in repelling an unsuccessful immigrant vis-à-vis the home country. Country #1 has a conditional advantage in repelling an unsuccessful individual vis-à-vis country #3 if $r_1 < R_3 < R_1$ and an unconditional advantage if $r_3 < R_3 < r_1$.

Finally, country #1 does not have an unconditional advantage in attracting individuals who possess human capital unless the home country option is removed. Obviously, the last result is very important in our consideration of forced migration.

The results of the foregoing model regarding forced migration can be summarized as follows. For an individual who has migrated and is successful, the removal of the home country option is of no consequence -- the individual remains in the new country. For an individual who has migrated and is unsuccessful, the removal of the home country option has two major consequences. First, the new country has an unconditional disadvantage in repelling the individual vis-à-vis a country paying lower professional wages that are lower than the new country's labor wages (i.e., country #3 if $r_3 < R_3 < r_1$). Second, the new country has a conditional advantage in repelling the individual vis-à-vis a country paying lower professional wages that are higher than the country's labor wages (i.e., country #3 if $r_1 < R_3 < R_1$).

For an individual who has not yet migrated, a target country has an unconditional advantage in attracting the individual only if the option of remaining in the home country is not available, i.e., only if migration is forced.

## ACKNOWLEDGEMENTS

I wish to thank Lisa Carlson and Robert Muncaster for their technical assistance, Konrad Talmont-Kaminski for his helpful insights on the experiences of immigrants, and two anonymous referees for their constructive suggestions. As always, the remaining errors are mine.

# REFERENCES


Borjas, G.J., ed. 2000. *Issues in the Economics of Integration*, Chicago: University of Chicago Press.

Borjas, G.J. 1999. *Heaven's Door: Immigration Policy and the American Economy,* Princeton: Princeton University Press.

Byerlee, D. 1974. "Rural-Urban Migration in Africa: Theory, Policy, and Research Implications," *Internal Migration Review* 3.

Chang, H.F. 2000. "The Economic Analysis of Immigration Law." In *Migration Theory*, C.B. Brettell and J.F. Hollifield, eds., New York: Routledge,.

Dacey, R. 2003. "The S-Shaped Utility Function, *Synthese* 135:243-272.

Dacey, R. 1998. "Race, Rationality, and Behavior." In M. Wolfson, ed., *The Political Economy of War and Peace*, Dordrecht, Holland: Kluwer Academic Publishers,.

Faist, T. 2000. *The Volume and Dynamics of International Migration and Transnational Social Spaces*, Oxford, UK: Clarendon Press.

Grossman, G.M. and E. Helpman. 1991. *Innovation and Growth in the Global Economy*, Cambridge: MIT Press.

Helpman, E. 1981. "International Trade in the Presence of Product Differentiation, Economies of Sale, and Monopolistic Competition," *Journal of International Economics* 11:305-340.

Helpman, E. and P.R. Krugman. 1989. *Trade Policy and Market Structure*, Cambridge: MIT Press.

Helpman, E. and P.R. Krugman. 1985. *Market Structure and Foreign Trade*, Cambridge: MIT Press.

Kahneman, D. and A. Tversky, eds. 2000. *Choices, Values, and Frames*, Cambridge, UK: Cambridge University Press.

Krugman, P.R. 1992. "Does the New Trade Theory Require a New Trade Theory," *The World Economy* 15:423-441.

Krugman, P.R. 1991. *Geography and Trade*, Cambridge: MIT Press.

Krugman, P.R. 1990. *Rethinking International Trade*, Cambridge: MIT Press.

Kwok, V. and H. Leland. 1982. "An Economic Model of the Brain Drain," *American Economic Rreview* 72:91-100.

Luterbacher, U. and J.-P. Theler. 1992. "Some Game Theoretical Considerations on Negotiations About Migrations," presented to the Peace Science Society (International), Erasmus University, Rotterdam.

Petersen, W. 1958. "A General Typology of Migration," *American Sociological Review* 23:255-266.

Rogers, R. 1981. "Incentives to Return: Patterns of Policies and Migrants' Responses." In M.M. Mitz, C.B. Keely and S.M. Tomasi, eds., *Global Trends in Migration: Theory and Research on International Population Movements*, New York: The Center for Migration Studies.

Romer, P.M. 1990. "Endogenous Technological Change," *Journal of Political Economy* 98:S71-S102.

Romer, P.M. 1987. "Growth Based on Increasing Returns Due to Specialization," *American Economic Review* 77:56-62.

Romer, P.M. 1986. "Increasing Returns and Long-Run Growth," *Journal of Political Economy*; 94:1002-1037.

Thomas, B. 1973. *Migration and Economic Growth*, Cambridge, UK: Cambridge University Press, (second edition).

Todaro, M.P. 2000a. *Economic Development*, New York and London: Longman (seventh edition).

Todaro, M.P. 2000b. "Urbanization, Unemployment, and Migration in Africa: Theory and Policy." In D. Ghai, ed., *Renewing Social and Economic Progress in Africa*, New York: Macmillan,. (Also available as paper 1997/#104 at www.popcouncil.org/pdfs/wp/104.pdf.)

# ENVIRONMENTALLY INDUCED POPULATION MOVEMENTS: THEIR COMPLEX SOURCES AND CONSEQUENCES

*A Critical Review*

Nurit Kliot
*University of Haifa*

## 1. INTRODUCTION

This chapter deals with population movements which are induced by environmental forces – the latter are broadly defined. The passing of the bipolar world gave rise to increasing concern by the international community for non-military sources of instability – environmental degradation, rapid population growth, growing un- and underemployment and poverty, ethnic tensions, human rights violations, transnational terrorism and large-scale international migration. The fear of mass migration of environmental refugees – people driven from their homes as a result of ecological destruction – has become a major issue in the international community. The mutual dependence of the peoples of the world on a single common planetary biosphere means that the environmental decline of one country or region is a problem for the entire community of nations (Swain, 1939). The New World Order encourages an international solution to other existing global problems hitherto swept under the carpet. The reckless abuse of the human environment is one of such problems (Ezeonu and Ezeonu, 2000: 41-48). Though major research attention has been attributed to the South-North migration and East/West migration, most of the movements have been from rural areas to urban areas inside the developing countries or from one developing country to another (Swain, 1993).

The world's interest in environmentally induced migration is also reinforced by the assumed "security threats" attributed to such migrations. Movements of persons across borders affect security in international relations at three levels:

- The national security agendas of receiving and transit countries which perceive massive international population movement as a threat to their economic well-being, social order, cultural and religious values and political stability
- Relations between states, as movements tend to create tensions and



*J.D. Unruh, M.S. Krol and N. Kliot (eds.),*
*Environmental Change and Its Implications for Population Migration*, 69–99.

impair, bilateral relations, thus impacting upon regional and international stability.

- Irregular migration flows and involuntary population displacement have significant implications for the individual security and dignity of migrants and refugees (Lohrmann, 2000: 4-5).

Most of the current literature on environmental migrants or environmental refugees points to rapid population growth, poverty, shortage of nature resources and their degradation as the main causal forces of environmental migration. It is, therefore, useful to present the major streams of thoughts on the complex relations between the human population and resources.

Since Thomas Malthus published his *Essay on Population* in 1798, the notion that humankind will sustain hunger and disease if population growth is not adjusted to the limited resources of the planet, was adopted by many scientists and policy-makers in the western world. One way in which science and philosophy coped with issues of "over-population" and immigration is ideological and valuation vectors lenses. An ideology in the Marxist orthodoxy is the system of false consciousness which obscures the recognition of capital processes, class conflict and other related social dynamics. The valuation vector goes far beyond the class consciousness of a given group or the belief systems which inform its assessment of population size: the vector incorporates their entire approach to a phenomenon that is the means through which perceptions, intentions, expectations and values are mediated, altered and expressed (Petrucci, 2000: 327).

Probably the most prominent valuation factor is the New-Malthusiaian "limits to growth" and resource scarcity as has been advocated since the 1970's. Accordingly, the *Club of Rome* "Limits" school is the belief that the basic behavior mode of the world's system is exponential growth of population and capital, followed by resource depletion and collapse. New Malthusianism became linked, in some quarters, to repressive or neo-imperialist policies. One of the most well-known expressions for that is the famous formula depicted by Ehrlich and Holdern (1971) in which $I = P \times A \times T$ where "I" is Human Ecological Impact which is a function of "P" (population) and "A" – Affluence. (A=per capita consumption) and a technological factor T relating to impact per unit resource use (Ehrlich and Holdern, 1979). This neo-Malthusian algebra was hugely influential is casting environmental degradation entirely in terms of consumption and its associated parameters (Petrucci, 2000; Graeger and Smith, 1994; Goodland, 1994).

The environmentalists and "Greens" vary in their attitudes to the population-environment matrix: most saw population control as best negotiated but intolerant "Greens" positions favored measures in which anthropocentric concerns should lose absolute priority and the need for immigration controls in order to protect resources (Petrucci, 2000: 330).

A Marxist overview of the population/resources nexus is quite different. The argument is that the planet's 'ultimate limits' have not yet been reached and so cannot be held responsible for poverty and environment damage. Marxists believe that poverty is endemic to commodification and would occur (to some degree) at any population size by virtue of capital transformation of society's needs, organization and means of production. Capitalism tends to alienate individuals and communities from their environment.

Any attention to population size must, therefore, be balanced by thorough considerations of the ecological effects of social inequity and the associated means by which capital has been and is extracted and concentrated in global free market systems (Petrucci, 2000). Similarly, views on immigration are also presented in contrasting perspectives.

Chapman (2000) examined relations between immigration and the environment along two very different paths: that of utilitarianism vs. rights-based approaches. Within the framework of utilitarianism and according to the principle of equity, if two million migrants enter the USA, they should be divided between *all* USA families, namely, we import and spread out the misery equally (Chapman, 2000: 194).

Another path to solve a problem of migration is from the perspective of rights. Rights represent a belief that persons deserve respect and one solution to their problems would be to create the conditions necessary to provide meaningful aid to those countries generating migration.

A right to immigrate does not appear to be a basic right. Only in cases of political persecution or severe economic deprivation so excessive, and environmental degradation so comprehensive, could immigration be considered a *basic right* (Chapman, 2000: 196). Subsistence as a human right is basic. A healthy environment is just one of the conditions necessary for human well-being, but one that provides the foundation for many other rights. Environments are often degraded because of violations of other human rights like access to productive land, economic opportunities and medical resources, but they can also be degraded by supporting rights that promote unsustainable practices (Chapman, 2000: 198). In order to honor the right to a livable environment for necessitous immigrants – certain changes in First World Countries would need to take place, particularly amelioration of those political and economic conditions that make migration necessary (Chapman, 2000). In a similar vein, the various sections in this chapter will present at least two major "Explanations": "mainstream" and "critical" or "alternative explanation". This approach is adopted based on the assumption that the theoretical framework of environmental migrations is still in its formative stages.

The first part of the chapter reviews all the definitions of such migrants and the difficulties which accompany attempts to isolate this type of migration. The second section of the chapter deals with the basics of migration and circulation as these comprise the foundation of "environmental migration". The third part explores the relationship between environmental degradation, population growth and migration. The fourth section discusses the concepts of resilience and vulnerability in their relationship to environmentally-induced movements. The fifth part presents and discusses a few types of environmental "push" forces such as natural hazards, disasters, and development projects. The sixth and last section deals with the environment-migration-security nexus. The final section presents the policy implications of the former discussed sections.

## 2. THE BASIC DICTIONARY OF ENVIRONMENTALLY-INDUCED DISPLACEMENT: HISTORICAL (EVOLUTIONARY) APPROACH

The most frequently coined phrase for environmentally-induced population migration is "environmental refugees". El-Hinnawi, who is often quoted as the source of this definition, described a process in which land degradation, mostly in Third World countries, has been the main factor in the migration of subsistence farmers into the shantytowns of major cities, producing desperate populations vulnerable to disease and natural disasters (El-Hinnawi, 1985: 4). Tolba adopted the term and stated that "throughout the world there is copious evidence that the carrying-capacity of many life support systems is being overloaded to the breaking point, and where such systems have collapsed, the options for the poor are either to flee or stay put and starve" (Tolba, quoted in Lazarus, 1990: 14). Jacobson (1988) extended the term "environmental refugees" not only to victims of natural disasters and land degradation but also to people who are forcibly evacuated from their homes due to large development projects or because of industrial accidents or disasters such as Bhopal (Jacobson, 1988). Lonergan (1995) added a fifth category of environmental migrants which is related to conflict and warfare.

More recent definitions for these two types of population movement were summarized by Lonergan (1995, 1998) as follows: "Environmentally induced population movements are caused by a) environmental stress such as an earthquake, cyclone or other natural disaster; b) Development projects which create a permanent change in the habitat, thus forcing people to leave their homes; c) Cumulative changes or "slow-onset" changes such as deforestation, land degradation and climate warming; d) Industrial accidents such as the nuclear accident at Chernobyl; e) Conflict and warfare: environmental degradation is considered by many to be both a cause and effect of armed conflict" (Lonergan, 1988; Lonergan, 1995).

In addition to "environmental migrants", "ecological migrants" and "environmental refugees", another important concept is "Internally Displaced People" (IDPs) who could be, sometimes, refugees, or sometimes "environmental migrants." It would seem that IDP's are "persons or groups of people who have been forced to flee or leave their homes or places of habitual residence as a result of or in order to avoid the effects of armed conflict, situations of generalized violence, violations of human rights; natural or human made disasters and who have not crossed an internationally recognized state border (McClean, 1998: 10-12). The distinction between IDP's and refugees is becoming increasingly unclear (*Refugees*, *106*, 1996: 3). In addition to the above "fuzziness" or complexity of the concept a few more confusions of this nature can be depicted. *First*, and most important, there is criticism of the "taken-for-granted" adoption of the linkage between environmental degradation and population movement. "There is too often an uncritical acceptance of a direct causal link between environmental degradation or environmental stress as a possible cause for population displacement, that can be separated from other social, economic or political causes" (Lonergan, 1998: 81).

Lack of employment, poverty and general deprivation is closely connected with political situations; it is, therefore, difficult to delineate the tenuous frontier between political and economic factors. Political crises that have rocked many developing countries have left them unable to harness and adequately manage their resources (Castillo, 1995; quoted in Afolayan, 2000: 15).

The seven main determinants of emigration dynamics as observed from many studies in developing countries are: the demographic factor, the vulnerable economy and the debt problem, the socio-cultural factor, ecological disasters, social networking, government migration policies and regional economic integration (Afolayan, 2000: 10). In a case-study from Morocco the authors stipulate this ambiguity as follows: "Although Morocco faces a wide range of environmental and political problems, Moroccan immigrants can be defined neither as political nor environmental refugees. If Moroccans leave for Spain for socio-economic reasons emigration is also influenced by geographical proximity of the two countries and by the large network of relations and friends already in Spain" (Carella and Pace, 2001: 71).

This is a reflection of course on the research of migration movements in general which is "an extremely varied and complex manifestation and component of equally complex economic, social, cultural, demographic and political processes operating at the local, regional, national and international levels" (Castles and Miller, 1993). People are on the move for such a variety of reasons that even other concepts such as "refugee" are less and less capable of capturing their experiences, which are likely to involve a complicated mixture of compulsion and choice.

In addition to the genuine difficulty in the determination of all relevant variables and their relation to each other, a critic in a similar vein states that the present concept of environmental migrants deals with the "environment" as a single cause of migration, whereas the process (of migration) is based on a set of reasons explaining why an individual or family may be forced to leave their home or relocate (Black, 1998). The use of the term "environmental" implies a false differentiation between overlapping and interrelated categories of sources for migration (McGregor, 1993). Kibreab (1994) claims that the definition of environmental refugees lies in the agenda of policymakers in the North who wish to further restrict asylum laws and procedures and could lead to the withdrawal of asylum from those who currently receive it (Kibreab, 1994). Other problems that were raised are concerned with the use of "refugees" to depict environmental migrants. This has to do with the entitlement of refugees as against "simple" environmental migrants.

Not only does this conflict with the standard definition of refugees which was codified in the 1951 Convention on the Status of Refugees and 1967 Protocol, but it may undermine current work towards using broader human rights criteria to determine refugee status (Lonergan 1998 a). The current refugee definition can be used to limit refugee status to those outside their country of origin with a well-founded fear of persecution, the latter being defined in narrow political terms.

In 1994 at the Ditchley Conference, Report No. D94/10, S. Collins stated: "Although there was some support for applying the term "environmental refugee" to people forced to move in such circumstances (i.e., environmental degradation, concentrated population and poverty combine with civil strife) the overall consensus seemed to be that the concept "refugee" should be reserved for people whose migration was forced by political or man-made factors" (*RPN*, January 1995: 6).

In practice, UNHCR and other aid agencies have long assisted a broader group than those included by the narrow 1951 Convention (McGregor, 1995: 7). In addition, the Accord of the Organization of African Unity (1969) extended its definition of "refugee" to include also environmental refugees (Lonergan, 1998). The official journal of UNHCR, the periodical *Refugees*, expressed this new trend well. "We (UNHCR) have abandoned traditional refugee assistance based on individual registration of refugees and individual refugee entitlement. We have done so because we have found that, particularly in areas of instability and underdevelopment, traditional assistance creates dependency among refugees and complicates assistance to the host community which is equally in need. Under (UNHCR) cross-mandate approach, the needs of the entire *Environmentally Induced Population Movements* community are assessed, including the needs of refugees, internally displaced, drought victims, demobilized soldiers and returnees" (*Refugees,* December 1992: 12).

Environmental refugees are of course part of the large number of persons who are living outside their countries of origin which were estimated in the late 1990s at 150 million (Lohrmann, 2000: 5). They include migrant workers and their families, refugees and permanent migrants.

Developing countries in general are said to be generating close to 55 percent of the world's millions of migrants by the early 1990s. Some of these countries host a high percentage of emigrants. However, an increasing proportion of emigration is to the developed countries of the North and the Middle East. The magnitude of Africa's refugee population is related directly to warfare, persecution, poverty and environmental calamities (Afolayan, 2000). The number of migrants in some parts of Africa is staggering: estimated 1.76 million Sahelians (many probably environmental migrants) lived in other African countries (Findley and Traoré, 1995: 485).

If there is ambiguity, vagueness and contradictory interpretation of the term "environmental refugees", it is not surprising to learn that estimates of the magnitude of environmentally-induced migration are also contested and debated. According to Myers (1995) there are now at least 25 million "environmental refugees" and the International Organization for Migration goes further, providing that by the turn of the century there may be one billion persons who have been environmentally displaced from their original homes" (IOM, 1992).

These numbers were vigorously criticized by Black (1998) and Lonergan (1998), basically because they are not based on strict empirical research, inaccurate statistical data, vague definitions and figures which are "educated guesswork" (Lonergan, 1988: 8; Black 1998: 24-5). Estimates of environmental refugees vary between 15 and 64 million – depending on the various sources of data (Myers, 1995; IOM, 1992; Lohrmann, 2000).

Putting the numbers aside, there is a general agreement that environmental degradation and environmental stresses play an increasingly important role in population displacement. These roles will be examined in the following sections.

In this paper, the broadest definition of environmental migration is adopted in order to discuss, in depth, all possible conceptual avenues in their relation to that phenomenon and to focus on shortcomings and weaknesses in the current theoretical debate of environmentally-induced displacement.

# 3. THE ESSENCE OF MIGRATION AND CIRCULATION

## 3.1 Migration: Conceptual framework

Migration, even in the context of environmental stresses, is a form of mobility. The two categories of mobility are circulation and migration. Circulation has been defined as "a great variety of movements, usually short-term, repetitive or cyclical in character, but all having in common the lack of any declared intention of a permanent or longstanding change of residence" (Zelinsky 1971). Included in these type of movements are: nomadism, transhumance pilgrimage, tourism, and commuting. Many environmental migrants, particularly those victims of natural and man-made disasters, are moving in a circulatory fashion: When the calamity ends, they return to their place of origin to re-build their homes.

*Migration* is normally used only to describe a movement from one administrative unit to another that results in a permanent change of residence. Migration could be *internal* (within the border of a state) and *international*, from one country to another. Migrants can be in-migrants or out-migrants when they move across internal administrative boundaries. They are immigrants or emigrants when they cross international boundaries (Hornby and Jones, 1993; Clarke, 1973). Migrations are embarked upon from an area of origin and are completed at areas of destination. Many environmental migrants are internal migrants but environmental migrants may also be refugees, as is the case in Africa, and then they become international migrants.

## 3.2 Processes and Patterns of Migration

The start of migration studies is usually traced back to Ravenstein's 1885 "Laws" in which the relations between distance and the volume of migration was central, emphasizing reasons for migration and with the selection of migration ("who moves"). The most basic classical model of migration is the push-pull concept. This model states that for any individual the decision to migrate results from the interplay of two forces: pressures at the permanent place of residence ("pushes") and inducements from a number of destination ("pulls"). Examples of push factors are low wages, unemployment, political racial and religious oppression and natural disasters such as drought, famine and floods. Pull factors include employment offers and opportunities, better medical and social provision, political and religious tolerance and in the case of environmental migrants – "environmental security". It is clear that in some cases only "pushes" will be of major significance. This is true for unforeseen disasters of all kinds. For many environmental migrants this is the most important force of movement.

Migration is to be selective or differential, namely, age, gender and socio-economic status determine who will move. But for many environmental migrants migration is not selective – whole communities are forced to leave. In other cases only the relatively rich can afford to move. This is the case in many poor areas in India, for example, a person needs to have some resources if he/she wants to leave their area of origin. Major forms of migrations are classified as rural to urban and this is also true for many environmental migrants who move to the major cities after a disaster in which they have lost all their property. There are also urban to urban migrations, migrations between core and periphery (in both directions) and from more "traditional" areas to the most modern ones. As for the distance of movement, we differentiate between *local* migration, intra and inter-spatial components of migration, inter-district, and intra- inter-provincial migration. International migration may be: First World ("North") to other Third World countries ("South"), First World – to Third World ("South to South") and from First World to Third World countries (Parnwell, 1993). There is no accurate account of environmental migrants and even their number is disputed, but most of the case studies on environmental migrants come from the poor developing countries in Africa, Asia and Latin America.

The temporal dimension of migration is also crucial: people can move temporarily for a few weeks, months, or a year, and they can move for a longer period or when they emigrate permanently to another destination. As stated before, many environmental migrants are temporary migrants, but there is not enough information from field research and case-studies to generalize on this type of movement.

The abovementioned "classical" models of migration are based on economic reactions. De Haan (1999) argues that patterns of migration are determined by social and cultural institutions, embedded in local customs and ideologies. Many circulations in the developing world which are perceived as "forced environmental migration" are institutionalized mechanisms to cope with resource scarcity (de Haan 1999).

# 4. POPULATION GROWTH, ENVIRONMENTAL DEGRADATION, AND MIGRATION: "PUSH" AND "PULL" FRAMEWORK OF ANALYSIS

## 4.1 Population Growth as a "Push" Factor

Population growth which leads to environmental degradation or population growth which is accompanied by poverty are perceived as major "push" forces for environmental refugees. However, some migration movements are motivated mainly by the "pull" forces of cities or employment opportunities in the immigration destinations.

Population growth, and more particularly the further expected and inevitable rapid population growth, prevails in almost every explanatory matrix for environmental degradation and disaster pressures. In the year 2000 the population of the planet surpassed 6.0 billion, and a further 2 billion more people are expected in the following twenty-five years (Blaikie, Cannon, et al., 1994). Global population growth is currently estimated at 1.3 percent, representing an annual addition of just under eighty million (Benedick, 2000: 7). The U.N. estimated that in fifty more years, by 2050, the world's population will probably lie within a range of 7.3 billion (low variant) to 10.7 billion (high variant). Developing nations will comprise eighty percent of the global population.

Ninety percent of future population growth is predicted to occur in low and middle income nations, so population control is needed there. Indeed, rapid population growth within subsistence economies compounds environmental degradation such as soil erosion, depletion of natural capital deforestation, indirect effect on watersheds and irrigation and so forth, thereby increasing the impact of human activities on local environments. Thence, people leave those environmentally abused areas, becoming environmental refugees (Goodland, 1994).

In the transition toward environmental sustainability, population stability is the only acceptable goal. Most of these countries already face problems of health and deteriorating environments, many of which are hazard-prone, with large numbers living in ecologically fragile areas and now confronting combined environmental impacts brought about by both poverty and early stages of industrialization (Benedick, 2000: 7).

The high population growth in the developing countries has multiplied pressure on all renewable resources. At the same time most of the developing countries are extremely dependent on their renewable natural resource base to sustain their economic activity, to which the scarcity tendency has been shifted. Due to environmental destruction, there may be a reduction in the availability of cultivable land, forests, fresh water, clean air, fish resources for the consumption of humankind. Greater demand for these resources is expected due to population growth, urbanization and developmental initiatives (Swain, 1993). Furthermore, environmental change can lead to dramatic reduction in agricultural output, lack of water for irrigation, and massive deforestation. In the developing countries, where agriculture is the most important source of subsistence, its decreased production might result in the loss of people's livelihood. The loss of means of living due to environmental degradation could lead to the migration of affected populations (Swain, 1993: 12-14). The continent suffering most from land degradation that is the outcome of population growth is Africa. In sub-Sahara Africa deserts are fast encroaching southward, forcing populations to abandon their traditional base and to migrate to coastal regions. An estimated 50 percent of the poor live on marginal land characterized by low productivity and highly susceptible to environmental degradation. Many landless marginalized farm laborers, deprived of the means to improve their living conditions and alleviate poverty, abandon rural areas in search of survival in the cities. Drought, desertification and cyclical famines in 1973-4, 1984-5, 1990-91 and 1994 have triggered waves of environmentally displaced persons from parts of the region (Adepojo, 1995).

One of the outcomes of the accelerated population growth is the increased concentration of population in urban areas. Cities encroach on farmland, ecologically sensitive wetlands or watersheds, increase overcrowding which bring greater vulnerability to epidemics, eruptions, storms and flooding (Blaikie Cannon et al., 1994: 35; Benedick 2000: 15).

It is difficult to object to the idea that population growth is a significant global pressure contributing to increasing vulnerability. However, the linkage remains uncharted. Because of many mediating and intervening factors such as income levels, consumption patterns, technological infrastructure and the economic and political institutions in a particular society, it is difficult to establish, with scientific precision, clear correlations between population pressures and environmental degradation.

Numerous examples are presented to substantiate the link between environmental change and population movements but the most frequent come from Africa (Rwanda, South Africa, the Sahel), Central America (El Salvador, Haiti, Chiapas, Mexico), the Indian sub-continent (Bangladesh, Pakistan) (Homer-Dixon and Blitt 1998; Myers, 1995; El-Hinnawi, 1985). There is little doubt that each of these regions has experienced significant environmental stress: droughts, deforestation, soil degradation and flooding. But it is also clear that social, economic and institutional processes are involved and rapid population growth, and inequitable land distribution, civil war, and extreme poverty all play their role in resulting population displacements (Lonergan, 1998: 10).

## 4.2 Population Growth, Poverty and Environmental Degradation as "Push" Factors

One of the most contentious issues is that of population growth in relation to poverty. The key factor is that certain populations are becoming more vulnerable to environmental change because of other factors; primary among these are poverty and resource inequality (Lonergan, 1998: 11).

The link between poverty and environmental degradation is twofold: the poor have not only suffered disproportionately from environmental damage, but have become a major cause of ecological stress themselves. Increasing poverty in developing nations places stresses on environmental sources and sinks. Poverty forces people into short-term environment mismanagement (Goodland, 1994).

One-third of the impoverished population in the world are Africans. Twenty-nine African countries are classified as Least Developed Countries. Among the many factors that contribute to the region's persistent poverty are low agricultural production, low non-form income, low education, poor health and nutrition and rapid population growth (Adepojo, 1995: 323). The effect of poverty on emigration dynamics can be exemplified in a case study from the Philippines.

Poverty in the Philippines grew steadily through the 1970s and 1980s until it reached 50 and 60 percent in rural and urban areas respectively in 1988. Displaced urban workers and landless agricultural workers migrated to the upland frontier areas in search of better opportunity and converted forests to subsistence agricultural production. These lands are effectively open to deforestation by improper logging and illegal homesteading. The environmental results have been unsustainable agriculture on land that is poorly suited to agriculture in any event, plus erosion and degradation of the upland forest and downstream losses due to heavy sediment deposition on commercial agricultural land, in water catchments intended for hydroelectric power production and in the coastal fishery (Amacher Cruz et al., 1998).

At the international level, the incidence of poverty and quest for employment increase migration from poor countries to rich, seeking to leave environmentally damaged, unproductive or unhealthy lands. Poverty and environmental abuse such as polluted forests and soils, unhealthy water contribute to the refugee flow from Eastern Europe into Germany and Sweden, and from El Salvador to the USA (Goodland, 1994).

But there is a considerable debate about whether population growth is a cause or consequence of poverty in the Third World or a complex interaction of both (Blaikie Cannon et al., 1994). An even more controversial question is the possible connection between poverty and voluntary or non-voluntary migrations. Gosine (1996) for example, stated that since Thomas Malthus' *Essay on Population* was first published in 1798, overpopulation propaganda has been used as an effective tool to blame the poor and excuse the rich for the former's state of poverty. The neomalthusian ecologist Paul Ehrlich warns that too many people leads to too little food and a dying planet (Ehrlich and Holdern, 1974). Gosine claims that evidence from the experiences of indigenous peoples in Central and South American reveals that manipulation of resources by external forces, not resources scarcity, is responsible for the apparent creation of environmental crisis (Gosine, 1996: 27-8).

Gosine went even further, and stressed that the purpose of overpopulation propaganda is to reinforce a global system of economies which allows the rich North to have its own cake and eat that of the South, as well. Not only are natural resources and wealth unequally distributed in Latin America but so, too, is the experience of citizenship. Refugees (including environmental refugees) are created by national and international state policies which reject the notion of equal citizenship or equal environmental rights for all people.

The critical overviews on environment–resource-population linkages focus on two major issues: depletion of resources by the North (namely the industrialized nations) vs. the South (the developing countries) and the issue of unequal distribution of wealth between nations and within countries. The world is hurtling away from environmental sustainability at an unprecedented rate. The combination of unbridled consumerism, waste and the growth in scale of industrial economies results in the increased use of the global environment as sources of raw materials and as sinks for wastes (Goodland, 1994).

Environmental problems worldwide are ultimately the result of an adverse global tendency: although the resources and ecological durability of our planet are finite, humans conduct themselves demographically and economically as though this were not the case. The faith in free market forces in which Northern economic welfare will eventually percolate into Southern economy was found to be disappointing. From environmental and economic perspectives it is a problem that the prescription to raise Northern income, as a means to promote growth in the South, fails to alleviate poverty, worsens income distribution within and among nations and reduces environmental sustainability (Goodland, 1994: 28).

Inequitable income distribution also has an impact on the environment. The distributional profile of a country significantly determines whether economic development will have either a positive or negative effect on the rate of forest loss. In countries where levels of inequality are high, development will tend to exacerbate deforestation rates while in countries where distributional profiles are nonegalitarian, the negative effects of growth and development on forest cover will be ameliorated. The more egalitarian a country is – the more deforestation it is likely to have (Koop and Tole, (2001: 193-196).

## 4.3 Migration as a Coping Mechanism

Another criticism, on "mainstream" explanations about the relations between poverty and population movements, highlights migration as a coping mechanism or as a voluntary adaptation to poverty. De Haan (1999) pointed to the fact that there is a research focus on population movements as a result of crises – environmental, economic or demographic. Yet, migration is also a "normal" element of most, if not all, societies. Migration is the only option for survival for many poor people, after alienation from the land. Migration is a significant factor in agricultural or rural development. Views about migration are often based on an assumption of sedentarism. There is much evidence to challenge this sedentary bias and to view population movement as the norm rather than the exception. Examples of high rates of mobility as survival strategies in a risky environment come from the Sahel, from southern Africa and from Asia (de Haan, 1999: 8).

Seasonal rural to urban migration of young men is becoming an increasingly important part of the rural economy of the West African Sahel. The Fulani, in common with other Sahelian pastoralists are highly mobile and are associated with large-scale transhumance, moving substantial distances with their herds in search of pasture and water. Since the major droughts of the early 1970s when herd numbers were drastically reduced, the Fulani of Northern Burkina Faso have been engaging increasingly in temporary migration to cities to seek alternative sources of livelihood to supplement an agropastoral mode of subsistence (Hampshire and Randall, 2000).

Apart from drought related migrations, the Sahel has always been a region of important migration movements because of the significance of agriculture. In the Sahel population dynamics, including migration, must be considered from the ecological context as the consequence of an imbalance between human and land-environment systems. Sahelian migrants also move to Europe. By 1990 the number of Africans registered in France was 77,000, but many more are clandestine migrants (Findley and Traoré, 1995). Lock, Adger and Kelly (2000) showed that voluntary migration in Vietnam, which is circular in its nature, is a coping measure and a fundamental component of household coping strategies which contributes to livelihood security and resilience at the household level (Lock, Adger and Kelly, 2000: 30).

A counter perspective is provided by Nord (1998), who found that the combined migration of the poor (and non-poor) tends to maintain, and even exacerbate, pre-existing patterns of poverty concentration in rural America. He also found that the poor respond to a different set of migration pushes and pulls than do the non-poor (Nord 1998: 349).

To summarize, there is evidence that the relations between population growth, poverty and migration is complex and, again, under-researched in relation to many facets of environmental resources. In some societies migration is not forced and does not constitute a calamity but a way of life. In many cases in the Third World the migration is circulatory and temporary, while in other cases it is permanent, and very often it is to the urban agglomerations in the Third World countries, where the poor remain poor or become poorer. One explanation of the variation in patterns of reaction to environmental forces lies in the concepts of resilience and vulnerability. These concepts were evolved within the theoretical framework of disasters and natural hazards, but are often used in a broader connotation to depict environmental migration in general.

# 5. THE OTHER "PUSH" FORCES: DISASTERS, NATURAL AND MAN-MADE HAZARDS, DEVELOPMENT PROJECTS AND DISPLACEMENT OF PEOPLE

## 5.1 Disasters

Natural hazards have been defined in four ways:

a) "A naturally occurring or man-made geologic condition or phenomenon that presents a risk or is a potential danger to life or property."
b) "An interaction of people and nature governed by the co-existent state of adjustment of the human use system and the state of nature in the natural events system.
c) "Those elements in the physical environment which are harmful to man and caused by forces extraneous to him."
d) "The probability of occurrence within a specified period of time and within a given area of a potentially damaging phenomenon" (Alexander, 1993: 4; Blaikie Cannon et al., 1994).

Disasters occur not only because natural hazards harm people, but they are also the product of the social, political and economic environments – mainly because of the way those environments structure the lives of different groups of people. Disasters are often discussed in relation to natural hazards but we are reminded that some definitions of disasters include wars, civil strife, and famine (Blaikie Cannon et al., 1994: 3-4). Many factors affect disaster victims: social factors such as economic situation, income distribution and poverty, and living in areas susceptible to natural hazards. Thus, many disasters are usually a complex mix of natural hazards and human action (Abbot, 1996).

Vulnerability is the key concept in relation to disasters. Vulnerability is the capacity of a person or a group to anticipate, cope with, resist and recover from the impact of natural hazards (Blaikie Cannon et al., 1994). Some groups in society are more prone than others to damage, loss and suffering. The most vulnerable groups are those that also find it hardest to reconstruct their livelihood following disaster.

Natural sources of hazards constitute plate-tectonics (earthquakes, volcanic eruptions and mass movements such as landslides, mudslides, avalanches and catastrophic subsidence); climate (cyclones, tornadoes, hurricanes, drought and famine, floods); ocean waves (and tsunami) and fires (Abbott, 1996). Hazards have varying degrees of intensity and severity. A disaster occurs when a significant number of people experience a hazard and suffer severe damage and/or disruption of their livelihood.

The root-causes for vulnerability (to disasters) are a set of well-established widespread processes within society and the world economy. The most important root causes that give rise to vulnerability are economic, demographic and political processes. Such root causes are normally a function of economic structure, legal order, the functioning of states (peace and war, for example) and the distribution of power. Dynamic pressures "translate" the effects of root causes into the vulnerability of unsafe conditions. Dynamic pressures include reduced access to resources as a result of the way regional or global pressures such as rapid population growth, epidemic disease, rural-urban migrations, rapid urbanization, war, foreign debt and structural adjustment, etc., work through to localities (Blaikie Cannon et al., 1994: 24). People are vulnerable and live or work under unsafe conditions. Their presence in hazardous places is often the result of broader political economic pressures which may displace weaker groups.

## 5.2 The Concepts of Social Resilience and Vulnerability

The concept of resilience is widely used in ecology but its meaning and measurement are contested. Social resilience is an important component of the circumstances under which individuals and social groups adapt to environmental change (Neil Adger 2000: 347). Ecological and social resilience may be linked through the dependence on ecosystems of communities and their economic activities. The question is whether societies dependent on resources and ecosystems are themselves less resilient and whether institutions themselves are resilient to change (Institutes in this case are defined in the broadest sense to include habitualized behavior and rules and norms that govern society, as well as the more usual notion of formal institutions with membership, constituencies and stakeholders). Social resilience is a community's or an individual's ability to withstand shock and stress without significant upheaval. These shocks and stresses could be caused by significant changes in social structure and life-styles brought about by such occurrences as government politics, civil strife or environmental hazards (Lock Adger et al., 2000).

The concept of resilience is directly related to the concepts of social vulnerability and criticality which also refer to the environment-society relationship. Vulnerability broadly defined as the potential for loss is an essential concept in hazards research. Vulnerability assessments are used to determine the potential damage and loss of life from extreme natural events/disasters (Cutter, 1996). Cutter (1996) made the important differentiation between three meanings of vulnerability:

1) *Vulnerability as pre-existing condition*, namely, the distribution of some hazardous condition, the human occupancy of this hazardous zone such as flood plains, coastal areas, seismic zones) and the degree of loss associated with the occurrence of particular event (Cutter, 1996).
2) *Vulnerability as a tempered response*: This facet of vulnerability focuses on coping responses including societal resistance and resilience to hazards. The nature of the hazardous event or condition is usually taken as a given or at the very least viewed as a social construct, not a biophysical condition. This perspective highlights the *social construction of* vulnerability, a condition rooted in historical, cultural, social, and economic processes that impinge on the individual's or society's ability to cope with disasters and adequately respond to them (Cutter, 1996: 533).
3) *Vulnerability as hazard of place:* In this perspective, vulnerability is conceived as both a biophysical risk as well as a social response, but within a specific area or geographic domain (Cutter, 1996: 533).

Another concept, distinct from vulnerability, is environmental criticality which "refers to situations in which the extent or rate of environmental degradation precludes the continuation of current use systems or levels of human well being, given feasible adaptations and societal capabilities to respond" (Kasperson et al., 1995: 25). This *criticality* is a state of an area or region which incorporates various attributes including environmental degradation and some aspects of human adaptation (Kasperson et al., 1995). The concept of environmental criticality is therefore applied to geographical areas or to resource systems (Neil Edgar, 2000).

In the context of this paper, our focus of interest is to the possible conceptual linkages between resilience, vulnerability, criticality and migration. Mobility and migration are (among other indicators) important indicators of resilience. However, resilience or changes in resilience cannot simply be inferred from the presence or absence of migrants in any area of community. Significant population movement can be evidence of instability or could be a component of enhanced stability and resilience, depending on the type of migration. Migration and circular mobility occur for a plethora of reasons. Displacement migration may be caused by a deleterious state of affairs at the home locality and often has negative impacts on social infrastructure in both sending and receiving areas.

Circular and seasonal migration are important aspects of social stability. In many parts of the agrarian world seasonal migration contributes to livelihood security and resilience at the household level through remittances providing opportunities for diversification and reduction of resource dependency (Neil Adger 2000: 357). But in the face of significant external stress, population displacement is often an indicator of the breakdown of social resilience. Coping strategies of households which face extreme food insecurity (which might be caused by diverse factors ranging from climatic extremes to wars) are short-term adjustment and adaptation to extreme events. They are usually involuntary and almost invariably lead to a different subsequent state of vulnerability to future famine situations (for examples of coping strategies see Corbett, 1988; Jodha, 1975). This discussion is needed to replace our perception of environmental migrants not necessarily as simply a group of victims that is wholly and completely helpless and subject to the whims of nature, but for the understanding of the more complex behavioral patterns in which migration is a function of social-resilience and vulnerability.

## 5.3 Disasters, Vulnerability and Migration

Table 1 shows the estimated number of people affected, but not killed by natural disasters 1980-1990. Not surprisingly drought (and often famine) claim about half of the victims of natural disasters followed by floods.

Many environmental migrants, caused by disasters, are found in the developing countries, which are extremely susceptible to natural catastrophes.

The disaster areas of developing countries often show common features: productivity is low, infrastructure is inadequate and rural areas are politically and economically marginal. In South Asia, floods, cyclones and in some areas drought and desertification have led to both internal and international migrations (Shah, 1995; Afaloyan, 2000).

Bangladesh is particularly vulnerable to floods: floods in August/ September 1988 affected 48 million people destroyed at least one million homes and brought the economy to a standstill for six months. A flood in 1991 left 200,000 persons dead and millions homeless (Myers, 1993).

*Table 1: People Affected by Natural Disasters 1980-1990*

| Disaster Type | Approximate number affected (thousands) |
|---|---|
| Droughts | 952,200 |
| Floods | 524,600 |
| Windstorms | 150,300 |
| Earthquakes | 28,400 |
| Landslides | 3,100 |
| Volcanic eruptions | 620 |
| Wildfires | 610 |
| Tsunami | 1 |
| Total | 1,659,831 |

Based on Chester et al., 2001: 93

In Latin America and the Caribbean land shortage, lack of viable resource base and depleted soil fertility reinforce people's vulnerability to disasters such as cyclones and other storms, and to earthquakes. The earthquake of February 1976 in Guatemala affected sixty-four percent of the national population who were left homeless as a result of damage or destruction (Alexander, 1993: 496). In Africa, persistent drought and famine triggered migrations in the Horn of Africa and in the Sahel, the most fragile part of Africa (Oucho, 1995: 36-7; Afoloyan, 2000). War, repression and tyranny frequently blend with the effects of natural disasters (Alexander, 1993: 503).

The forecasts for the future are bleak. The climate change forecasts for the two periods 2050's and 2080's show an increase in the number of people at risk in millions due to higher temperature. Larger and more vulnerable exposed population in the 2080's than in the 2050's is due to the greater number of people, more than from increases in temperature or changes in precipitation and sea level rise (Parry, Arnell, McMichael et al., 2001: 181). Global warming, and the more rapid melting of the Himalayan glaciers could create 30 million environmental refugees in India and 125 million in Bangladesh by the year 2050 (Myers, 1993).

A final note in this subject relates to the direct (causal) relationship between disasters and migration which is often assumed but not accompanied by documented evidence. Many of the disaster-affected people are, most likely, temporary migrants and when the calamity is over, they try to return to their home, but many disasters are not documented and do not provide accurate information on disaster victims' movement *after* the disaster (Alexander, 1993; Blaikie et al., 1994). There is more exact information on people who are killed in disasters on damaged property. The few detailed case studies from Ethiopia (the 1972-3 famine) and the 1980s wild fires in Kalimantan, Indonesia confirm the above generalization and point to the need for in-depth studies into the migratory pattern which follows disasters (Seamen et al., 1977; Shaker, 1987).

## 5.4 What Lies between Environmental Migrants Asylum Seekers and Economic Migrants?

Economic migrants are often classified as people who are motivated by economic incentives to immigrate but are, in one form or another, environmental migrants or pretend to be asylum seekers (refugees) who face life-threatening danger if they return to their country of origin. Most of the refugee asylum seekers and economic migrants in the world move from one developing country to another (Lohrmann, 2000: 12). Refugees and migrants looking for work are rapidly becoming the driving force for many government decisions and policies and often strain relations between neighboring states (India-Bangladesh and Honduras-El Salvador are examples of such a situation).

As the EU (and to a lesser extent, the USA) virtually halted all primary immigration except for purposes of family unification, more and more people, from the less developed realms in Europe (Albania, C.I.S. countries) try to enter Europe illegally or as asylum seekers. Most of the Albanians who arrive in Italy are poor peasants looking for jobs. But most of the asylum seekers come from Third World countries – some of them are truly refugees needing protection – but mostly they are victims of natural calamities, underdevelopment poverty, socio-economic problems and ecological disasters. With few opportunities for authorized migration they try to enter European countries legally or illegally, often endangering their lives (*Refugees*, No. 85, May 1991: 16-17). Many incidents of trafficking those migrants to Europe end in death (*Refugees*, No. 111, 1995: 26). It is important to emphasize that it is almost impossible to discern the causes of migration waves such as the ones from former C.I.S. countries which include refugees from the conflicts in the Caucasus, Crimean Tatars, ecological migrants from Chernobyl and Kazakhstan or Russian-speaking people leaving Central Asia en masse because they do not feel at home there any longer (*Refugees*, No. 103, 1996: 18).

The temptation to reach Europe as legal or illegal immigrants is great because often those people are not returned to their countries of origin and remain illegally in their destination, often employed as cheap labor with no rights, and no social protection (*Refugees*, No. 85, May, 1991). European and North American needs for cheap labor are still great and illegal migrants or economic migrants often fill manual low-paid jobs which the local population will not take. Europe and North America have become a magnet to the impoverished people of the East, Southeast and South, some of whom are environmental migrants and others who are "just economic migrants in desire of a better job" (*Refugees* 87, October 1991; *Refugees* 88, January 1992).

The combination of push and pull forces, the difficulty of differentiating among refugees, asylum seekers or economic migrants, have brought some experts to call for a reassessment of the way the international community deals with refugee problems. They say that instead of waiting for people to be displaced, international organizations should be concentrating on removing root causes such as violation of basic human rights and extreme poverty within the countries of origin (*Refugees*, No. 88, January 1992). However, doing so would require vast amounts of foreign development aid in a world of shrinking resources. The impoverishment of the South, through debt financing alone, has itself been a factor in the enrichment of the North. To pay that debt, more and more of the environment of the poor countries is being sacrificed: hardwoods, fish, forests for grazing land – all are being pawned or virtually given away.

"The poor escape our notice almost entirely unless war or an exotic disease breaks out on yet another program with IMF is signed (when their national debt brings their economies to bankruptcy). The challenge is that of mobilizing global science and technology to address the crises of public health, agricultural productivity environmental degradation and demographic stress confronting these countries" (Sachs, 1999, *The Economist* August 14th: 17-19). That this observation puts our behavior in a good light is reflected by another excerpt from the same issue of the *Economist*. "For many migrants from Honduras and Nicaragua, the devastation visited upon their countries by Hurricane Mitch has brought an unexpected blessing in its wake: those living illegally in the USA when hurricane Mitch struck. They have been given time to apply for a temporary stay. Costa Rica offered a far more open-armed amnesty to illegal migrants from the Central American neighbors (*The Economist*, August 14, 1999: 46).

## 5.5 Development Induced Displacement

Among the categories of environmental migrants are included those displaced by diverse development activities. The most obvious examples are the many thousands who were evicted due to the construction of large dams such as Sardar Sarovar in India or the Three Gorges Dam in China (Stein, 1998).

There are two forms of displacement: direct displacement which takes the form of eviction by the inundation of a valley or by legal prohibition of the right to collect forest products. Indirect displacement occurs through more circuitous chains of development impacts, such as erosion resulting from destructive logging practices that lead to landslides and to river flooding, or chemical pollution that kills fish and thus undermines the livelihood of fishing communities (Penz, 1997: 3-4).

There are many less conspicuous cases of development induced displacements. Among these are the impact on forest dwellers from reserving forests for industrial and commercial logging, and from converting forests to agricultural uses. The introduction of mining, and the chemical pollution accompanying it, affects rivers, fishing and paddy farmers. In many of these cases indigenous people are involved and their land rights are not respected. Though most of the development- induced displacement takes place in the poorer countries, the processes are not absent from developed countries such as the hydro project in Quebec, Canada, which displaced indigenous people who lived in the area for many generations. A detailed case study from Haiti shows severe environmental degradation, either as a direct consequence of a large project (a dam), or an indirect consequence of certain agrarian development strategies which induced a significant number of people to migrate into Haiti's urban centers (Howard, 1997).

The Sardar Sarovar Dam Project, one of the more controversial development projects which raised world-wide opposition, when completed will displace at least 100,000 people in 245 villages (Uytterlinde, 1997: 12-17). India has a very bad record in resettling and compensating people displaced by large dam projects and the same applies to China. An estimated 20 million persons have been displaced in India by official projects such as dams, mines, forest preserves and irrigation projects (Shah, 1995: 574). Finally, it is worthwhile mentioning that even policies which are adopted for the establishment and management of parks and protected areas often circumscribe or ban livelihood activities of many rural communities who have lived in these areas for a long time. This process, which constitutes a form of displacement, takes place in Belize, Botswana, Cambodia, the Dominican Republic, Ecuador, Malawi, Namibia, Panama, Senegal, Sri Lanka and Tanzania (Vandergeest 1997: 28-31). IUCN and WWF are working hard to involve local communities in conservation efforts, so that local people will not be alienated from these parks and will benefit from their establishment.

## 6. ENVIRONMENT, ENVIRONMENTAL SECURITY AND MIGRATION

"Environment and security", "Ecoviolence – Links among Environment Population and Security", "Environment and Conflict", "Environmental Change and Security" – are some of the titles of a research focus of the last decade, which attempts to provide a theoretical framework which ties types of environmental change to conflicts on various levels.

In its most basic form the above-mentioned studies investigate transformation or changes in the environment, mostly anthropogenic, which play a role in conflicts or lead to violent encounters (Bächler, 1995; Homer-Dixon and Blitt, 1998). The central point in many of these studies is that rapid population growth, particularly in developing countries, leads to depletion of local/regional renewable resources like cropland, forests and fresh water supplies. Scarcity of renewable resources is linked or plays a role in many conflicts, nicknamed eco-conflicts. Thus, the environment, and environmental concerns are integrated in the concept of security. Environmental security was used to depict situations in which internal instability and international conflict were the possible results of resource scarcity – real or anticipated – such as the oil crisis of the 1970s (Dabelko and Dabelko 1995: 3-5). Various authors have tried to demonstrate the links between environmental change and acute conflict. Homer-Dixon found evidence of environmental scarcity serving as an underlying yet strong cause of intrastate conflict in detailed case-studies from Chiapas, Mexico, Gaza, South Africa, Pakistan and Rwanda (Homer-Dixon and Blitt 1998). But Bächler (1995, 1998) found that international eco-conflicts resulting from degradation of a transboundary ecoregion also took place in many parts of the world. Many of these conflicts were not violent and revolved around shared water sources: the Farakka barrage on the Ganges (India/Bangladesh) Gabcikovo hydropower station on the Danube (Hungary/Slovakia), Lake Chad shrinking water (Nigeria/ Cameroun, Chad) (Bächler, 1995: 25).

In this chapter we would like to concentrate on the place of population displacement in these studies. Bächler (1995, 1998) identified few categories of population movements in his typology of environmentally induced conflicts. First, heavy environmental degradation can be the source of ethnic tension when ethnic groups with different socio-economical traditions share a sensitive ecoregion. Rwanda-Burundi, Zaire and Kenya are a few examples of that type. A second type of conflict is the periphery-center conflict in which the pressure is put by the modern center of a country on certain regionally limited habitats of the rural population, pushing these populations into even more precarious circumstances. Large agricultural mining or industrial projects are usually the underlying causes, such as the copper mines in Bougainville (Papua New Guinea), Narmada Dam project or the overuse of renewable resources in the Amazon region of Brazil.

A third type of environmental conflict is the displacement conflict which arises from the massive resettlement of people to other eco-geographic locations within their own country. The desertification of the Sudan-Sahel or Central China caused internal migrations to ecologically better regions or to urban agglomerations. But massive resettlements are often forced by government measures in connection with large technical projects (Bächler 1995: 14-15). All the above were intrastate conflicts but Bächler also found internationalized intrastate conflicts in cases where 'environmental refugees' cross national borders. Examples are drawn from the border between Bangladesh and India, or the environmental migrants from degraded northern El Salvador who crossed the border to Honduras. According to Homer-Dixon (1994, 1995) population movements are among the intermediate social effects which, as their name indicates, mediate between environmental scarcity and instability and conflict.

A cautious and sometimes critical approach to the role of the environment in migration movements is adopted by Lonergan (1995, 1998). As both migration and environment are complex entities and processes from the social, economic, political, and institutional aspects, drawing a linear deterministic relationship between environmental degradation, migration and security is not only inappropriate but impossible. The advocates for the relationship between environment and population movement claim that environmental degradation is a root cause of population displacement. But many indicators of migration potential such as population growth, economic restructuring, increasing economic disparities, and increased refugee flows, have only indirect relationship with the environment. Lonergan criticized the confusion over the definitions of environmental refugees, the uncritical acceptance of a direct causal link between environmental degradation and population displacement. Lonergan showed that the links between "push" environmental forces and displacement are difficult to ascertain or have a minimal impact (Lonergan, 1998: 10). Industrial accidents have had relatively little impact on migration, with the exception of Chernobyl. Most accidents have resulted in a short-term relocation, but very few people have been displaced permanently. As for development projects, they do displace large populations, but in theory these projects include a resettlement component and are unlikely to produce waves of environmental refugees. The links between natural disasters and population displacement are also problematic. The number of homeless, as a result of these disasters, is significant, but it does not imply that these people migrated to different regions or countries (Lonergan 1998: 10). Lonergan, though, relates to the category of cumulative or slow-onset environmental change as the most important force in population displacement. Similarly, Black (1998) challenged many of the statements on migration as a result of drought and desertification (Black, 1998: 26-28). Black provided empirical evidence from the Sahel to show that migration is a survival strategy of rural Sahelians. He even provides evidence that emigration from the Senegal River in Mali actually declined during the drought of the mid-1980s. Black also scrutinized many of other premises and predictions related to the environment migration matrix which are not powerful enough to provide satisfactory explanations or "predictions".

Duedney (1991) also amplifies some skeptical arguments in relation to the assumed relations between environmental degradation and national security. Interstate violence has little in common with either environmental problems or solutions and environmental degradation is not very likely to cause interstate wars (Deudney, 1990).

Environmental-induced mobility is often a circulation – not migration – and many of these movements are for short-distances and are only temporary, but not enough empirical data exists on these movements, and information on these migratory cycles from Third World countries is still lacking. There are also dualistic messages from the studies of poverty in its connection to environmental displacement. More research is needed on migration as a measure of adaptation of vulnerable poor people against the vagaries of the environment. Such research may show that there are voluntary displacements side by side with involuntary displacement in many environmentally degraded regions of the world.

There is a prominent gap in our knowledge of environmental displacement which is an outcome of natural or industrial hazards. That lack led Lonergan to conclude that population movements of these types of environmental forces are not significant (Lonergan, 1998). However, this conclusion has not been ascertained (proven). Integrating concepts of resilience, vulnerability, and criticality in the research agenda of detailed studies of Third World environmental displacement will benefit the current theoretical framework. Finally, the research on the possible links between environmental stress, population movements and conflict have probably led to more criticism and "alternative" explanations than any of the other topics which concern the relations between environment and migration. Many case-studies have been explored of people who were forced to leave their homes as a result of flooding and climate change, and the same applies to the assertion that refugee movements are the cause for environmental degradation and conflict (Black 1998: 30-32). It is appropriate to conclude the discussion by quoting from Lonergan: "Generalizations about the relationship between environmental degradation and population movement mask a great deal of the complexity which characterizes migration decision-making. It is extremely difficult to isolate the specific contribution of environmental change in many forms of population movements, especially those which are more "voluntary" in nature" (Lonergan 1998: 11-12).

# 7. POLICY DIRECTIONS

The policy implications of environmentally induced migration are many and relate mainly to population policies, immigration policies, social and economic policies, and finally, environmental policies.

## 7.1 Population Policies

Direct population policy measures that curb population growth without taking the long "detour" through the development of the societies concerned (Wöhlcke, 1999: 90). Though this, blunt policy is not often stated openly, many social and economic policies aim, indirectly at the same goal.

## 7.2 Social and Economic Policies

Many of the policies which are advocated in order to "solve" environmental migration are economic and social and have to do with poverty alleviation and wealth creation in developing countries (Graeger and Smith, 1994). In recent years, states of the First World have targeted the countries of origin (mostly Third World countries with policies which advocate trade liberalization, foreign investment and development assistance which can reduce over time emigration pressures through creating income-generating opportunities reducing unemployment and improving wages in countries of origin (Lohrmann, 2000: 14). Upper income nations should see that it is in their own interest to invest in the South to reduce inequality to alleviate poverty to protect and improve the global environment (Goodland, 1994).

There are also calls for major changes in management of the debts developing countries owe to their creditors. The present policies were found to "create serious problems in terms of both conflict and environmental degradation" (Graeger and Smith, 1994: 115-117). The integration of Third World countries in the global economy has often increased indebtedness and impoverishment. Economic structural adjustment led to many negative changes mainly in the rural sector, transforming many rural people to "environmental migrants". A major transformation in policy is needed in order to change environmental mismanagement. Finally, a reduction in per capita consumption in high-income nations and a decrease in environmental abuse is required in the First World (Goodland, 1994).

## 7.3 On the Political Level

On the political level, perhaps the most important recommended policy concerns empowerment and participation in electoral democracy. The empowerment of women and their education is essential priority for stabilizing population size by reducing fertility rates. Measures to eliminate or resolving environmental conflicts are also promoted. Sometimes armed intervention has taken place in countries affected by large population displacement.

## 7.4 Immigration Policies

Many industrialized and developing countries contend that the increasing flow of economic migrants, asylum seekers, refugees and environmental migrants had become an unbearable burden and many turned to adopt policies of "closed doors".

Finally, there is room for just a brief account of migration policies, particularly international. Most governments in almost all the countries of the First World adopted strict controls on entry and exit of both aliens and citizens though implementation of the existing policies and regulations are not fully enforced (Kritz and Zlotnik, 1992: 10). States which restrict the entry of aliens for permanent resettlement may tolerate refugees – for humanitarian reasons. This applies often also to environmental refugees. While the gates for permanent migrations are closed, foreign (cheap) labor still plays an important role in the international flows of migration (Kritz, Lim and Zlotnik, 1992).

## 7.5 Environmental Policies

Environmentally sustainable future is the wish of all and, for those countries which have a choice, environmental sustainability costs less in the long run. Continuing a policy of "business as usual" will be massive migration and conflicts. Among the most prominent policies towards a more sustainable resource use is the policy to reduce "Greenhouse" gases which was adopted by the community of nations or the international agreements concerning the ozone layer depletion. However, policy operates within constraints. How well societies, communities and groups adapt to change and pressure ultimately depends on factors beyond the control of policymakers, such as a society's ability to organize itself, its previous experience of change, its capacity for social innovation and the degree of participation in political process.

## 8. CONCLUSIONS

The research nexus which links environment and migration made significant progress in empirical studies, but it still struggles for a sound/solid theoretical foundations. The current studies are characterized by dual nature which provide "mainstream" perspectives on the environment-resource-population matrix and "alternative" explanations – mostly using Marxist ideology and context.

Much of the current literature is in agreement that environmentally-induced migration has multiple and changing causation and that constitutes the major difficulty in the analysis of this process. The factors which are associated directly or indirectly with environmentally-induced migrations are population growth, poverty, mismanagement of resources, environmental degradation, nature disasters and complex global and political powers and systems.

Though Malthusianism and "neo-Malthusian" concepts and the enclosed "limits on growth" perspectives still play an important and central role in shaping our thoughts and policies concerning environmental migrants, there is a growing body of alternative directions in which explanations and recommended policies tend to blame world capitalist systems and imperialistic past and present policies for many of the environmental migration waves in Third World countries.

Because of the ambiguity in the definition of what constitutes an environmental migrant, there is a great variation in the estimated number of environmental refugees. Furthermore, as this chapter has shown, it is becoming more and more difficult to isolate a single "push" force for migration in the developing world which propels most of the immigrants in the world – including environmental refugees. Among the "push" forces which often were attributed to environmental migration are development projects and industrial disasters which are questionable as "environmental" forces even if a perspective of multi-causation to environmental migration is taken for granted.

Most of the current studies also forecast a rapid growth in environmental migration mainly due to climate change trends and the accelerated pace of resource degradation due to the present state of population growth. Most of the recommended policies also point to curbing population growth as the most important policy to impact resource and environmental management. However, there are more and more recommendations which call for a drastic transformation in the North resource consumption patterns, a conviction that only poverty alleviation and heavy investments in the South will, eventually curtail environmental mismanagement. As usual, implementation of these truly positive policies is slow and insufficient and hence, we may expect a constant growth in the number of environmentally induced migration.

## REFERENCES

Abbott, P. 1996. *Natural Disasters*. Dubuque, IA: W.M. Brown.
Adepojo, A. 1995. "Emigration dynamics in sub-Saharan Africa." *International Migration* 33(3/4): 315-390.
Adger, W.N. 2000. "Social and ecological resilience. Are they related?" *Progress in Human Geography* 24(3): 347-364.
Afolayan, A.A. 2001. "Issues and challenges of emigration dynamics in developing countries." *International Migration* 39(4): 5-36.
Alexander, D. 1993. *Natural Disasters*. New York: Chapman & Hall,.
Amacher, G., W. Cruz 1998. "Environmental motivation for migration: Population pressure, poverty and deforestation in the Philippines." *Land Economics* 74(1): 92-101.
Bächler, G. 1995. "The anthropogenic transformation of the environment: A source of war?" In *Environmental Crisis: Regional Conflicts and Ways of Cooperation,* K. Spillman, G. Bächler, eds. Zurich: Center for Security Studies and Conflict Research (ENCIOP); 11-27.
Bächler, G. 1998. Why environmental transformation causes violence: A synthesis. Environmental Change and Security Report Project; 4, Spring: 24-44.
Benedick, R. 2000. "Human population and environmental stresses in the twenty-first century." *Environmental Report* 6: 5-18.
Black, R. 1998. *Refugees, Environment and Development*. Harlow, Essex: Longman.
Blaikie, P., T. Cannon, I. Davies, B. Wisner 1994. *At Risk – Natural Hazards. People's Vulnerability and Disasters*. London: Routledge.
Carella, M., R. Pace 2001. "Some migration dynamics specific to southern Europe – South-North and East-West axis." *International Migration* 39(4): 63-100.
Chapman, R. 2000. "Immigration and environment: Settling the moral boundaries." *Environmental Values* 9(2): 189-210.
Chester, D.K., M. Degg, A.M. Duncan 2001. "The increasing exposures of cities to the effects of volcanic eruptions: A global survey." *Environmental Hazards* 2: 89-103.
Dasgupta, P. 2000. "Population and resources and exploration of reproductive and environmental externalities." *Population and Development Review* 26(4): 643-690.
de Haan, A. 1999. "Livelihood and poverty: The role of migration – A critical review of the migration literature." *The Journal of Development Studies* 36(2): 1-47.
Deudney, D. 1990. "The case against linking environmental degradation and national security." *Millennium* 19(3): 461-476.
Dessler, D. 1994. "How to Sort Causes in the Study of Environmental Change and Violent Conflict." In *Environment Poverty and Conflict*, N. Graeger, D. Smith, eds. Oslo: International Peace Research Institute, Report 2/94, 91-112.
Ehrlich, P., G.P. Holdern 1974. "Impact of population growth." *Science*; 171: 1212-1217.
El-Hinnawi, E. 1985. *Environmental Refugees*. Kenya: UN Environmental Programme.
Ezeonu, I.C., F. Ezeonu 2000. The environment and global security. *The Environmentalist* 20(1): 41-48.
Findley, S., and S. Traore 1995. "Emigration from the Sahel." *International Migration* 33(3/4): 469-520.
Goodland, R. 1994. "Environmental Sustainability: Imperative for Peace." In *Environment, Poverty and Conflict,* N. Graeger, D. Smith, eds. Oslo: International Peace Research Institute, 19-46.
Gould, W.T.S., A.M. Findlay eds. 1994. *Population Migration and the Changing World*

*Order*. New York: Wiley.
Graeger, N., D. Smith eds. 1994. *Environment, Poverty and Conflict*. Oslo: International Peace Research Institute, Report 2/94.
Hampshire, K., S. Randall 2000. "Pastoralists, agropastoralists and migrants: Interaction between fertility and mobility in Northern Burkina Faso." *Population Studies* 54(32): 247-261.
Homer-Dixon, T. 1994. "Environmental scarcities and violent conflict: Evidence from cases." *International Security* 19(1): 5-46.
Homer-Dixon, T. 1996. Project on environment, population and security: Key findings. Environmental Change and Security Project Report; 2, Spring: 45-49.
Homer-Dixon, T., J. Blitt 1998. *ECOVIOLENCE – Links among Environment, Population and Security*. Boston: Rowman & Littlefield Publishers.
Homer-Dixon, T., M. Levy 1996. Environmental scarcity and violent conflict: A debate. Environmental Change and Security Project Report; 2, Spring: 49-71.
Hornby, W.F., M. Jones 1993. *An Introduction to Population Geography*. Cambridge: Cambridge University Press.
Howard, P. 1997. "Development-induced displacement in Haiti." *Refugee* 16(3): 4-11.
IOM/RPG 1992. *Migration and the Environment*. Geneva: International Organization for Migration, and Washington D.C.: Refugee Policy Group.
Jacobson, J. 1988. *Environmental Refugees: A Yardstick of Habitability*. Washington D.C.: Worldwatch Institute.
Jodha, N.S. 1975. "Famine and famine policies: Some empirical evidence." *Economic and Political Weekly* 10: 1619.
Kasperson, R., J. Kasperson, B. Turner 1995. "Critical Environmental Regions." In *Regions at Risk: Comparisons of Threatened Environments,* J. Kasperson, R. Kasperson, B. Turner, eds. Tokyo: United Nations Press, , 1-41.
Kibreab, G. 1994. "Migration, Environment and Refugeehood." In *Environment and Population Change*, B. Zaba, J. Clarke, eds. Liège, Belgium: International Union for the Scientific Study of Population, 115-129.
Koop, G., L. Tole 2001, "Deforestation, distribution and development." *Global Environmental Change* 11(3): 193-202.
Kritz, M., H. Zlotnik 1992. "Global interactions, migration systems, processes and policies." In *International Migration Systems,* M. Kritz, Lin Lin, H. Zlotnik, eds. Oxford: Clarendon Press, 1-18.
Lock, C., N. Adger, W. Kelly, P. Mick. 2000. "Changing places: Migration's social and environmental consequences." *Environment*. September: 25-35.
Lohrmann, R. 2000. "Migrants, refugees and insecurity. Current threats to peace." *International Migration*; 38(4): 3-22.
Lonergan, S. 1995. Population movements and the environment. Refugee Participation Network (RPN), 18, January: 4-7.
Lonergan, S. 1998. The role of the environmental degradation in population displacement, Environmental Change and Security Project Report; 41, Spring: 5-15.
McClean, J. 1998. National response to internal displacement. Forced Migration; 1, January-April: 10-12.
McGregor, J. 1993. "Refugees and the Environment." In *Geography and Refugees: Patterns and Processes of Change,* R. Black, V. Robinson, eds., London: Belhaven.
McGregor, J. 1995. Are there environmental refugees? *Refugee Participation Network* (RPN); 18: 7-9.
Milozi, D. 1995. "Emigration dynamics in southern Africa." *International Migration* 33(3/4), 521-556.

Myers, N., J. Kent 1995. *Environmental Exodus: An Emergent Crisis in the Global Arena*. Washington D.C.: The Climate Institute.
Nord, M. 1998. "Poor people on the move: County-to-county migration and the spatial concentration of poverty." *Journal of Regional Science* 38(2), 329-351.
Oucho, J.O. 1995. "International migration and sustainable human development in eastern and southern Africa." *International Migration* 33(1): 31-54.
Parnwell, J.J.G. 1993. *Population Movements and the Third World*. London: Routledge.
Parry, M., N. Arnell, T. McMichael 2001. "Viewpoint: Millions at risk: Defining critical change. Threats and targets." *Global Environmental Change* 11(3): 181-183.
Penz, P. 1997. "Development induced displacement." *Refuge* 16(3): 1-3.
Petrucci, M. 2000. "Population: Time bomb or smoke screen." *Environmental Values* 9(3): 325-352.
Refugees, 81, 1990. The environment, *Refugees 81*, December: 14-16.
Refugees, 85, 1991, *Refugees, 85*, May.
Refugees, 87, 1991. Refugees, 87, October
Refugees, 91, 1992. Special Issue: Emergency, Refugees, 91, December.
Refugees, 103, 1, 1996. Special Issue: The internally displaced, *Refugees, 103*
Sacks, J. 1999. Helping the world's poorest. *The Economist*; August 14: 17-20.
Seaman, J., J. Holt. 1980. Markets and famines in the third world. *Disasters* 4(3): 287-297.
Shah, Nasra M. 1995. "Emigration dynamics from and within South Asia." *International Migration 33*(3/4): 559-625.
Shaker, M.T. 1987. *An Analysis of Squatter Settlements in Dhaka, Bangladesh*. Ph.D. Thesis. The University of Liverpool.
Smith, D. 1994. "Dynamics of Contemporary Conflict: Consequences for Development Strategies." In *Environment, Poverty and Conflict*, N. Graeger, D. Smith, eds. Oslo: International Peace Research Institute, Report 2/94, pp. 47-89.
Stein, M. 1998. The Three Gorges: The unexamined toll of development-induced displacement, *Forced Migration*; 1: 7-10.
Swain, A. 1993. *Environment and Conflict*. Uppsala, Sweden: Department of Peace and Conflict Research, Report No. 37.
Tolba, M. 1990. Quoted in Lazarus, P., 1990. New strangers at the door, *Refugees* 81, December: 14-15.
Uytterlinde Flood, L. Sardar 1997. Sarovar Dam: A case study of development-induced environmental displacement. *Refuge* 16(3): 12-17.
Vandergeest, P. 1997. Protecting nature and displacing people. *Refuge*; 16(3): 28-31.
Zelinsky, W. 1971. The hypothesis of the mobility transition. *Geographical Review* 61: 219-49.

# MIGRATION INDUCED LEGAL PLURALISM IN LAND TENURE

*Implications for Environmental Change*

Jon D. Unruh
*Department of Geography, Indiana University, Bloomington, IN 47408 USA*

## 1. INTRODUCTION

Increasingly pervasive migration in the developing world is emerging as an important force in global environmental change scenarios (e.g., Amacher et. al., 1998; McGregor, 1994; Southgate, 1990; Myers, 1997; Doos, 1994; Ghimire, 1994; IUCN, 2000). Whether by forced dislocation or self-selected migration, a wide variety of causal factors related to resource conditions, from food insecurity, conflict, and water resources, to political, social and economic disruptions, are increasingly leading to larger scale migrations with significant impacts on the environmental resources needed to sustain livelihoods. Such impacts can then lead to additional dislocation and migration due to resource scarcity and conflict. The impact on natural resources and resource use systems in temporary and permanent destination areas extends beyond direct environmental degradation, leading local (non-migrant) communities to change resource use arrangements due to the presence of significant migrant populations, and governments to search for appropriate policy and enforcement instruments.

Given that a large proportion of migrants in the developing world are rural inhabitants who seek rural destinations (Ghimire, 1994), the environmental impact of migrant presence in destination areas operates within the domain of interaction with land based resources and the local communities who have pre-existing claim and use rights to those resources. In this context rural land resource rights (land tenure) play a primary role in how migrants intersect with destination resources and communities, and the resulting environmental consequences. A great deal of valuable work has contributed much to our understanding of the important role of land tenure with regard to how humans interact with the environment (e.g., Katon, et. al., 2001; Ostrom, et. al., 1999; Amacher, et. al., 1998; Thesihuesen, 1991; Southgate, 1990). And in aggregate land tenure plays a primary role on land cover change at various scales (Unruh, 1995a). In a migration context often the first and most important interaction between migrants, and between migrants and local communities, is over access to resources, and most often,



*J.D. Unruh, M.S. Krol and N. Kliot (eds.),*
*Environmental Change and Its Implications for Population Migration,* 101–118.

the land resources needed for near term food security (e.g. McGregor, 1994; Unruh1995a, 1993).

One of the more important emerging aspects of migration in the developing world is the increasing diversity of those who migrate. (McGregor, 1994; Schmeidl, 1998). McGregor (1994) reviews the literature on migrant and refugee livelihoods, noting that the economic and ecologic changes that take place in destination locations for migrants result in very diverse experiences in different places. Not only are migrants emerging from a wider variety of states and regions within states, but from a wider variety of livelihood systems, religions, ethnic groups, and socioeconomic strata (Schmeidl, 1998). As well, the array of specific reasons for dislocation, and the variety of experiences during migration, add to the diversity of migrant characteristics, and importantly, to the diversity of their approaches to resource access in temporary and permanent destination locations. Diversity in tenurial constructs in a migration context comes about both as migrants carry with them notions about property rights arrangements that are familiar, and seek or are compelled to pursue new constructs in new locations.

Due to the diversity of tenurial constructs coming into contact with each other through migration, and the primary role that land tenure issues play in impacts on the environment and environmental change, what is the aggregate land tenure pattern which is likely to emerge as migrants and migrations become more common, complex, and conflictive? And what are likely to be the primary environmental repercussions of this pattern? This paper draws on a literature review and the author's research and policy experience with land tenure, migration, and environmental change conditions in East, Central, and Southern Africa to examine the increase in forms of 'legal pluralism' with regard to land tenure which result from migration and the interaction between migrants and local communities. The argument is made that the primary land tenure issues that will develop with increased rural migration in the developing world will be associated with a significant increment in legal pluralism in land tenure--or, multiple operative notions about resource use, access, and claim. Land tenure related impacts on environmental change will to a significant degree be associated with this pluralism, and as a result how states are prepared to contend with pluralism will become an important concern. The availability of legislative tools along with the ability and willingness of states to employ these, will impact how conflictive situations of legal pluralism over land resources become, and their overall impact on the resources needed for food security, development and conservation.

Subsequent to a brief introduction to legal pluralism, the bulk of the paper examines several of the more important ways that legal pluralism in land tenure develops in a migration context. This is followed by a brief discussion of the environmental impacts associated with tenurial pluralism, and finally the constraints involved in national level legislative response.

## 2. LEGAL PLURALISM AND LAND TENURE

Legal pluralism as a domain of legal studies has developed quickly in recent years, with issues of land tenure occupying a significant place in this progress (e.g., Benda-Beckmann, 1995; Galanter, 1981; Merry, 1988; Moore, 1973; Griffiths, 1986; Lund, 1998; Prill-Brett, 1994; Vel, 1992). Important definitions of legal pluralism range from pluralism in the juristic sense, when a state brings about different laws for different groups of people (Merry, 1988), to claiming that all societies are legally plural especially in less formal ways, in that making rules, obtaining compliance, and punishing rule breakers exists in a wide variety of societal subgroups (Abel, 1982; Henry, 1985). Much important research has focused on this informal domain, which recognizes legal pluralism as multiple, nonjuristic, alternative, forms of "normative ordering" (e.g., Merry, 1988; Griffiths, 1986). Griffiths (1986) notes that the law "which is actually effective on the 'ground floor' of society is the result of enormously complex and usually in practice unpredictable patterns of competition, interaction, negotiation, isolation, and the like," such that "the legal organization of society is congruent with its social organization."

Of particular utility in a migration context, is the semi-autonomous social fields approach described initially by Moore (1973), in which separate social fields of 'legality' interact over time and space. This interaction can take a number of paths from eventual reconciliation between fields, to increasing separation or multiplication of fields, depending on the nature of the interaction and associated resource-related, socio-political, and economic forces. Moore (1973) further articulates the role of semi-autonomous social fields in the way individual and group behavior and the processes of interaction within and between fields determine what 'law' is operative in a given space and time. Griffiths (1986) builds on Moore's (1973) work, noting that it is to a large degree the network of rights and obligations in human relationships which constitutes a 'legality' within a social field.

Land tenure in a very fundamental way is a system of rights and obligations in human relationships (Migot-Adholla and Bruce, 1994). Legal pluralism in land tenure signifies the different sets of rights and obligations regarding land resources, as these reside within multiple social fields or normative orders. One of the most common examples of legal pluralism regarding land occurs in much of the developing world, where, due to the existence of both customary and formal tenurial systems, legal pluralism in land is an approach to practical governance. One the other hand one of the most aggravated examples of incompatible and confrontational legal pluralism regarding land exists in the Middle East, where the Israeli-Palestinian lands issue has vexed attempts at peacemaking for some time.

In a migration context, tenurial pluralism emerges as features of pre-migration tenure systems are shaped by the experiences of dislocation, migration, and attempts at settlement in new locations. This provides for a wide variety of approaches to accessing land resources in new locations with different political, social, cultural, and physical environments. At the same time non-migrant local communities that experience large influxes of migrants can abandon features of their own land tenure systems that others (migrants) are not following, and likewise pursue alternative modes of land access, use, claim, and dispute resolution. Important here is the role of tenure security. The search for tenure security in a physical, political, and socio-economic environment that is new and unfamiliar is an important driver in the search for alternative ways of configuring property rights; and an important factor in the development of legal pluralism. This search can be particularly robust when food insecurity and conflict are compounding influences--an important consideration given that most migrants are poor and lack the means to provide for near-term food security (Ghimire, 1994). Tenure insecurity can be significantly aggravated by the lack of dispute resolution institutions able to operate between different notions of tenure, and seen as legitimate, effective, and enforceable by the parties concerned (Unruh, 1997a, 1995b).

# 3. DEVELOPMENT OF LEGAL PLURALISM IN LAND TENURE

## 3.1 Rights and Obligations in Land Tenure

Dislocation and migration profoundly change relationships between people--especially dislocation that results from severe circumstances (e.g., extreme food insecurity, resource scarcity, etc.) and migration experiences which are of long duration. Because of the importance of the spatial aspect in migration, established rights and obligations regarding land, property and resources are at the forefront of change. The socio-spatial repercussions and the subsequent reduction in the relevance of specific administration, enforcement, and other property-related institutions and norms, result in altered relationships between people, land areas, land uses, production systems, and population patterns. In essence, migration and its repercussions reconfigure the network of social relations upon which all land tenure systems--and hence agricultural and other land-based resource use systems--depend. Such changes in social relations as they affect land tenure become operative through specific interrelated processes. While these can be numerous, this section examines five of the more pervasive processes that change rights and obligations in land tenure and result in the development of multiple normative orders regarding tenure.

## 3.2 Dislocation

### 3.2.1 Physical Separation and Status Change

Population dislocation plays a primary role in the development of legal pluralism with regard to land. Physical separation of people from established home areas and ways of land use and tenure, can be the first and most dramatic step toward the development of a changed approach to land rights. Physical separation changes, terminates, or puts on hold prevailing social rights and obligations regarding land and property, especially where actual occupation, or social position forms the basis or a significant aspect of claim and use rights. This change occurs primarily because community members are no longer physically present to exercise rights and obligations. For those who are not dislocated or choose not to migrate, alternative ways of viewing land can move forward quickly, especially if marginalized sectors within a community see an opportunity to enhance rights. For migrants, land must usually be sought elsewhere especially for agricultural populations, but with an approach to access and claim to land resources different from what prevailed in a home area. This comes about with a change in status as community members become dislocatees, migrants, squatters, female-headed households, and refugees in new locations. Most African populations for example rely on relatives and other community members for security of person and property and for assistance in dispute resolution involving these. In circumstances involving dislocation and migration however, many variables serve to rework or reconfigure this construct, especially if the destination area has little or no community entry. Affected customary populations (both migrating and receiving) can move to establish alternative land tenure arrangements that follow newly emerging situations and norms, or pursue variations of old arrangements which will work under prevailing circumstances. The direction that this emergence takes and how rapidly it occurs can depend to a significant degree on experiences associated with dislocation, migration, and reception.

### 3.2.2 Return to Area of Origin

Return to a home area subsequent to dislocation can provide for additional opportunities for the development of legal pluralism. Relevant to the idea of 'return' is the potentially significant resistance and animosity toward returnees by community members who did not migrate. Such animosity can be pronounced when alternative ideas regarding land tenure are brought back with returnees, particularly when these ideas involve a perceived reduction in rights and/or power for those who stayed. Also relevant to 'going back' are the presence and activities of other actors--squatters, large landholders, commercial interests--all of whom may seek access to lands thought to be unoccupied or abandoned. Krznaric (1997) observes how dislocation and migration influenced the development of legal pluralism over land within returning groups of Guatemalan migrants versus those who stayed, due to the migrant's raised political awareness during their exile in Mexico. In Guatemala such raised awareness, itself different for different groups of dislocatees, was the result of the opportunity to advance certain interests suppressed under the predislocation community political arrangement--such as those of women and lower socioeconomic strata. An organizational capacity also emerged within some sectors of the Guatemalan refugee returnee community that adhered to a transnational language of rights (human rights, refugee rights) that was appropriated and used by groups of returnees. Hammond (1993) notes similar contrasts for Nicaragua and El Salvador, and Unruh (1997b) describes for Mozambique the difficulties in re-accessing home areas occupied by squatters and large commercial cotton interests.

## 3.3 Change in Penetration of the State

### 3.3.1 State Lands Administration

Dislocation and migration necessarily result in a change in the influence of state institutions and state law on populations, with the overall effect spatially variable. Subsequent to a dislocation event, the state's land administration institutions in areas of dislocation can be rendered much less influential, and rules unenforceable as pre-existing relationships with the state are disrupted. This can occur in several ways: (1) land records pertinent to affected areas of the country (both dislocating and receiving) become outdated as land and property transactions take place and go unrecorded; (2) a general recognition emerges by the dislocated population (and frequently the host population) of the unworkability of state land administration as a national institutional endeavor which serves their interests as migrants (or hosts); (3) in the context of a resurgence in the use of certain traditional norms in substate groups (frequently tied to identity); and, (4) increasing numbers of migrants and members of local host communities abandon the state tenure system in favor of alternatives, which then acts to further subtract adherents to state law in a 'momentum effect.' If dislocation is due to conflict and involves the state, then the influence of the state can be reduced considerably as the legitimacy of the national formal land tenure system is compromised. In such a case the derivation and pursuit of alternative forms of property rights arrangements can occur with considerable energy in the context of opposition to the state. At the same time the weakening or subordination of the state may have the effect of encouraging land grabbing by urban based elites. This can Take place as state controls diminish and those in a position to do so use the instruments of the state with greater ease to obtain land--in a form of legal pluralism derived from a changed position of the state with regard to property rights.

In Somalia, the complete reduction of state power and the search for order has led to the installation of Islamic Shari' a courts for administration and enforcement of a number of aspects of society, including land and property rights and land dispute resolution (UNDP-EUE, 1999). During Mozambique's RENAMO war, the dislocation and resulting migration of approximately half of the national population, together with forces associated with the war, resulted in a considerable reduction in the capacity of the state to administer land. This allowed not only the RENAMO opposition, but a variety of groups to exert alternative approaches to land access, claim, and use. At the same time, many non-migrant communities in Mozambique refocused their attention on their own traditional ways of land tenure, dropping any recognition of state land administration that existed prior to the war. In some cases this allowed the occupation, or re-occupation, of lands formerly seized by the state or commercial interests (Unruh, 1997b).

In a different effect on the state Mbembe (1999) observes that the efforts of the international conservation community have been significantly successful in moving large areas of land from effective state control, to extra-territorial status. This influences the development of legal pluralism both in the dislocation of local inhabitants, and the duality of state versus international territorial control.

### 3.3.2 Land Related Grievances and the State

Pre-dislocation ideas of the 'unjustness' in the way the state dealt with land rights for a population can constitute an important force in the reduction of state penetration in land issues during migration scenarios. Such ideas can range from simple disappointment in or distrust of the state and its ability, willingness, or bias in handling land issues, to the perception of the state as the enemy. The latter can be especially powerful if there exists an accumulation of land-related grievances against the state brought on by land alienation and discrimination, corruption, state intervention in agricultural production, dislocating agricultural or population programs (e.g., villagization), and heavy-handed approaches to enforcement of state decisions and prescriptions regarding land issues. In aggregate, this can result in what Ranger (1985) calls a "historical consciousness of grievances" with regard to land rights issues; these can become especially pronounced if grievances merge with other issues not necessarily related to land. In these cases dislocation and migration become an opportunity to derive and/or pursue alternative normative orders regarding land access and use. Such alternatives, once developed, can persist with considerable tenacity, justifying themselves by appeals to perceived historical wrongs done to certain groups (Merry, 1988). De Soto and del Castillo (1995) note that land grievances had been at the core of the Salvadoran friction since the colonial era, and were some of the primary causes of the conflict in the 1980s. This was also the case in Zimbabwe's liberation war regarding land expropriations by the Rhodesian state (Ranger, 1985), and in both Mozambique's RENAMO war and Ethiopia's Derg war as a result of government villagization programs. Variants of such conditions also prevail for more recent problems in southern Mexico, and in the way the land issue has been handled over the course of the conflict in the Middle East.

### 3.3.3 Interventions

Government and international donor interventions in migration scenarios can contribute significantly to the emergence of tenurial legal pluralism. In attempts to mitigate environmental degradation associated with migration, intervention programs can encourage migrants to plant trees (Ghimire, 1994) or construct anti-erosion and water catchment installations, etc.--all of which can be seen by local non-migrant populations as attempts to claim land, and can in fact be used as such. Ghimire (1994) observes however that interventions such as tree planting can be more successful if migrants are first provided with secure access to land. Such an arrangement would be an explicit promotion of legal pluralism by an intervention intended specifically to both assist migrants and mitigate resource degradation.

### 3.3.4 Flight from the State

In broader fashion, migration plays an operative role in what Mbembe (1999) calls, for Africa, the current "flight from the state." Mbembe observes that a pronounced reconfiguring of the African continent is underway due to the effects of demography, urbanization, military, economic, and religious factors in which migration plays a primary functional role in bringing members of different groups into contact. In this pervasive scenario, different forms of political and social orders are emerging which bypass the state system. For land tenure, this means that forms of formal tenure connected to the state may increasingly be only one of several approaches to tenure rights which emerge through processes of migration. Nigeria provides an important example, where localities answer to a wide variety of jurisdictions and authorities that span a spectrum of government, ethnic, traditional, and religious authorities (Mbembe, 1999). Migration is the operative process which brings subjects of these authorities and notions of legitimacy into close contact, resulting in a variety of conflicts. Often these are explicitly over land and resource access, with often no dispute resolution institutions operable between different and often competing notions of tenure.

Mbembe (1999) as well notes that the increasing number of migrants to different locations in Africa is providing for an "extraordinary rise in xenophobia." Such an effect can translate readily into a land tenure context, where attachments to land derive legitimacy from notions of insider - outsider in a variety of configurations. This, together with a reduction in state power, can bring pluralism to the fore in a variety of ways.

### 3.3.5 Identity Change

For considerable numbers of people who find themselves in migration scenarios, identity can become intricately bound up in land occupation, access, or perceived rights to specific lands in very powerful ways. Smith (1988) notes, that "the ethnic 'self' remains the fundamental territorial 'self' in Africa." In many cases the existence of ethnic, religious, geographic, or other identities to which primary attachments persist, can be based on connections to land, home area, or territory (Unruh, 1998). Dislocation from home areas can result in a relative rise in the influence of identity-based attachments to land, especially if there is an identity component to the dislocation event, and destination locations for migrants become problematic for reasons involving identity (Ibrahim, 1998).

Ibrahim (1998) notes that "the assertion of a person's identity is intended to give meaning to an encounter with others." The definition of identity in a context of migration can be predicated on how an individual or group see themselves with respect to other groups. Thus, approaches to land employed by one group in a migration context can be purposefully rejected by another, leading to a situation of opposed legal pluralism over land in destination areas. As the identities of those involved in migration develop and take on enhanced definition, or redefinition (Ibrahim, 1998), approaches to land issues will reflect this and can become a prominent feature in pluralistic approaches to tenure.

Notions of identity can also involve land claim justification based on earlier historical occupation of areas that migrants can then seek out as destination locations, supported by oral histories that can be traced back through time into mythologies about how various peoples came to exist in an area and in the world (Comaroff and Simon, 1977). Such justification can gain renewed strength during migration, and the pursuit of a 'return' to historical lands or territory--from which groups were expelled or departed, recently or long ago--can become a priority in a migration event. In some cases, such a situation can be seen as a singular opportunity to regain historical lands.

### 3.3.6 Changes in Legitimacy and Evidence

Claims to land, property, and territory are based on the perception of legitimacy in various forms. In a migration context legitimacy in land tenure becomes significantly problematic, as migrant groups emphasize, modify and derive notions of legitimacy which will allow access and use of land resources. The possession of evidence to prove and support rights of access and claim to land resources is a fundamental feature of land tenure systems. While formal tenure regimes hold the document to be the primary form of evidence, customary tenure systems and normative orders regarding land contain a wide variety of informal evidence that derives legitimacy and authority from relevant customary social and cultural features. During the course of dislocation and migration, evidence and legitimacy of evidence is subject to considerable change, primarily due to the role that community plays in determining what evidence is regarded as legitimate (Unruh, 1997a). In one sense the boundary of a legal field (Griffiths, 1986) with regard to a set of normative rules regarding land, can exist between those who do and do not regard certain forms of evidence as legitimate.

Shipton (1994) notes that within the administration of land (both formal and customary) the question of who controls the "language," and the "translations" of reality into legitimate evidence and how land is dealt with (demarcations, transfers, inheritance, access, etc.) becomes critically important (also Murphy, 1990). Such control over what is or becomes evidence, legitimizes or de-legitimizes units of aggregation, kinds of rights, transactions, rituals, and ways of land use (Shipton, 1994). Thus competition and confrontation over who exercises this control with regard to a specific land area, or specific rights within an area, can influence the development of legal pluralism. This occurs as some claimants find themselves with evidence different from that considered legitimate or possessed by others as dislocation and migration scenarios reconfigure the important aspects of evidence (e.g., community, occupation, relevant cultural-ecological features, customs, ways of administration, etc.). The same effect can manifest itself in a more nuanced fashion as the relative value of pre-existing evidence can shift to reflect changed circumstances (Unruh, 1997a). This was the case in Mozambique, where, because land rights for the customary tenure system were bound up in historical interconnections of community interaction regarding land, and locally legitimate evidence of rights, migration resulted in a shift of legitimate evidence for many affected groups (emergence of alternative norms). This occurred as changes in specific evidence and community interaction regarding this evidence meant that shifts in the availability and legitimacy of evidence occurred. As a result 'social' customary evidence such as testimony, community and lineage membership, and history of occupation were significantly devalued due to widespread dislocation and migration. At the same time, the existence of permanent, physical investments in land, such as agroforestry trees, greatly increased in value as evidence (Unruh, 2001).

### 3.3.7 Processes Associated with Destination Areas

#### 3.3.7.1 Tenure Insecurity as a 'Pull' Factor

The character of destination areas can exert considerable influence in the development of legal pluralism involving migrants--beginning with the relative degree of attraction to a destination area. In a study by Amacher et. al., (1998) land tenure insecurity was found to be a primary factor in selection of destination sites by migrants in the Philippines, with similar examples elsewhere in Asia. Migration destinations in this context can often be: (1) at the frontier of development, (2) in marginal lands, (3) where the state has neglected development or land tenure, or (4) where political instability has likewise caused tenure insecurity (Amacher, et. al., 1998; Myers, 1997). Such situations are seen as attractive opportunities for migrants in that they represent the possibility for pursuing various forms of claim for themselves, however temporary or unofficial (Amacher, et. al., 1998).

Pre-migration tenure insecurity in such locations can often be the result of forms of tenurial pluralism to begin with. This can be the case when lands are claimed by both the state and local communities, or there is confusion over what rights different groups have. Conversely the opportunity for migrants to pursue pluralistic forms of claim is much reduced in locations where the in-place local community has strong and secure tenure rights that are effective and enforceable.

An initial search for areas of tenure insecurity by migrants can then be followed by the search for tenure security once claims are made; and this can significantly fuel the development of legal pluralism. But with no specific way to obtain tenure security in such areas, a number of different approaches can be attempted by migrants. Local communities--tenure insecure themselves--can be quick to abandon or change ways of tenure in the face of significant migrant arrival, because there can be little reason to continue with rules that others are not following.

### 3.3.7.2 Relationships between Migrant and Host Communities

The nature of the intersection between local and migrant communities is one of the more important sources of tenurial legal pluralism in a migration context. Although there is a significant literature concerning the interaction between migrant and host communities on a variety of topics (e.g., Kok, 1989; Wilson, 1992; Unruh, 1993; Clay and Stokke, 1991) land tenure repercussions remain largely unexamined. McGregor (1994) notes how migrant and local societies themselves are not homogeneous. As a result, how migrants interact with local communities will not be homogenous. Interaction can depend on a wide variety of factors, including the nature of arrival (timing, seasonality, and magnitude); the socio-economic status of migrants vis a vis the local community; food security; and similarities or differences in ethnicity, land use, and production systems. Ostrom et. al., (1999) make the point that if resource users are added rapidly such as through migration, similar understandings of resources, and resource use and access are unlikely to be shared between local communities and migrants. As well there is variability in assistance migrants can receive as internally displaced persons (IDPs) or refugees from local, regional, national, or international organizations, with such assistance frequently contentious with regard to local communities.

How migrants settle in destination areas can have a significant influence on the derivation of legal pluralism for both migrant and local communities. Different patterns can emerge from settlement in refugee camps or other concentrated areas, versus settling scattered within the host community, versus scattered in a wide rural area in and around the host area. Concentrated settlement of migrants due to security, humanitarian, or resource availability reasons can encourage forms of tenure that are more competitive and contentious (Ghimire, 1994), and lead to spatially acute forms of resource degradation (land, fuelwood, timber, water, etc.) (McGregor, 1994). Decisions about where government and donors locate concentrations of refugees are rarely made with local community consultation, compensation, or coordination, especially with regard to how land is accessed. Legitimacy in tenure and tenure security for migrants in such a situation is often derived from the presence of a government or international entity seeking to service concentrations of refugees or IDPs via the formal national tenure system (McGregor, 1994). Meanwhile local communities often continue with customary tenure. Differences in tenure security between local and migrant communities in such a case can be significant.

Settlement of migrants in more dispersed fashion within a local community results in differentiated approaches to land tenure. While a good relationship between host and migrant populations may result in secure access to land based resources (Wilson, 1992) this is frequently difficult to achieve. Migrants can derive their own forms of land access in reaction to rules of exclusion by local communities, or exclusion from the institutions which facilitate an equitable place in the local land tenure system. Often such rules of exclusion are constructed and enforced in an attempt by the local community to avoid resource degradation and a 'tragedy of the commons' scenario (Ostrom, 1990). Such exclusion can often be facilitated by the weak position of migrants vis a vis local communities (socially, economically and politically) (Ghimire, 1994). However with needs and desperation frequently high among migrants, alternative ways, reasons, and legitimacy can quickly be configured in reaction to exclusion, competition, and confrontation with local communities. Where migrants are able to engage local resource access and use arrangements, these can often be a subset of local rules, or special rules set up for migrants which provide for more limited rights. This is the case with refugee communities in western Tanzania, and internal migrants in Zambia.

## 4. ENVIRONMENTAL IMPLICATIONS

While tenurial legal pluralism alone does not necessarily lead to resource degradation, the combination of migration and legal pluralism provides for situations that lack mitigating institutions and rules that can prevent environmental degradation. In part this can be due to the lag time between the arrival of migrants and any derivation of institutions and rules able to mitigate degradation. But as noted above, to a significant degree much about migration and the effects it has on local populations, regions, and countries effectively discourages the derivation of the necessary institutions, rules, and laws that would serve to mitigate degradation.

Environmental degradation in this regard involves four primary interrelated processes. The first comprises the repercussions of tenure insecurity. Without secure tenure, migrants and the local communities affected tend to pursue more extensive cultivation and extractive practices. When this occurs on more marginal and fragile lands, degradation can be much more pronounced. Lands that are arid or semi-arid, have steep slopes, and are otherwise marginal for agriculture or pastoralism are especially susceptible to such degradation, and as noted above can be attractive to migrants (Amarcher, et. al., 1988).

Second, migrants often face tenure regimes in developing countries that encourage deforestation and land clearing (Southgate, 1990). The removal of trees and other vegetation is often a prerequisite for establishing claim and rights to land, and coupled with tenure insecurity in destination locations and a local institutional ability to effectively deal with tenure problems associated with migrants, can lead to significant deforestation (Southgate, 1990).

Third, when use rights are a significant aspect of claim for migrants, continuous use of agricultural lands via cultivation and grazing in order to maintain rights within various tenurial constructs can lead to significant degradation (Southgate, 1990). In a modeling effort, Southgate (1990) notes that significant land degradation occurs in parts of the developing world where migrants arrive to engage in agriculture, specifically because acquiring property rights to 'idle' land entails both removing vegetation and continuous use of land resources. In such a case migrant non-access to local and/or formal ways of access, use, and claim can encourage continuous use of land.

Fourth, land resource degradation due to the effects of migration can itself have a significant role in the continued development of legal pluralism in land tenure. This is because as land resources in a given location become increasingly degraded and scarce, and competition and confrontation emerge as significant themes of interaction, different forms of tenure will be more vigorously attempted in order to improve tenure security. As well, any existing forms of legal pluralism can become more pronounced as competition along with the absence of mediating institutions, can serve to drive different tenurial normative orders into increasingly opposed domains.

## 5. CAN NATIONAL LEGISLATIVE EFFORTS KEEP PACE?

What are the prospects for national legislative efforts in the developing world to keep pace with the need to derive and implement formal legal constructs which can effectively manage emerging pluralism in land tenure due to migration, and hence mitigate resource degradation? Recent historical scenarios do not paint a promising picture (e.g., Odoul, 2001). Migrants can find that even if lands are available, significant rights to resources--and hence tenure security--are unavailable for those not from the host country, region, or local community (Ghimire, 1994; Kabera, 1987). And legal pluralism in land caused by or aggravated by migration--much of which will be taking place across international borders--will pose particular problems in legislative terms. Significant incentives exist which discourage the derivation and implementation of legislation for resolving the dilemmas of tenurial legal pluralism in migration contexts. Host communities, regions, and nations most often receive migrants with significant reluctance, and usually view their residence as temporary (Ghimire, 1994). As a result there can be little enthusiasm (locally or nationally) for deriving the legislative means to deal with land conflicts between different forms of tenurial normative orders involving migrants, especially if the state views alternative forms of tenure as detracting from the national land tenure project (Thiadens, 2001). And international bodies can have very limited influence encouraging nations toward such legislation (Thiadens, 2001).

In reality legislative trends in many developing countries are moving in the direction of laws which increasingly restrict formal legal options for migrants regarding important aspects of land tenure; including dispute resolution, access, and use rights. Such reluctance or inability by national governments to derive and implement legislation and institutions able to effectively deal with legal pluralism in land tenure will significantly compromise the 'adaptation' between formal and non-formal forms of tenure which Bruce (2000) highlights as a significant pattern in coming decades in the developing world.

However there are examples where problematic pluralism did not develop in migrant situations, providing an example where conditions can vary with different results. Tanzania has had varying experiences with refugees, some of these include allowing refugees to become citizens with full tenurial rights to lands in largely uninhabited regions of the country (Christensen, 1985). As well some Burundi refugees who settled in what is now Congo were able to rent land from local landlords, in an example where host tenure rules were able to be of use to migrants (Ghimire, 1994). And in Zambia internal migrants from the Gwembe valley to the general area east of Kafue National Park were able to work with local chiefs in the derivation of an alternative rights system which allows access to land, including inheritance. These examples illustrate that migration need not necessarily result in resource degradation and conflict. But unfortunately these cases are the exception and not the rule.

## 6. A LOOK FORWARD

Migration will increasingly be used as a primary coping and adaptation strategy in response to the effects of global environmental change--including climate change. As patterns and processes of rural to rural migration come to operate in an increasingly pervasive manner, especially in the developing world, patterns of legal pluralism in land tenure will become an important feature. Environmental repercussions will to a significant degree be linked to important aspects of this pluralism. Lands most susceptible to the effects of migration and the resulting legal pluralism will include marginal areas and lands that are for a variety of reasons already tenure insecure. While the derivation of the necessary institutions and legal means can show promise, current trends are not encouraging. Much in the way of policy research is needed in order to adapt to the effects of global environmental change in its different manifestations. Policy tools able to cope with the migration - tenurial pluralism - environment nexus will be part of this need, and a significant challenge to the international community.

## REFERENCES


Abel, R.L. 1982. *The Politics of Informal Justice,* 2 Vols. New York: Academic Press.

Amacher, G.S., W. Cruz, D. Gerbner and W.F. Hyde 1998. "Environmental motivations for migration: population pressure, poverty and deforestation in the Philippines." *Land Economics* 74: 92-101.

Benda-Beckmann, F. von, 1995. "Anthropological approaches to property law and economics." *European Journal of law and Economics* 2: 309-336.

Bruce, J.W. 2000. "African tenure models at the turn of the century: individual property models and common property models." *Land Reform* 1: 17-27.

Clay, J. and O. Stokke 1991. *Food Aid Reconsidered: Assessing the Impact on Third World Countries.* London: Frank Cass.

Comaroff, J., and R. Simon 1977. "The invocation of norms in dispute settlement." In *Social Anthropology and Law,* editor I. Hamnet. London: Academic Press.

Doos, B.R. 1994. "Environmental degradation, global food production, and risk for large-scale migrations." *Ambio* 23: 124-130.

Galanter, M. 1981. "Justice in many rooms: courts, private ordering and indigenous law." *Journal of Legal Pluralism* 19: 1-47.

Ghimire, K. 1994. "Refugees and deforestation." *International Migration* 32: 561-569.

Griffiths, J. 1986. "What is legal pluralism?" *Journal of Legal Pluralism* 24:1-52.

Hammond, J. 1993. "War up-rooting and the political mobilization of Central American refugees." *Journal of Refugee Studies* 6: 104-122.

Henry, S. 1985. "Community justice, capitalist society and human agency: the dialectics of collective law in the cooperative." *Law and Society Review* 19: 303.

Ibrahim, F. 1998. "Editorial: identities in sub-Saharan Africa." *GeoJorunal* 46: 77-78.

IUCN, 2000. "IUCN-CEESP Environment and Security Task Force Briefing." IUCN. Presented at the World conservation Conference, Amman.

Kabera, J.B. 1987. "The refugee problem in Uganda." In *Refugees, a Third World Dilemma,* editor J.R. Rogge. New Jersey: Rowman and Littlefield.

Katon, B., A. Knox and R. Meinzen-Dick 2001. "Collective action, property rights, and devolution of natural resource management." Policy Brief Number 2, CGIAR Systemwide program on Collective Action and Property Rights, International Food Policy Research Institute.

Kok, W. 1989. "Self-settled refugees and the socio-economic impact of their presence in

Kassala, Eastern Sudan." Journal of Refugee Studies 2: 419-441.
Krznaric, R. 1997. "Guatemalan returnees and the dilemma of political mobilization." *Journal of Refugee Studies* 10: 61-78.
Lund, C. 1998. "Struggles for land and political power: on the politicization of land tenure and disputes in Niger." *Journal of Legal Pluralism* 40: 1-22.
Mbembe, A. 1999. "Migration of peoples, disintegration of states: Africa's frontiers in flux." *Le Monde Diplomatique,* November.
McGregor, J. 1994. "Climate change and involuntary migration: implications for food security." *Food Policy* 19: 120-132.
Merry, S.E. 1988. "Legal pluralism." *Law and Society Review.* 22: 869-896.
Migot-Adholla, S.E., and J.W. Bruce 1994. "Introduction: are indigenous African tenure systems secure?" In *Searching for Land Tenure Security in Africa*, editors J.W. Bruce, and S.E. Migot-Adholla. Dubuque Iowa: Kendall/Hunt Publishing.
Moore, S. 1973. "Law and social change: the semi-autonomous social field as an appropriate field of study." *Law and Society Review* 7: 719.
Myers, N. 1997. "Environmental refugees." *Population and Environment* 19: 167-182.
Murphy, A.B. 1990. "Historical justifications for territorial claims." *Annals of the Association of American Geographers* 80: 531-548.
Odoul, R. 2001. "Ethiopian refugees need not fear harassment in camps." *The East African,* July 2-8.
Ostrom, E. 1990. "Governing the Commons: The Evolution of Institutions for Collective Action." Cambridge: Cambridge University Press.
Ostrom, E., J. Burger, C. Fields, R.B. Norgaard and D. Policansky 1990. "Revisiting the commons: local lessons, global challenges." *Science* 284: 278-282.
Prill-Brett. 1994. "Indigenous land rights and legal pluralism in Highlanders." *Law and Society Review* 28: 687-697.
Ranger, T. 1985. *Peasant Consciousness and Guerrilla War in Zimbabwe.* London: James Currey.
Schmeidl, S. 1998. "Comparative trends in forced displacement." In *Internally Displaced People: A Global Survey*, editor J. Hampton. London: Earthscan Publications.
Shipton, P. 1994. "Land and culture in tropical Africa: soils, symbols, and the metaphysics of the mundane." *Annual Review of Anthropology* 23: 347-377.
Smith, A.D. 1988. "Toward a geography of peace in Africa: re-defining substate determination rights." In *Nationalism, Self Determination and Political Geography, editors* R.J. Johnston, D.B. Knight, E. Kofman E. New York: Croom Helm.
de Soto, A. and G. del Castillo 1995. "Implementation of comprehensive peace agreements: staying the course in El Salvador." *Global Governance* 1: 189-203.
Southgate, D, 1990. 2The causes of land degradation along "spontaneously" expanding agricultural frontiers in the Third World." *Land Economics* 66: 93-101.
Thesihuesen, W. 1991. "Implications of the rural land tenure system for the environmental debate: three scenarios." *Journal of Developing Areas* 26: 1-24.
United Nations Development Programme - Emergencies Unit for Ethiopia (UNDP-EUE) 1999. "Horn of Africa the Monthly Review - Somalia: power of Islamic courts and Shari'a militia increasing." Addis AbabaU: NDP-EUE.
Unruh, J.D. 2001. "Land dispute resolution in Mozambique: institutions and evolution of agroforestry technology adoption." Washington DC: CAPRi Working Paper 12. IFPRI.
Unruh, J.D. 1998. "Land tenure and identity change in postwar Mozambique." *GeoJournal* 46: 89-99.
Unruh, J.D. 1997a. "Post-conflict recovery of African agriculture: critical resource

tenure in Mozambique". Ph.D. Dissertation, University of Arizona

Unruh, J.D. 1997b. "The role of land conflict and land conflict resolution in a peace process: Mozambique's return to agriculture." *Refuge* 16: 28-33.

Unruh, J.D. 1995a. "Agroforestry, reforestry and the carbon problem: the role of land and tree tenure." *Interdisciplinary Science Reviews* 20: 215-228.

Unruh, J.D. 1995b. "The relationship between indigenous pastoralist resource tenure and state tenure in Somalia." *GeoJournal* 36: 19-26.

Unruh, J.D. 1993. "Refugee resettlement on the Horn of Africa: the integration of host and refugee land use patterns." *Land Use Policy* 10: 49-66.

Vel, J. 1992. "Umbu hapi versus umbu vincent: legal pluralism as an arsenal in village combats." In *Law as a Resource in Agrarian Struggles,* editors F. von Benda-Beckmann, and der Velde Mvan. Wageningen: Wageningse Sociologisch Studies 33. Agricultural University.

Wilson, K. 1992. "Enhancing refugees own food acquisition strategies." *Journal of Refugee Studies* 5: 226-247.

# SEMI-ARID NORTHEAST BRAZIL: INTEGRATED MODELLING OF REGIONAL DEVELOPMENT AND GLOBAL CHANGE IMPACTS

Maarten S. Krol[1], Dagmar Fuhr[2] and Andreas Döring[2]
[1] *Water Engineering and Management, Faculty of Engineering Technology, University of Twente, PO Box 217, 7500 AE Enschede, the Netherlands.*
[2] *Center for Environmental Systems Research, University of Kassel, 34109 Kassel, Germany.*

## 1. INTRODUCTION

Societies in semi-arid areas in developing regions are amongst those most vulnerable to climate variability and potentially most vulnerable to climate change. The vulnerability to climate variability emerges from a combination of the level of availability of natural resources and the human dependency on these resources. In semi-arid regions, the limited availability of water and the low reliability of this availability pose strong restrictions on the use of natural resources. On the other hand, population often has a high density, and is strongly dependent on natural resources with little short-term options to reduce the dependency.

Reasonable conditions in the wetter years support the persistence of population in the area; marginal or poor conditions in dryer years often have strong impacts as social disruption and population migration. The adverse conditions and arrears in development hamper significant improvements in the quality of life.

Regional tradition and adaptation to local conditions serve as experiences to cope with droughts. This may help to survive emergency situations, but does not really reduce vulnerability, as it does not reduce development deficits (including education, access to markets and credits). These deficits inhibit local population to improve their own living conditions.

In many semi-arid developing regions, such as Northeast Brazil, population in the rural hinterland relies heavily on agricultural production and thus depends directly on the natural resources land and water that are available. Present-day climate conditions already are marginally capable of fulfilling the societal demand for water; therefore climate change poses a serious potential threat. The interrelationships between climate, water



*J.D. Unruh, M.S. Krol and N. Kliot (eds.),*
*Environmental Change and Its Implications for Population Migration,* 119–144.

availability, agriculture, economic self-support, and the quality of life shape the sensitivity and vulnerability of the region.

In the Northeast of Brazil social structures were established over the last century under difficult natural and climatic conditions. The social structures are not very capable of dealing with social and ecological crises in the region. Coping mechanisms, which were developed for such problematic natural and social living conditions, are highly questionable as individual and social concepts for solution. This is especially the case for migration activities, which are no longer traditional compensatory migration or seasonal labour migration.

*Figure 1: The 'drought polygon' with a semi-arid climate in Northeast Brazil. The study area, indicated by the red contour, consists of the States of Ceará and Piauí.*

It is the goal of the present paper to understand the relationships between the variable availability of water and societal impacts in the context of global change. Here, climate change is one of the important components of global change, but other tendencies affecting regional development may be more important.

Many aspects of global change are highly uncertain, especially in their emergence at the regional scale. Climate change projections at the global scale are increasingly showing consensus. However large uncertainties exist at the regional scale, e.g. even the sign of precipitation changes over Northeast Brazil are unclear (see section 4).

Realistic projections of climate change impacts in semi-arid developing regions cannot be made without consideration of the development path of the region. This development may have important implications on the dependency of the society to natural resources. Growth of economic activity will increase demands, also the demand for natural resources, tending to increase the dependency. On the other hand, the relative importance of the most sensitive, agricultural sector may cease, reducing dependency.

This calls for an internally consistent integrated assessment of regional development and climate change. In this assessment, the approach to integration will be a key ingredient. In this paper, pathways of regional development are studied using integrated scenario techniques, while climate impacts are represented using integrated modelling.

Northeast Brazil shows characteristics that make it specifically vulnerable. The region is semi-arid, with precipitation averaging 900 mm per year, but on large stretches annual precipitation is below 600 mm. It is not the annual amount of precipitation that creates problems, but rather its distribution and cycle. The mean annual cycle shows a rainy period with reasonable to ample rainfall amounts, and a mostly 6 month long dry period with extremely low or no rainfall. The annual precipitation amounts show large fluctuations, with irregularly recurring droughts that may last for several years. Droughts do correlate with the El Niño phenomenon and with unfavourable conditions in the pressure systems over the Atlantic Ocean. Spatial distribution of precipitation is very heterogeneous, with spatially limited droughts occurring even in years with median rainfall amounts at the sub-regional scale. Potential evapotranspiration is above 2000 mm per year.

Surface flow is largely restricted to the rainy period and the following months. Most of the river network is not perennial, or was perennial only after reservoir construction. Reservoirs are a regionally common strategy to extend water availability into the dry period (smaller reservoirs) or even into a forthcoming dry year (larger reservoirs), mostly aiming at safeguarding urban water supply. In most of the region, groundwater does not offer a reliable alternative for surface water, as its availability is spatially very heterogeneous, and many wells supply saline water only.

Most of the area is crystalline, with shallow soils. Many of the regionally common soil types have low nutrient content. Alluvial soils close to the river network are best suited for agriculture. The rainfall regime and soil quality combine into relatively unfavourable conditions for rain-fed agricultural production. Yet, a large share of the rural population depends on subsistence farming. Inequity in the distribution of land is extreme in Northeast Brazil; subsistence farmers generally have small plots and few options or competencies to improve their situation.

In severe drought years, the regular functioning of society in the interior is disrupted. Local water and food supply systems collapse and have to be sustained by governmental emergency programs. Water is rationalised and its supply is arranged using transport by trucks over longer distances, food packages are distributed and emergency employment programs are initiated. Population from the rural areas moves into the villages and regional centres Mostly, these aid programs are triggered by societal tension, resulting in e.g. assaults on food transports or blockades of city halls by suffering population, demanding emergency aid.

In a regional study on Northeast Brazil, Magalhães and Rebouças (1988) make integrated considerations on the causal chain from climate to societal impacts, focussing on the sensitivity of agriculture to climate variability. They conclude that 'there evidently is a lack of integrated climate-economy-society studies in Brazil. Given the magnitude of climatic impacts on the Brazilian economy and society, especially in North-eastern Brazil, the development of further research in this field is recommended, with particular emphasis on research that will lead to policies that increase the resilience to drought.'

Latest assessments of available climate observations conclude that global climate has been warming over the last century and that human influences play a major role in climate change (Houghton, et al., 2001). For the 21st century, an enhanced warming is projected to take place under a broad variety of assumptions on global developments. Global precipitation is projected to increase, but at low latitudes, where most of the semi-arid regions in developing countries are situated, both regional increases and decreases are projected. Where precipitation increases, an increased inter-annual variability is assessed to be very likely.

Scientific consensus on climate change projections for specific semi-arid regions is not available at present. This may partly be due to the small scale of these regions, compared to the global dimension represented in the climate models supporting the present assessments. Both the spatial resolution of global climate models and the modelling skill at the regional scale hamper a direct application of results. Regionally specific detailed interpretations of climate change projections are thus required.

The research area of integrated modelling in the context of climate change studies is developing rapidly. Integrated models initially originating from natural science models (energy balance, carbon cycle, crop yield), where models for greenhouse gas emissions from energy use and representations of climate change impacts were included, like IMAGE (Rotmans, 1990), IMAGE 2 (Alcamo et al., 1994). In an alternative approach, macro-economic models were extended with cost estimates for impacts of climate change (like GCAM, Edmonds et al., 1994). Later on, integrated models emerged with a more centrally defined integration approach (like ICLIPS, Toth et al., 1997 or TARGETS, Rotmans and de Vries, 1997).

Regional climate impact assessments also tend to include ever more integrative approaches, extending from agricultural impacts (MINK study, Rosenberg et al., 1993) to representation of interests of various socio-economic sectors (MBIS study, Cohen, 1997). Earlier regional studies on North-eastern Brazil (Magalhães et al., 1988) were broad and rich in information but did not arrive at an integrative concept as adopted in the present (Bronstert et al., 2000).

The implementation of this concept in a fully coupled integrated model, forms a very challenging task, see Section 2. The resulting model dynamically describes the relationships between climate forcing, water availability, agriculture and selected societal processes, allowing the analysis of possible climate change impacts or simulation of scenarios of regional development (Döll et al., 2000).

## 2. INTEGRATED MODELLING

The basic aim of the integrated model is to give a dynamic representation of the relationships between water availability and societal impacts in the context of climate variability and global change. This representation reflects the existing quantitative understanding of the major processes. The integration itself generates a value added in various perspectives. First, it serves to evaluate the multidisciplinary information and modelling, assessing the sufficiency and identifying knowledge and data gaps. Second, it can be used to assess regional sensitivities and development pathways, based on the understanding represented in the model.

Although there does not exist a generally accepted approach to integrated modelling, some characteristics are commonly found. Basic characteristics of an integrated model are its focus on a problem rather than on individual processes, its focus on dynamic behaviour rather than on exact replication of the current situation.

Integrated modelling starts with a systems analysis of the problem, the assessment of relationships between water availability and quality of life and migration in rural semi-arid North-eastern Brazil at the meso-/macro-scale in the context of global change, especially climate change. A top-down analysis of the problem identifies which are the basic variables, processes, and external forcing that should be accounted for in the integrated model (Fig 2).

In this analysis, the focus is explicitly on internal features of the most relevant dynamic behaviour rather than on what is best understood, as for instance the influence of water scarcity on water consumption. The specific challenge is to find a good compromise between the clarity in the representation and the comprehensiveness regarding the main dynamic processes. Special attention should be paid to cross linkages and feedback processes, which already are important in the present functioning of the system (Bronstert et al., 2000), but can especially influence long-term dynamics, as for instance the effects of regional development on the agricultural sector and on water extraction.

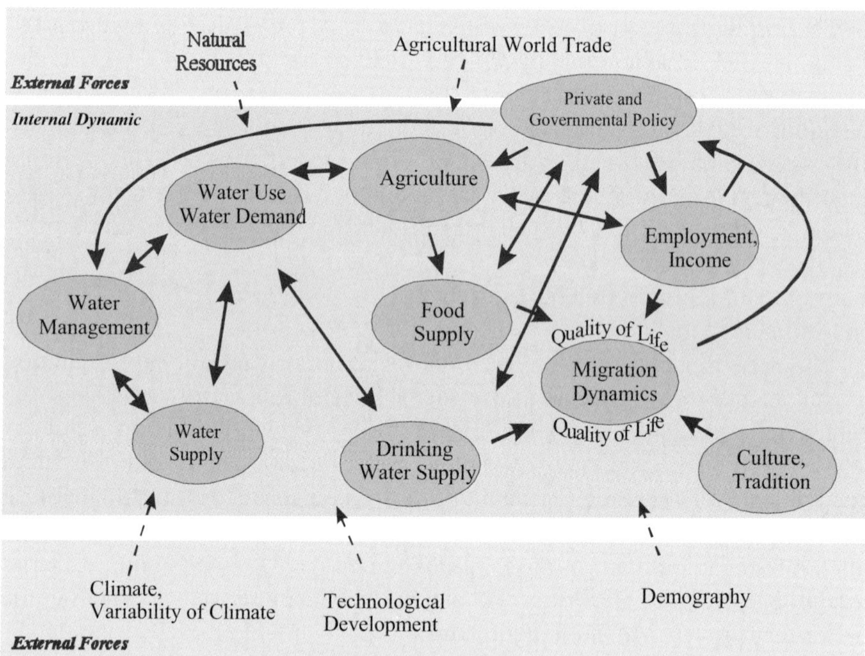


*Figure 2: Conceptual model of dynamic relationships between water availability, agriculture and migration in North-eastern Brazil, as influenced by global change.*

A specifically interesting feature of the dynamics is the ability of the system to dampen or enhance the variability resulting from the variability in the precipitation. Here e.g. water infrastructure and water policy can reduce the variability in the availability of water resources, but increased water demands nearing the availability may amplify variability of impacts as the insufficiency of water supply.

The context of global change forces the analysis to consider long-term behaviour of the system. A reasonable dynamic understanding is preferred over an excellent static understanding, and processes influencing long-term changes are more in focus than processes just explaining heterogeneity at the micro-scale. Clearly, only (aggregated) process-based models can simulate both the effects of individual influences of global change and policy interventions. In this sense, integrated models generally show a mix of deductive and inductive approaches.

In the analysis, geographically explicit integrated modelling is chosen as the most explicit way to approach consistency considerations and regional validity. The Semi-arid Integrated Model (SIM) was constructed for this regional integrated modelling at the meso-/macro-scale (Krol et al., 2001), and is extending the commonly found integrated modelling further into human dimensions.

SIM is built up in a modular way, where each module consists of a disciplinary contribution. Integrative tasks include guaranteeing the consistency between modules, defining interfaces between modules and filling up gaps with plausible and clear basic assumptions. Modules of SIM are the large scale hydrological model WASA, the large scale water use model NoWUM, FAO crop models (LSC and water sensitivity model), the agricultural economy-model RASMO and the model of demography and migration MigFlow. Next, climate scenario data are important inputs to the model

- Climate input: Scenarios for future climates or a historic climate reconstruction are input to SIM. The scenarios result from long-term daily observations combined with climate trends from General Circulation Models (Werner and Gerstengarbe, 1997).
- WASA. This large-scale water balance model (Bronstert et al., 2000) describes run-off, storage in water reservoirs and soil moisture. The model is based on a hydrotope-approach, resolving for quasi-homogeneous units within municipalities. Hydrotopes are defined by their soil, topographic characteristics and vegetation, with a typical size of 10 to 50 km2. The units were determined using a GIS-based soil information system. On these units, vertical processes as evapotranspiration, infiltration, and percolation are represented. Lateral processes couple the units, transporting the generated runoff to the river

network, which is digitised at the scale of municipalities. River flow is routed through the region. Within each municipality, lumped water balances represent small reservoirs; large reservoirs (> 50 Mm3) are described explicitly.

- NoWUM. This water use model simulates water withdrawal and consumption at the municipal scale, considering the various water use sectors (Hauschild and Döll, 2000). Sectors included are private sector (households), agriculture (irrigation and animal water use), industry and tourism. For each sector, indicators for the sectoral activity are combined with water use intensities. Effects of reduced water availability and water price on water use intensity are represented. Part of the extracted water generates return flows. Extraction- and return-flows are accounted for in WASA.
- Agricultural production. Crop yields are simulated using FAO schemes for yield response to water stress and to soil quality. Yield responses to water insufficiency are modelled, accounting for sensitivities in different phenological stages in crop development, as in CROPWAT (FAO, 1979), and put through for 14 regionally important crops (including maize, bean, rice, cashew). Water stress is simulated as non-optimal evapotranspiration, as calculated by WASA at the hydrotope scale, aggregated over the phenological periods. Impacts of soil-quality are statically represented, based on the LSC procedure (FAO, 1983), using data from the soil information system.
- RASMO. The agro-economic model has two important goals. It should represent the economic output of the agricultural sector under the variable climatic conditions, using the yields as simulated using the FAO models. But it should also represent long-term changes in land use due to trends in prices of commodities, climate, and population. Economic output is simulated by bookkeeping production costs, use of produced goods for on-farm consumption as food or feed, and trading on the market. Long-term changes in agriculture are represented as an optimisation of farm income under mean recent climate conditions, varying cropping and husbandry production activities and intensities, under the restrictions of available land, technical, and financial opportunities (Hinterthür et al., 2001). Simulations are performed at the municipal level for three farm size classes.
- MigFlow. The model of population development describes natural processes (birth and death) and migration at the municipal scale. Migration amongst municipalities and to external destinies is modelled to occur when gradients in the quality of life exceed migration costs, see the next section. Quality of life is a composite indicator, where mean municipal income per head is the dominant influence.

The common resolution of all modules is the municipality and the year; the extent consists of the states of Ceará and Piauí (together 332 municipalities in year 1996) and periods of 30 to 100 years. Some modules have fines resolution, hydrotope or day. Modules exchange information at the finest common resolution.

# 3. UNDERSTANDING, DESCRIPTION OF MIGRATION

One prominent quality of the two Federal States (Piauí and Ceará), which were examined, is their juvenile age distribution resulting also in a high birth rate. In 1996, the birth rates were 25.3 births per 1,000 inhabitants in Piauí and 27.5 births per 1,000 inhabitants in Ceará, both substantially above the Brazilian average of 19.7 births per 1,000 inhabitants. However, the high fertility in the Northeast is compensated by migration. Thus the population growth in Piauí and Ceará between 1991 and 1996 was substantially lower than the total Brazilian average (Brazil 1.42 %, Ceará 1.35 %, Piauí 0.69 %).

Migration, however, cannot generally be interpreted as negative. In the context of the social system, it is an adaptation mechanism to high demographic pressure resulting from the continuous disequilibrium between high birth rates and comparatively slower regional economic development.

For the actor, migration is a function of weighing up different adaptation strategies to achieve certain goals in life (i.e. a certain quality of life). It is the expression of the expectation that certain goals in life could better be realized at another location because even innovative action strategies at the location of origin do not promise success. The discrepancy between expectations and 'fulfilment potential' can be the consequence of worsening living conditions, or increased or modified expectations. The investigation of the ‘objective’ living conditions therefore only gives limited insight in the actually experienced ‘subjective’ quality of life. Subjective testimonies of the affected population about their expectations to their environment as well as about their abilities to deal with modifications give information about the extent of the fulfilment of their expectations regarding their actual living conditions.

Changes in the employment structure represent one of the main causes for changes in direction and extent of migration streams in Northeast Brazil. Different Brazilian studies (Wajnmann et al., 1998; Scudele, 1998) and also the results of interviews with the local population carried out in our investigation indicate that there is a strong connection between the insufficiency of employment in the rural area of Northeast Brazil and intensive migration. In our investigation we focus on labour migration since intensive work-related relocation of people from the rural areas towards the major cities gives rise to substantial problems in the coastal areas where most employment opportunities are found. Apart from the employment-driven migration, additional reasons for migration are missing infrastructure of public services, particularly in health care and education. These qualities are increasingly satisfied in the coastal cities.

Migration in our study is interpreted as a multidimensional phenomenon. Therefore, an integrative concept is needed to connect the different thematic spheres. Possible causes for migration are located in the environmental sphere (lack of natural resources), in the social sphere (missing social support) and in the personal sphere (education). The concept of "quality of life" integrates these different phenomenological spheres. Psychological impacts are reflected in the TCA model (see Figure 3), spatial aspects are integrated in the simulation part (MigFlow).

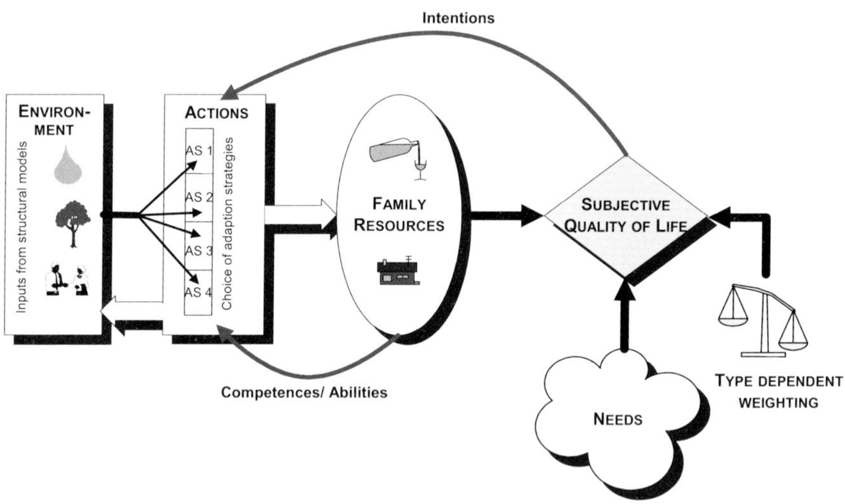


*Figure 3: Structure of TCA, showing the central role of „Subjective Quality of Life" for the prototype collective actor: The fitting of needs – dependent in their extend on the type of actor – with available family resources creates „Intentions" towards actions. The „Choice of Adaptation Strategies" is determined by „Competencies / Abilities" of the actor (which could be described as a subset of „Family Resources") and the „Intentions" to transform „Environmental Resources" (social and natural) into a new state of „Family Resources". These, in turn, are evaluated again in the next step of the new ratio of expectations to „Needs".*

## The Threat-Competence-Adaptation (TCA) model and its theoretical background

To reach the goals described above, it is necessary to formulate an actor model reflecting the dynamic interdependence of internal and external states.

Different adaptive action strategies have to be implemented in this model to realize feedback functions of behaviour on external (e. g. characteristics of the social surrounding) and internal variables (e. g. degree of satisfaction with social relations). The set of the variables used and their interrelations was derived from theoretical considerations based in social psychology. Survey data was used to check the validity of the proposed model.

With respect to the multidimensionality of the migration phenomenon, an integrative concept of quality of life was developed. This concept takes into account both the satisfaction of elementary needs and environmental conditions composed of the natural environment as well as the social environment.

On the basis of the framework presented above, concepts for objective and subjective quality of life can be derived. "Objective quality of life" describes the characteristics and developments of the physical-biological environment, the social surroundings and the economic and political systems. These variables are included in the action model as 'indicated potential'. In contrast, "subjective quality of life" is measured by the extent of the need-specific contentment. This need-specific contentment can be influenced by limitations in available resources or competences, shifting objective potentials outside of the range of desired improvements in living conditions.

The general concept of the action submodel of migration can be described in three steps:

1. fitting of needs and possibilities/opportunities (affordances),
2. selection of different adaptive action strategies with reference to a set of theory-guided deduced needs and empirically-derived competencies,
3. maximisation of changes in satisfaction of needs.

A set of needs (existence, security, freedom, flexibility, affiliation) was derived from Max-Neef's works (Max-Neef, 1991) who intended to put together a set of human requirements with inter-cultural validity. Furthermore, two resource types are differentiated: infra-structural (medical support, education) and agricultural resources (soil quality, temporal distance to latest drought, water availability). Different adaptive action strategies are identified (migration, innovation and persistence). The selection of different adaptive action strategies is based on the protection-motivation theory by Riptoe and Rogers (1987). It postulates that the probability of specific strategies depends simultaneously on both the perceived personal ability (the actor's competencies to act (e.g. education) and the perceived situational threat (described by the actual need state and the availability of action resources).

The satisfaction with the actual living conditions is the important result of that coping process. The postulation of a general goal to maximize this satisfaction leads to the actor's preference to select those strategies that are supposed to result in an optimal need satisfaction considering competencies and action resources.

The Threat-Competence-Adaptation model is based on the prototypical collective actor (e.g. family or household). According to farm size and family phase, different actor types can be differentiated. Each type has a characteristic profile of needs and their importance as well as of different competencies. The possibilities for the adaptive strategies described are implied for each actor type in each municipality.

**Implementation problems**

The migration model, formulated in the previous paragraph has a high requirement for input data. Especially for the analysis of migration motivation and population development, this is a big problem because many demographic and migration-related data only exist on the level of the federal states and aggregated over population subgroups with different vulnerabilities, e.g. as resulting from different conditions concerning land ownership or education.

The main data problems concern:

- availability and quality of data,
- coarse resolution of population data,
- underreporting of births and deaths,
- no accepted procedure to measure migration,
- missing time series.

The data input requirements for models explaining migration, such as TCA, can only be fulfilled using micro-scale questionnaires, additional to the readily available secondary data. Spatially explicit models covering larger areas, such as required for the integrated model, can rely only on the coarser, highly aggregated data. In order to develop a partly reliable migration model despite these problems, we decided to support the interpretation of the limited data by using knowledge gathered by interviewing experts and rural population in two selected municipalities. The municipalities were selected for covering the range from sites with a very dry climate, poor soils, low development, being a migration source (Tauá, Ceará) to a regionally attractive centre where economy is diverse (Picos, Piauí).

**Results of the Surveys in Picos and Tauá**

About 100 households (102 families in Tauá, and 104 families in the Picos area, including Picos and eight municipalities around Picos) were questioned in March 1999 with respect to their socio-economic situation, their experience with migration and their satisfaction with the farming situation.

One person was questioned as a representative of each household, in general the head of the household or his/ her partner. 87 men and 17 women were questioned in Picos and 54 men and 48 women in Tauá. The average age of the interviewees in Picos was about 48 years and in Tauá about 45 years. This is similar to the average age of the head of the households in this region.

Agriculture still plays an important role in municipal economy and is the principal occupation of most potential migrants and therefore chosen as a topic in the surveys. The main limitations to farming mentioned most frequently were (i) the absence or irregularity of precipitation, (ii) the desperate fight against pests, (iii) the lack of high-quality seeds, and (iv) the absence of a price guarantee for agricultural products.

Further, we were interested whether families were financially supported by family members who had already moved away. This is the case for about one third of all questioned households in both municipalities. Furthermore, about 15% of the questioned households in Tauá and Picos get financial support from the governmental retirement scheme. During the drought in 1998, there was additional governmental emergency aid available: These job programs or food aid were claimed by about two thirds of the questioned families in each of the municipalities.

We were further interested in the motivation to migrate and which town people preferred as a migration target. The main target is still the state of São Paulo since the main motivation for the majority of migrants is to find a new job. This is valid for 87% of the migrants who move to São Paulo. The second most important reason why people migrate is the reunification with family or plans to start a new family (16% of the motivation to migrate). The latter reason often leads to a migration within a smaller area when compared to migration due to economic motivation. As a third reason, people hope for an improvement of their general living conditions (including secondary or college education, making up 5% of the overall motivation to migrate). These improvements are mostly expected to happen in the South-East of Brazil.

It is remarkable to note from the questionnaires that neither the size of the farm nor the type of ownership seem to play an important role in the decision to migrate. Migration is practically equal from all sizes and types of farms, which can be explained by the fact that it is generally not the head of the family who decides to migrate but rather the sons and daughters.

Finally, we wanted to know where people prefer to live: in the countryside or the city. The answers show a clear opinion: people prefer living in the countryside since this is seen as more quiet and peaceful, and solidarity with others plays an important role. People are actually afraid to live in the city because of hustle and crime. Living in the big coastal cities therefore is not an objective in itself but can be seen as a coping mechanism due to the lack of regional development – particularly employment opportunities - in the countryside.

The results of the survey about the satisfaction with agricultural business suggest an upcoming of structural change, especially for future generations of the very small farm holders who do not seem content with farming as their sole profession. The lack of other jobs in rural area is the main reason for migration; climatic changes influence migration only indirectly (through economic impacts). Furthermore, quality of life in rural areas can only be improved through integrated measures, including agricultural assistance, improvement of infrastructure and creation of jobs in non-agricultural sectors in the rural area.

**Development of a space-related migration model (MigFlow)**

The TCA model presented in the previous section has an ambitious theoretical background based in psychological theory. The complex structure and feedback loops found in the model however lead to difficulties in the implementation. The insufficient database on the municipality level underscores the problems of comprehensive modelling, especially in the field of the subjective perception of the quality of life and the personal competence.

Hence, it was necessary to develop an applied submodel as a simulation environment to test hypotheses concerning interactions between the development of the quality of life and human action on a spatial level, the Migration Flow Model (MigFlow). Additional theoretical concepts from TCA are successively being integrated. MigFlow has a more pragmatic approach than TCA: it simulates migration without the comprehensive action model but directly related to the concept of quality of life. With respect to theory however, MigFlow has a limited scope. It cannot explain the migration phenomenon in general, but can clarify certain aspects, related to opportunities for potential migrants (e.g. innovation) based on differences in the quality of life in different municipalities in the Northeast of Brazil.

**Central Ideas**

The conceptualisation of migration processes in MigFlow follows the assumption that migration tends to equalize disparities of quality of life because migrants utilize opportunities to improve their living conditions. Therefore, migration is affected by opportunities in potential target regions and the competence to overcome obstacles of moving as well as to benefit from these opportunities for the potential actors.

These findings converge with the theoretical assumptions of "classical migration theory" (Ravenstein, 1885; Lee, 1966):

- The degree of migration inside a given territory varies directly with the diversity of regions included in that territory.
- The migration rate between two locations is inversely related to the obstacles for overriding the distance.

MigFlow consists of three main modules: demography, quality of life and migration.

The sub-model demography calculates the total population for each municipality in cohorts of 5 years based on age-specific and sex-specific rates of mortality and fertility (IBGE, 1997). The trends for mortality and fertility follow scenario assumptions: We assume a decrease down to 60% (fertility) and 80% (mortality) of 1996 rates in 2025.

Quality of life is modelled as an integrated indicator composed of

- the aggregated family income (HDI: PNUD, IPEA, FJP, IBGE, 1998);
- an indicator for educational infrastructure (secondary education and upwards) (SOLAM);
- and an indicator for health infrastructure (Costs) (SOLAM).

The selection of suitable indicators was accomplished by two main aspects: their prominence in the results of the survey and the availability of comparative official data on the municipality level. These indicators are further made dynamic by scenario assumptions (for the general scenario assumptions compare also Döll et. al., 2000).

MigFlow models Ceará and Piauí as a system of interacting spatial units - the municipalities - with specific properties. The communication with external territories outside the model boundaries is represented in one virtual unit (being represented by São Paulo which is the primary migration target outside of the survey region; see also the results of the survey).

Migration is being calculated in the following steps:

- computation of disparities between the quality of life of the different spatial cells;
- computation of migration costs based on spatial distances between the cells (linear distance based on geographic coordinates);
- computation of potential migration to units with a higher quality of life: potential profit in the quality of life (cost benefit calculation) multiplied with competence factor and the affected population (households with an income less than half a minimum wage - Salario Minimo);

- recalculation of migration based on emigration and immigration restrictions;
- integration of migration in the demographic model.

For a more detailed description of the model the reader is referred to Grebe et. al, 2000, S. 10ff.

# 4. RESULTS FOR INTEGRATED SCENARIOS

In studies on long-term developments, integrated scenarios can play diverse roles. Scenarios can be used to scope the future, by considering various diverging but plausible scenarios. Scenarios can also be constructed to serve as a background for sensitivity studies or intervention assessments. In sensitivity studies, some external factors or parameter values are varied and the effect on the simulated integrated scenario is analysed. In intervention assessments explicit policies are introduced in background scenarios to assess their efficiency.

Here, two 'Reference Scenarios' of regional development try to project plausible pathways into the future, stressing the possibility of different futures. The relative likelihood of the one or other scenario is not considered.

Both scenarios emerged from interpreting a coarse generalised storyline in the various areas, sectors and variables. The scenarios are:

Reference Scenario A. Coastal boom and cash crops. In this scenario, economic globalisation and orientation on economic values shape development. Development focuses on profit optimisation, thus on most profitable production activities and production regions. In Ceará and Piauí this is interpreted to prefer coastal touristic and industrial activities and an orientation of agriculture on cash cropping.

Reference Scenario B. Decentralisation. In this scenario, a decentralisation of development aims at more participation of the population, and a more even distribution of activities over a geographic area. Development occurs integrated (sector-wide) and takes place in the regional centres in the hinterland, not only in the coastal area.

The scenarios were originally defined for the period 2000-2025 (Döll et al, 2000), but here they are extended to 2050, the timescale on which climate change may become significant.

The central assumptions made in the scenarios, next to climate, concern demographic drivers, economic growth, availability of agricultural land and its distribution over farm classes, irrigation policy, and water infrastructure.

In scenario A, the global focus on economic values results in higher growth rates than in B; in B the regional distribution of growth is more balanced. In scenario B, the distribution of land ownership is getting more even; in scenario A the high-input irrigation grows faster. Regional distribution of improved water infrastructure and new irrigation areas is strongly focussed on sub-regions with highest economic potential in scenario A and more even in scenario B.

State of the art global climate models show an increasing ability to simulate present day climate as well as historic trends over the last centuries at the global to continental scale (Houghton, et al., 2001). Still, the ability of these models in representing climate at the scale of North-eastern Brazil (NEB) is modest. This lack in ability may be caused by the relatively coarse resolution of the climate models, or by an imperfect representation of regionally important physical processes or the availability of regional data. Either way, the lack in ability seriously affects the applicability of model results for impact assessments.

The recommended approach to critically review regional performance in selecting model results to be used in assessments (IPCC-TGCIA, 1999) implies that only very few global model results allow regional interpretation. Two models distributing projections via the IPCC Data Distribution Centre (IPCC-DDC, 1999), still reasonably allowing a regional interpretation of their results for North-eastern Brazil are ECHAM4 (Roeckner et al., 1996) and HADCM2 (Johns et al., 1997). For a continuing increase of greenhouse gas concentrations at present rates, projections of precipitation changes over NEB (2070-2099 compared to 1961-1990) are –50% for ECHAM and +20% for HADCM.

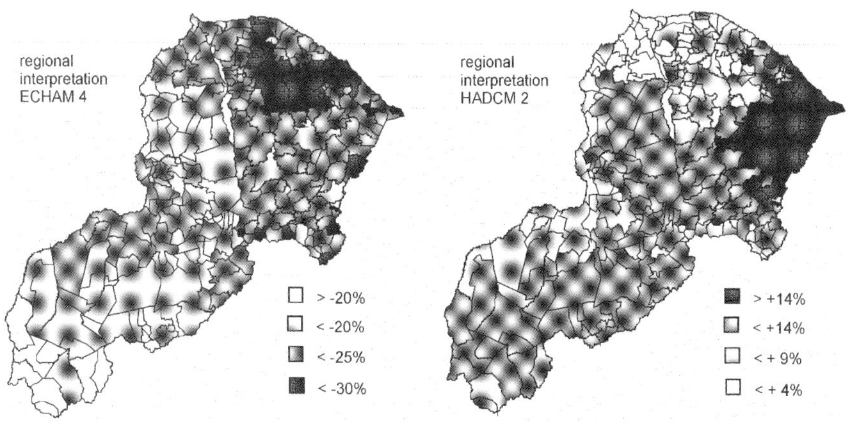


*Figure 4: Regional interpretations of GCM results for precipitation trends over the period 2001 to 2050, assuming continued growth of atmospheric greenhouse gas concentrations.*

From this, conclusions on likely precipitation changes in NEB, on median changes or probable ranges of precipitation change cannot be drawn. Still, in climate change studies for semi-arid North-eastern Brazil both the possibilities of a strong decrease in precipitation (-50%) and an appreciable increase in precipitation (+20%) should be considered as plausible to take place in the current century.

A direct use of the grid-scale results of the climate models is unattractive. The difference in scale between the global climate models and the hydrologic and agronomic models in SIM is too big to allow reasonable simulations at the model scale. Regional climate modelling may partly overcome this problem, but existing regional climate models for Northeast Brazil still give unsatisfactory simulations of precipitation, probably due to problems with physical parameterisations (Boehm, 2002). Downscaling methods exist to consistently couple climate trends, defined at a coarse scale with finer scale, regionally observed climate data, to yield a finer-scale interpretation of the climate model results. The downscaling method adopted combines observed daily historic climate data at the level of climate stations with long-term climate trends from GCM projections (Werner and Gerstengarbe, 1997).

In this paper, the two reference scenarios (A and B) and the two climate scenarios (ECHAM and HADCM) are combined into three scenarios, defined for the period 2000-2050: A+ECHAM, B+ECHAM and A+HADCM.

Over this 50-year period, precipitation shows a decrease by 20 to 30% in the ECHAM scenario and an increase by up to 15% in the HADCM scenario (Figure 4). The variability in precipitation (as coefficient of variation) is not changing much in the scenarios, reducing somewhat towards the end of the simulation period.

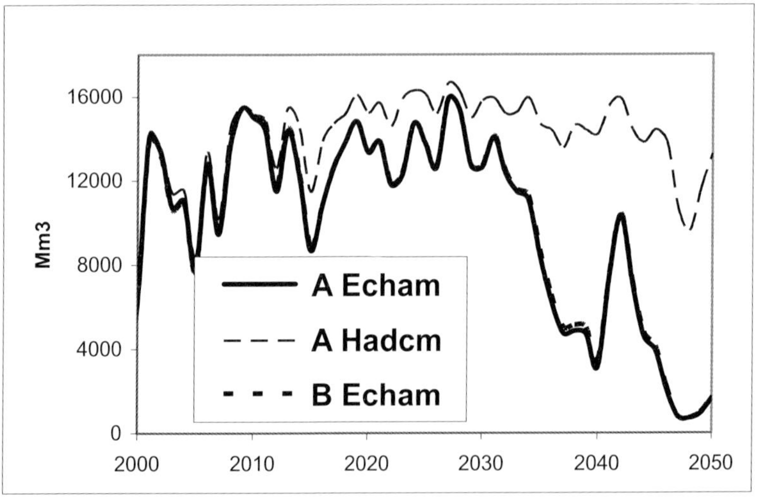


*Figure 5: Total volume of water stored in larger reservoirs at the beginning of the dry period.*

Surface water availability shows a sensitive reaction to this climate change. River discharge shows a dramatic decline in the second part of the simulation period for the ECHAM scenarios, whereas no significant change is found for the HADCM scenario. Differences between the A and B scenarios are small only, as the total annual river discharge is largely determined by natural processes. In this sense, river discharge is an indicator of climate change itself and not so much of the impact on society.

The total water volume stored in reservoir at the beginning of the dry period is another important indicator of surface water availability. This indicator is stronger linked to water management and water use, measuring impact on human resource use. Here we see (Figure 5) an initial rise in stored water due to increased storage capacity, assumed in both scenario A and B, followed by a clear downward trend for both ECHAM scenarios and a small trend for the HADCM scenario. For both the ECHAM scenarios, stored water volumes towards the end of the simulations are so low, that the reservoirs should be considered as fully inefficient for the climate conditions at that time. The differences between the A and B scenario are modest, with slightly higher values for the stored volume in scenario B. The difference is mainly due to a slower increase in irrigated area in scenario B than in scenario A effects of and on irrigation water use are much more pronounced in the dry period however.

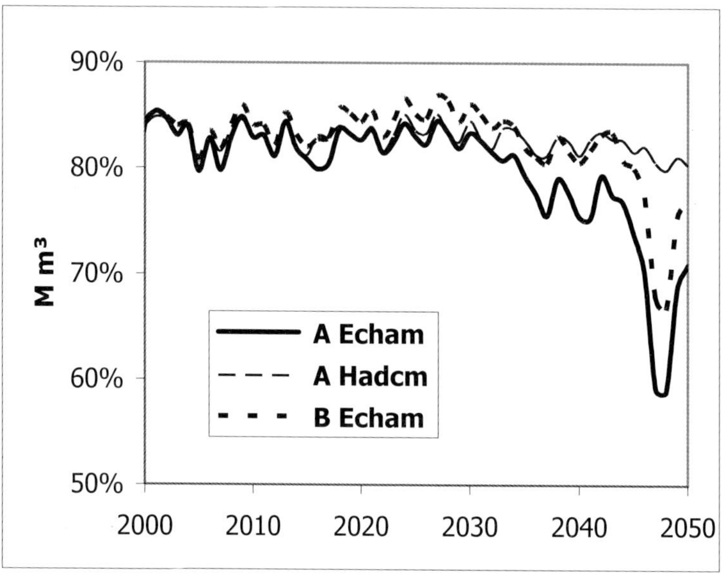


*Figure 6: Relative fulfillment of water demand; reductions are due to limited availability of water.*

The reduced availability of surface water, for the ECHAM scenarios, has a negative influence on the fulfilment of water demand (Figure 6). Whereas water availability decreases or remains about constant, water demand increases drastically in the next 50 years in the scenarios A and B. An initial increase of annual demand from 1500 Mm3 in 2000 to about 3000 Mm3 for scenario A and 2200 Mm3 for scenario B is followed by a small (about 10%) decrease in the demand, due to price-driven reductions in demand (water pricing policy is an emerging feature in Northeast Brazil). The fulfilment of total annual demand remains about constant in the HADCM scenario, but indicates increasing water stress in the ECAHM scenarios. Especially in the A-ECHAM scenario, where water demand is highest, shows a drop from the initial 85% to a mean 70% in the last decade simulated. In scenario B-ECHAM, fulfilment drops to 75%.

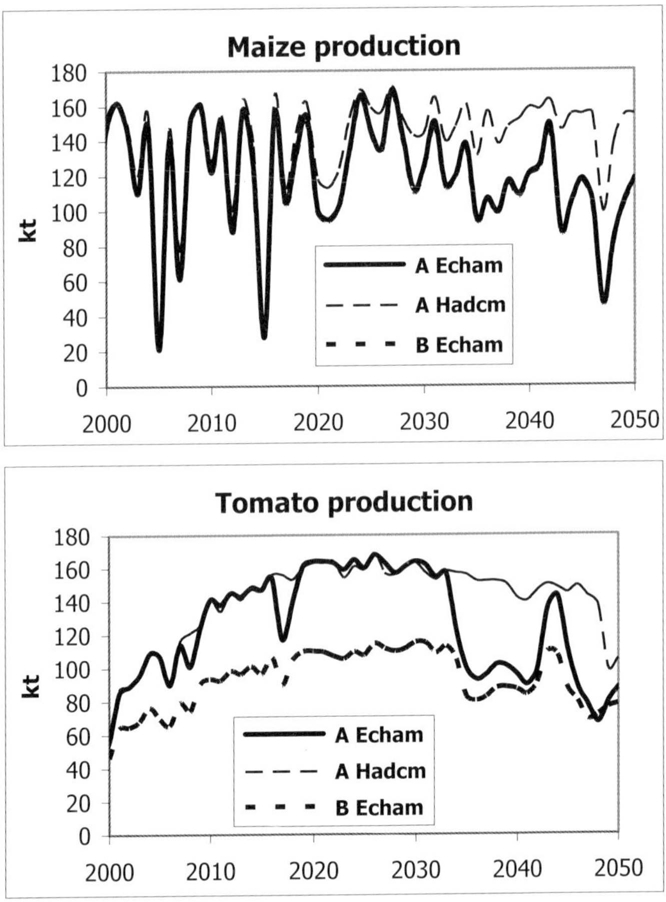


*Figure 7: Production of agricultural commodities in Ceará and Piauí in three future scenarios*

Agricultural production is still mainly rain-fed in Ceará and Piauí. This causes a high sensitivity to rainfall amounts, and long-term trends in precipitation may aversely affect regional production. Irrigated production can be less sensitive, as long as the irrigated area does not exceed the maximum area that can be reliably supplied with water. In both scenarios A and B, the cropping areas for the traditional food crops grown in subsistence farming (as maize) remain about constant, whereas the areas of irrigated crops (as tomato) increase by 100% (scenario B) to 200% (scenario A). The total production of Maize in Ceará and Piauí shows a high variability and a downward trend towards the end of the simulation period for the ECHAM scenarios and no clear trend for the HADCM scenario. Tomato production first increases with available irrigation area for all scenarios, the increase being faster for the A scenario. In the second half of the simulation period, water availability does not sustain irrigated production on all irrigatable areas in A-ECHAM, such that the production volumes are very similar for A-ECHAM and B-ECHAM (see Figure 7).

The neutral or downward trend in rain-fed productivity and modest growth of irrigated production (especially for the ECHAM scenarios) cause the agricultural sector to play a declining role in relative terms in regional economy. As population dynamics in the simulation model is mainly driven by economic development, this might cause a reduced influence of climate variability on migration in the model.

Total population increases strongly in all scenarios, starting from 10 million in 1995 reaching about 12 million in 2025 for all scenarios and 15 million in 2050 for the A scenarios and 14 million for the B scenario. The difference emerges from the higher assumed rates for economic growth in scenario A, reducing the tendency to migrate. Cumulative migration out of the region is simulated to reach about 5 million for scenario A and 6 million for scenario B. The mean annual rate of migration is reducing slowly in scenario B (-15% over 50 years) and strongly (-60% over 50 years) in scenario A towards the end of the simulation period.

The population in the (non-coastal) hinterland shows an even more marked difference between the two scenarios. In scenario A population in the hinterland decreases from about 5 million to 3.8 million, in scenario B it increases to 5.2 million. This differences results from the assumed regional distribution in economic growth, being more concentrated in the metropolitan and coastal areas in scenario A. This means that the continuing internal concentration of population in the region is very strong in scenario A and modest in scenario B.

Differences in population development and migration between the ECHAM and HADCM scenarios are small.

# 5. CONCLUSIONS / DISCUSSION

The integrated modelling of the causal relationships between climate variability, water availability, and agricultural and societal impacts proved useful in composing an internally consistent projection of future regional developments. The projections illustrate the implications for a possible future of plausible assumptions on development and formalised understanding of processes. It helps especially in analysing the consistency of assumptions, e.g. the consistency between the regional activities in irrigated production and availability of water resources.

In this respect, the present study extends existing regional long-term simulation studies of environmental (especially climate change) impacts in the direction of human dimensions.

Simulations clearly show that Northeast Brazil is very sensitive to plausible climate change.

For the more pessimistic scenario, the availability of river flow and surface water storage are alarmingly reduced; for the more optimistic scenario the present availability is about maintained. If the more pessimistic case would materialise, a rapid reaction in water management would be necessary to prevent severely worsening problems with water scarcity.

Agricultural rain-fed production shows a sensitive reaction on precipitation projections, whereas irrigated production may be seriously confined by future developments in water availability (stored water and river flow). This might have a large impact on the agricultural sector, as rain-fed low mechanised production of the primary food crops in subsistence farming already is economically marginal. In the more pessimistic scenario for precipitation in combination with strong growth of high-input agriculture (A-ECHAM), the expansion of irrigated production is not sustained by the projected water supply.

Both environmental impacts (water scarcity, harvest failures) are important in determining the living conditions of the local population, especially in the rural hinterland.

The formalised understanding of migration, represented in the simulations indicate that mean migration rates may remain high in future, albeit lower or even significantly lower than at present. The simulations suggest that developments in future migration will depend much more on integrated economic development of the rural areas than on climate, especially when the relative importance of agriculture will further decline. The effect of climate change on mean migration appears to be modest. Nevertheless, climate and its variability are simulated to have an influence on high migration occurrences in drought years.

Policies reducing the importance of rain-fed agricultural production or enhancing the buffering capacity in agricultural enterprises will have the positive effect of reducing migration peaks in drought conditions. Such policies include promoting diversification and improving access of farmers to credits and markets. As development alternatives are mostly envisaged for regional centres or sub-regions with less limiting water resources, gradual rural-urban migration may be expected to continue, and urban water demand will further increase. Policies aiming at a fixation of rural population in rain-fed production areas will lead to a continuation of the dependence on external support and possible high migration rates in drought years. Regional development policies focussing at integrated development of centres in the non-coastal areas where water resources can be made available might limit the concentration of population but still reduce the vulnerability to drought. Policies aiming at enhancing water use efficiency might seriously reduce the vulnerability to drought. These policies include reducing transport losses, promotion of water extensive production systems, both in industry and irrigation. In the more optimistic climatic scenarios, this would enhance water amount available for e.g. irrigation.

It should be noted that the simulation model only represents a partial understanding of migration processes. At this stage, migration is simulated to depend upon mean municipal income, and thus does not include the effects of income distribution. Additional data and research are needed to arrive at a more ample indicator for the quality of life, or to support models that describe migration motivation more intrinsically, including a better insight in the influence of additional variables, such as education and health care.

The indicator for the quality of life by now does not include a direct contribution from regional food or water supply, i.e. the effects of drought only enter through the reduced agricultural output. This must not lead to an underestimation of migration, as droughts directly cause a flow of temporary refugees rather than of permanent migrants. Drier circumstances could be interpreted be the trigger of migration rather than the root cause of it.

The topic of migration is deeply controversial, even its valuation. On the one hand, migration should be seen as a response to circumstances aiming at improving the current quality of life, and thus as a positive action, often aiming not only at drought relief, but also at lasting improvements of living conditions. On the other hand, large-scale migration is often connected to uncontrolled and hardly controllable actions as blockades of city halls or assaults on food transports, as a pressure in demanding drought relief aid. This and the public costs of emergency aid feed negative valuations of migration.

In the policy debate, the controversy is reflected in the possibly contradictory aims to both improve the fixation of population in the regions interior, and the aim to focus development programs on the sub-regions with highest potentials. The contradiction emerges from the fact that the regions with the severest drought problems, i.e. the main source regions for drought-related migration, are not amongst the sub-regions with the highest rate of economic return of development investment.

In this debate, the scenario analysis itself cannot deliver decisive information, as the valuation of aspects of the problem is the subject of the debate. But the results of the scenario analysis do give an important contribution to the debate by providing an ample set of information.

## ACKNOWLEDGEMENTS

The authors like to thank the German Federal Ministry of Education and Research, which supported the research as part of the WAVES program (Water Availability and Vulnerability of Ecosystems and Society) under grant number 01LK9713. During the period that the research was conducted and presented at the Wengen 2001 workshop, Krol was appointed at the Potsdam Institute for Climate Impact Research (PIK) in Gemany.

## REFERENCES

Alcamo, J., G.J.J. Kreileman, M.S. Krol and G. Zuidema. 1994. "Modeling the global society-biosphere-climate system: Part 1: Model description and testing." *Water, Air, and Soil Pollution* 76:1-35.

Böhm, U., F.-W. Gerstengarbe, D. Hauffe, M. Kücken, H. Österle and P.C. Werner. 2003. "Dynamic regional climate modelling and sensitivity experiments for the northeast of Brazil, in Global Change and Regional Impacts." Pp. 13-170 in *Water availability, vulnerability of ecosystems and society in the semi-arid northeast of Brazil*, editors T. Gaiser, H. Frischkorn, M. Krol and J.C. de Araújo. Springer.

Bronstert, A., A. Jaeger, A. Güntner, M. Hauschild, P. Döll and M. Krol. 2000. "Integrated Modelling of Water Availability and Water Use in the Semi-Arid Northeast of Brazil." *Phys. Chem. Earth, B* 25(3):227-232.

Cohen, S.J. (ed.) 1997. *MacKenzie Basin Impact Study (MBIS), Final Report*, Atmospheric Environment Service. Environment Canada, Toronto.

Döll, P., D. Fuhr and J. Herfort. 2000. *Szenarien der zukünftigen Entwicklung in Piauí und Ceará*. Statusbericht der ersten Hauptphase des WAVES Programms. Wissenschaftliches Zentrum für Umweltsystemforschung. Kassel, 2000 (http://www.usf.uni-kassel.de/waves/vorl_endbericht/ 3_szenarien.pdf ).

Edmonds, J.A., M.A. Wise and C.N. MacCracken. 1994. *Advanced Energy Technologies and Climate Change: An Analysis Using the Global Change Assessment Model (GCAM)*. Pacific Northwest Laboratory, Richland, Washington.

FAO. 1979. *Yield response to water*, FAO Irrigation and Drainage Paper 33, Rome.

FAO. 1983. *Guidelines: land evaluation for rainfed agriculture*, FAO Soils Bulletin 52, Rome.

Grebe, M., A. Döring, D. Fuhr and F. Matias. 2000. "Modeling Migration and Quality of Life. In: WAVES - Water Availability, Vulnerability of Ecosystems and Society in the Northeast of Brazil." Pp. 55-59 in *Proceedings of the Third International WAVES-Workshop in Freising - Weihenstehpan*, March 2000. Freising.

Hauschild, M. and P. Döll. 2000. "Water Management Modeling in Ceará and Piauí (Northeast of Brazil)." In *Proceedings of the German-Brazilian Workshop on Neotropical Ecosystems* Hamburg, September 3-8, 2000. (www.usf.uni-kassel.de/waves/english/vorl_ergebnisse/ hauschild.pdf).

Houghton, J.T., Y. Ding, D.J. Griggs, M. Noguer, P.J. van der Linden and D. Xiaosu. (eds.) 2001. *Climate Change 2001: The Scientific Basis*. Contribution of Working Group I to the Third Assessment Report of the Intergovernmental Panel on Climate Change. Cambridge University Press, UK.

Hinterthür, H. and H. Gaese. 1999. "Die Berücksichtigung des Produktionsfaktors Arbeit in einem Agrarsektormodell für einen bevölkerungsreichen semiariden Raum in Entwicklung: das Beispiel Nordost-Brasilien." *Göttinger Beiträge zur Land- und Forstwirtschaft in den Tropen und Subtropen* 133:99-105.

IBGE. 1997. *Contagem da População*. Rio de Janeiro.

IPCC-DDC. 1999. *GCM climate change scenarios and observed climate datasets*. http://ipcc-ddc.cru.uea.ac.uk.

IPCC-TGCIA (Intergovernmental Panel on Climate Change, Task Group on Scenarios for Climate Impact Assessment). 1999. *Guidelines on the use of scenario data for climate impact and adaptation assessment, Version1*. Prepared by T.R. Carter, M. Hulme and M. Lal.

Johns, T.C., R.E. Carnell, J.F. Crossley, J.M. Gregory, J.F.B. Mitchell, C.A. Senior, S.F.B. Tett and R.A. Wood. 1997. "The Second Hadley Centre coupled ocean-atmosphere GCM: model description, spinup and validation." *Climate Dynamics* 13:103-134.

Krol, M.S., A. Jaeger, A. Bronstert and J. Krywkow. 2001. "The Semi-arid Integrated Model (SIM), a regional integrated model assessing water availability, vulnerability of ecosystems and society in NE Brazil." *Phys. Chem. Earth B* 26(7-8):529-533.

Lee, E.S. 1966. "A Theory on Migration." *Demography* 3(1):47-57.

Magalhães, A.R., H.C. Filho, F.L. Garagorry, J.G. Gasques, L.C.B. Molion, M.S.A. Neto, C.A. Nobre, E.R. Porto and O.E. Rebouças. 1988. "The effect of climatic variations on agriculture in Northeast Brazil." Pp. 273-380 in*The Impact of Climatic Variations on Agriculture, Volume 2: Assessments in Semi-Arid Regions*, editors M.L. Parry, T.R. Carter and N.T. Konijn. Kluwer Ac. Publ.

Max-Neef, M.A. 1991. *Human Scale Development. Conception, Application and further Reflections*. New York, London.

PNUD, IPEA, FJP, IBGE. 1998. *Atlas do Desenvolvimento Humano no Brasil*. Belo Horizonte (Brazil).

Ravenstein, E.G. 1885. "Die Gesetze der Wanderung 1 und 2 (1885/1889)." Pp. 41-94 in reprint *Regionale Mobilität*, editor G. Széll, 1972.

Ripptoe, P.A. and R.W. Rogers. 1987. "Effects of components of Protection Motivation Theory on adaptive and maladaptive coping with a health threat." *Journal of Personality and Social Psychology* 52:596–604.

Roeckner, E., K. Arpe, L. Bengtsson, M. Christoph, M. Claussen, L. Dümenil, M. Esch, M. Gioretta, U. Schlese and U. Schulzweida. 1996. *The atmospheric general circulation model ECHAM-4: model description and simulation of present-day climate*. Max-Planck Institute for Meteorology, Report No. 218, Hamburg, Germany.

Rosenberg N.J., P.R. Crosson, K.D. Frederick, W.E. Easterling, M.S. McKenny, M.D. Bowes, R.A. Sedjo, J. Dramstadter, L.A. Katz and K.M. Lemon. 1993. "The MINK methodology: background and baseline." *Climatic Change* 24(1-2).

Rotmans, J. 1990. *IMAGE. An integrated model to assess the greenhouse effect*. Kluwer Academic Publishers, Dordrecht, Netherlands.

Rotmans, J. and B. de Vries. (eds.) 1997. *Perspectives on Global Change: the TARGETS approach*. Cambridge University Press/

Scudele, V.C. 1998. "O Mercado de Trabalho dos Eua e Alguns Indicadores Sobre a Primeira Inserção Sócio-Ocupacional do Fluxo Migratório de Governador Valadare." Pp 519-550 in *Encontro Nacional de Estudos Populacionais*, ABEP. Belo Horizonte.

Toth, F.L., T. Bruckner, H.-M. Füssel, M. Leimbach, G. Petschel-Held and H.-J. Schellnhuber. 1997. "The tolerable windows approach to integrated assessments." Pp. 403-430 in *Climate change and integrated assessment models [IAMs] – bridging the gaps*, Proceedings of the IPCC Asia-Pacific workshop on integrated assessment models, editors O.K. Cameron, K. Fukuwatari and T. Morita. Center for Global Environmental Research, Tsukuba, Japan.

Wajnman, S., B.L. Queiroz and V.C. Liberato. 1998. "O Crescimento da Atividade Feminina nos Anos Noventa no Brasil." Pp. 2429-2454 in *Encontro Nacional de Estudos Populacionais*, ABEP. Belo Horizonte.

Werner, P.C. and F.-W. Gerstengarbe. 1997. "A Proposal for the development of climate scenarios." *Climate Research* 8 (3):171-182.

# POPULATION MOVEMENTS, ENVIRONMENTAL CHANGE AND SOCIAL CONFLICTS IN THE BRAZILIAN AMAZON

Alexander López
*Institute for International Affairs, Unviersidad Nacional de Costa Rica*

## 1. INTRODUCTION

The relationship between population growth, environmental change and social conflicts in the Brazilian Amazon is very complex. Spite such complexity, the Brazilian Amazon shows that there is a geographical relationship between the areas of most intense frontier development and environmental change, the areas with the highest population growth and the main scenarios of social conflicts.

In addition, it is argued that population growth in the Brazilian Amazon is due basically to a migration process and not to high birth rates, and more important this migration process can be labeled as transformational migrations. The above means that the migratory process in the 1970s and early 1980s was a product of the opening up of the Amazon due to governmental activities. This explains why population movements in the Brazilian Amazon in the 1970s were more inter-regional, reflecting the opening of the Amazon and the development policy implemented in the region which made it very popular for people from other regions. In the 1990s it responds more to an internal dynamics of the Amazon, basically related to the creation of job opportunities in some Amazonian towns, consequently today population movements in the Brazilian Amazon are basically intra-regional.

I use two case studies (the states of Roraima and Pará) to show how the relationship between population movements, environmental change and social conflicts is constructed in the Brazilian Amazon. Moreover, these cases attempt to demonstrate that today the Amazon is more urban than rural, and what is more important these migrants are better understood as job seekers than traditional frontier pioneers.

Thus, population movement in the Amazon has to be understood in terms of a combination of factors of expulsion and factors of attraction linked to the whole country. These factors are not only environmental, but are also as I



*J.D. Unruh, M.S. Krol and N. Kliot (eds.),*
*Environmental Change and Its Implications for Population Migration,* 145–163.

will show linked to the distribution problems in Brazil and the misallocation of resources to the Brazilian Amazon.

Finally, the case studies of Roraima and Pará also show that the relationship between population movements and social conflicts has to be understood in terms of the incompatibility of different social groups sharing the same physical space, and the conflicting activities in terms of resource use which precipitate group identity clashes.

# 2. EXPLAINING ENVIRONMENTAL CHANGE IN THE BRAZILIAN AMAZON: SOURCES, EXTENT AND DYNAMICS

## 2.1 General

Currently there is an important body of literature linking environmental change to population movements. Therefore, the first variable being presented here is the one called environmental change understood as a large-scale natural or human induced destabilising interference in the environment. The sources of environmental change in the Brazilian Amazon are multiples and very complex, however there is a wide agreement that deforestation is the most important one. On a general level the main direct sources of deforestation in the Brazilian Amazon can be attributed to cattle ranching, colonisation and agricultural settlements, road building, mining, logging, dam construction, and urban development. Agropastoral activities in most of the Amazonian literature are considered to be the most significant. The category of private capital investment in cattle ranching through tax incentives, agricultural production through rural credits, and small farmer settlement in the Amazon region are the most important direct factors influencing the agropastoral activity and placing it as the most important source.

Fearnside (1986) has made an important classification of the causes of deforestation. He divides the present causes into two categories: proximate causes, and underlying causes. Proximate causes motivate landowner and claimants to direct their efforts to clearing forest as quickly as possible. The underlying causes are linked to wider processes in Brazil's economy (Fearnside, 1986: 42).

Among the main proximate causes of deforestation are, as Fearnside points out, land speculation, tax incentives, and negative interest loans. The first one is related to the fact that clearing establishes proprietary claims, and raises the resale value of land. Fiscal incentives allow businesses to avoid paying taxes owed on enterprises elsewhere in Brazil if the money is invested in Amazonian ranches. The third one has to do with the financing of government-approved ranching projects at nominal interest rates lower than inflation.

In addition, certain general macroeconomics policies such as income tax, the land tax and land titling regulations are providing economic incentives for deforestation. Land taxes were aimed to convert unused forestland into a more productive one, therefore farms containing forest were taxed higher than the ones containing only pasture and cropland. In this way there was a specific incentive for large landowners to convert their forestland. The policy on income tax refers to certain activities such as agriculture that were exempted from Brazil's income taxes, which provide other incentives for land acquisition in the Amazon by corporations and wealthy individuals.[1]

The other group of causes mentioned by Fearnside is comprised of the underlying sources, where he lists inflation, population growth, road building, and national politics. Inflation promotes speculation in real property, especially pasture land. Moreover, it increases the attractiveness of low-interest bank loans for clearing. Population growth increases the demand for subsistence production, the capacity to clear and plant both for subsistence and cash crops, and the political pressure for road building. Road building promotes immigration to the Amazon, and increases land clearing by persons already present. In terms of national politics there is a tendency of the Amazon interior residents to support incumbent governments providing incentives to increase political representation in these areas by creating new territories and states, justified by population growth achieved through colonisation programs and highway constructions (Fearnside, 1986: 45).

## 2.2 Extent of Deforestation

As deforestation is the most visible aspect of the transformation taking place in the Amazon, it is around this issue that public debate has been concentrated. Thus, an important discussion has taken place on the extent as well as the impacts of deforestation on the Amazon basin. In the debate some groups (political, and academic) may have exaggerated deforestation rates to further conservation aims, while others may have downplayed figures in order to encourage further development of the region.

The extent of deforestation has lead to academic and political debate, for two reasons: first, because of an inadequate understanding, even though the tool of remote sensing has been applied; and secondly, for its clear implications on policy making. For instance, the Brazilian government has been concerned about the empirical data published by studies on Amazonian deforestation, especially after the World Bank in 1988 financed a study on

Amazonian deforestation, which came out with a high rate of deforestation in the Amazon (close to 12 %).

Although deforestation is recognised as a problem, it is doubtful to present an exponential rate for the region that will mean that the cleared area could rapidly expand to encompass the entire region. Thus, rather than remaining exponential, forest depletion rates vary from year to year and vary from region to region.[2] Some of the trend analyses made in the early 1980s (exponential ones) indicate that the states of Pará, Mato Grosso, Maranhão, and Rondônia would be completely deforested by 1990. Data obtained from satellite imagery shows that in Rondônia, the most deforested of Amazonian states, no more than 12.6 % had been cleared in 1990.

According to May and Reis (1993), in the mid-seventies deforestation was practically restricted to the so-called Bragantina area, located on the eastern border of Pará with Maranhão, and to the north of Tocantins. During the late seventies and through the eighties deforestation rates within the region showed spectacular growth, most specifically in northern Mato Grosso, following a north-west path of expansion toward the states of Rondônia and Acre, stimulated by the paving of highway BR-364. This area also received a disproportionate share of economic activity, government investments, and regional development incentives.

In May 1994, the subcommittee of Western Hemisphere Affairs of the United States Congress conducted a hearing on the issue of deforestation in the Brazilian Amazon. Two scientists, Compton Tucker and David Skole, presented the result of an independent study commissioned by NASA. These results were published in the issue of March 25, 1993 of "*Science*" magazine.

Tucker and Skole stressed that the difficulty of monitoring changes in such a vast region as the Amazon gave rise to tremendous discrepancies with regard to the extent of deforestation. They specifically mentioned the erroneous figures published by the World Bank in 1988 that I have mentioned here. They found that 6 % of closed canopy forest had been cleared as of 1988, and around 15 % of the forested Amazon was affected by deforestation, caused by habitat destruction, habitat isolation, and edge effects.[3] For the entire Brazilian Amazon, they estimated that by 1988 around 230.000 $km^2$ had been deforested. These figures were close to but lower than the 280.000 $km^2$ estimated by INPE for the same year. The difference of 30.000 $km^2$ results from a different evaluation of the forest-*cerrado* boundaries in Mato Grosso and Tocantins.

Finally, INPE estimations show that there has been an important increment in the total area deforested. The total area deforested increased from 401.400 $km^2$ in 1989 to 517.069 $km^2$ in 1996.[4] However, when one takes the relative numbers it is possible to see that there has not been a constant increment in the annual rate of deforestation in the Brazilian Amazon. Figure 1 shows how the rate of deforestation during the period 1990-91 (0.30) is lower than during 1989-90 (0.37). The rate for the year 1995-96 (0.51) is also lower than the 0.81 during the year 1994-95.[5] That means clearly that the Brazilian Amazon is an open system ruled by internal and external forces that determine years of increasing rate of deforestation, but years of decrease too, of course in relative terms.

In conclusion there is an evident problem of deforestation. However the greening of the Amazon has contributed to slight exaggeration of the real situation. The deforestation problem has to be seen in context, which means understanding that it is strongly associated to socio-economic variables. The Amazonian deforestation must not only be presented as a problem in which the Brazilian society are cutting down the trees in the Amazon region, but rather as a problem which primarily has to do with the way of living of millions of Brazilian people, and in this way deforestation has an important incidence on population movements.

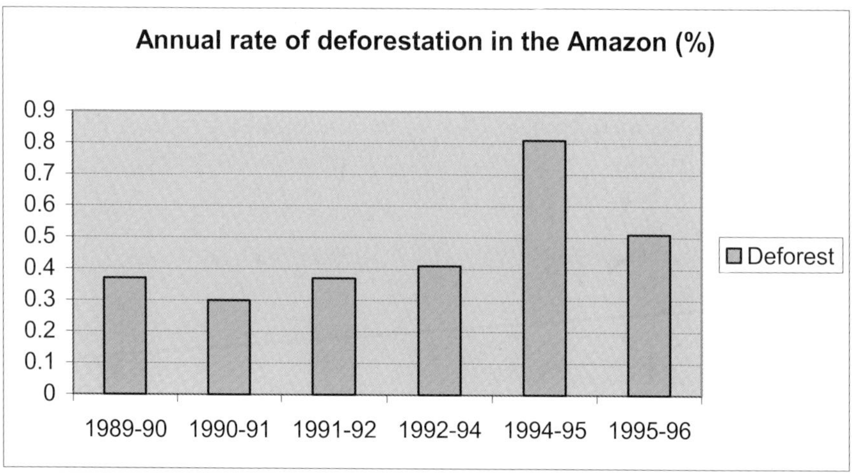


*Figure 1: Annual rate (%) of deforestation in the legal Amazon (1989-96). Source: INPE (1998)*

## 2.3 The Social Impacts of Deforestation and Its Relation to Population Movements

The process of environmental change and particularly deforestation is having a profound impact on the way people move around the Amazon basin. For instance, the process of removing tribes from the forest areas they still occupy has accelerated as new areas are targeted for mining and hydroelectric dams. The process of deforestation and ranching activities in general has had a direct effect on the lives of the forest-dwellers. The most evident conflict has been the expropriation of the customary lands of forest peoples. This situation has to do with the surviving strategies of several groups such as Indians, rubber tappers, nut collectors, and *Quilombos*, whose way of living is strongly related to nature, and whose social organisation is based on the communal use of natural resources. This way of living is in opposition to the private exploitation of these resources by miners, large landowners, mining companies, and logging enterprises etc.

## 3. DISTRIBUTION AS A SECOND VARIABLE EXPLAINING POPULATION MOVEMENTS

Beside the process of environmental change I argue that two core elements explaining population movements in the Brazilian Amazon are land and income distribution and allocation of resources. Distribution refers to, "the division of income and wealth among individuals or families. Distribution is by ethical criteria, just or unjust, or by statistical criteria, skewed or even. Both allocation and distribution can interact in a number of ways. For instance even efficient resource allocation, when coupled with a skewed land distribution, may result in the colonisation of land unsuitable for agriculture by poor farmers with no land" (Foy & Daly, 1992: 50-51).

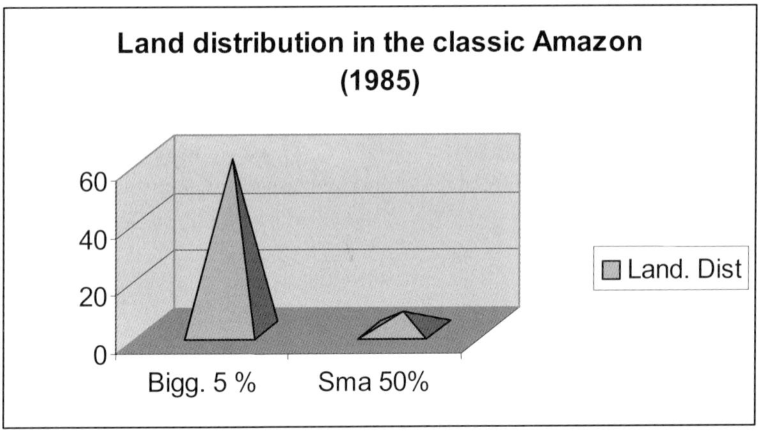


*Figure 2: Percentage of land occupied by the largest 5 % of the total farms and the smallest 50 % of the total farms. Source: IBGE Censo Agropecuário 1985. Includes data for the state of Rondônia, Acre, Amazonas, Pará, Roraima, and Amapá.*

Brazil has a skewed distribution of income and land. In the *Censo Agropecuário* (1985) farms under 100 hectares (representing more that 85 %

of the total) covered only 13 % of the total land area. The inequality in land distribution is also illustrated in Figure 2. There, using the same data from 1985 it is showed how in the classic Amazon the biggest five percent of the total farms occupy 59,6 % of the total land, while the smallest fifty percent account for only 6,6 % of the total land. This situation has not changed so much as demonstrated in the 1996 census.

Maldistribution has several repercussions. In the first place, its contribution to the enlarging of social gaps has had a direct effect on the level of conflict, because it is well-known that social gaps often act as a trigger of conflict. Secondly, maldistribution has also had a direct impact on the migration process to the Amazon, partially because of the land distribution inequalities elsewhere. It is well-known that the motive for the Trans-amazonian highway programs was to help defuse political tension in the north-east caused by great poverty, landlessness and the periodic droughts.

## 4. MISALLOCATION AS A THIRD CONTRIBUTOR OF POPULATION MOVEMENTS

Allocation of resources refers to the division of a given flow of resources among alternative uses. Allocation can be efficient or inefficient. Misallocation in the Brazilian Amazon has been witnessed above all in the areas of cattle raising, logging, and ownership rights. For instance, SUDAM allowed companies to invest up to 75 % of their income tax in approved Amazonian ranching development projects. In the same way the Brazilian Central Bank's rural credit has subsidised ranching through low interest loans. The consequences of misallocation are perceived in deforestation and soil erosion. The problem of allocation is notable by virtue of the inefficient outcomes from most of the economic incentives implemented by the Brazilian government. In the logging industry the government provided timber subsidies which encouraged deforestation without regard for sustainability of wood supplies.[6] As indicated by Repetto (1987), from 1981 to 1985, Amazonian forest-product producers and traders could borrow up to 100 % of their prior year's exports at interest rate as much as 30 % below Brazil's rapid inflation rate. Regarding ownership rights, Brazilian law encourages deforestation by allowing squatters rights. This implies that land clearing is considered an improvement to land and evidence of ownership if the peasant has been on the land for five years. Then, land was allocated under this system.

A study indicates that general tax policies, special tax incentives, the legislation linked to land allocation, and the agricultural credit system, all accelerate deforestation in the Amazon (Binswanger, 1989: 20). Poor allocations have also important impacts on population movements in the Amazon. For instance the capitalisation of these tax incentives into the price of land makes it impossible for the poor to acquire land by paying for it out of future production, since the capitalised value of the land is now greater than the capitalised value of the future products of the land by an amount equal to the capitalised value of the tax benefits. The above situation led the poor to perceive that the only access to land ownership was to squat at the frontier, and deforest rapidly in order to consolidate their claim to the land.

When taking into account both the Brazilian situation of land concentration, and the idea of economic growth formulated by the Brazilian government, one can clearly understand the conflict between *posseiros*, who responded to the government's call for small farmer colonisation, and the investor attracted by the profits that could be made by acquiring land with subsidies from the fiscal incentives program. The confrontations have been in many cases resolved outside the legal system. As Schmink and Wood (1992) state, hired gunmen resorted to a variety of violent means such as burning, beating, torture or murder, to persuade the *posseiros* to move on. In many cases the threats of violence was sufficient to convince *posseiros* that it was in their interest to accept whatever offer they could get for the improvement they had made. The result is that it became common practice for migrants to occupy a plot, clear it of trees, and then sell out. Known as the *industria de posse* (land claims industry), this procedure was one of the few avenues open to migrants to earn cash.

## 5. POPULATION GROWTH IN THE BRAZILIAN AMAZON

During the late 1980s and early 1990s the Amazon has been the object of special attention. However, as I have noted, this interest basically has to do with the deforestation problem, which in many ways has overshadowed part of the real problem in the Amazon, which is the social condition of the different social groups living in the basin. In many ways this "*greening of the Amazon*" has distorted perceptions of what is really happening in the region, and what the nature of the problem is. Thus, in many international as well as Brazilian circles, Amazonian problems have been reduced to the fact that the inhabitants and the Brazilian government are cutting down and burning all the forest. For that reason it has been more difficult to mobilise western public opinion on behalf of murdered peasants than on behalf of trees.

Contrary to popular belief, most of the Amazon population is currently urban. As can be seen in Table 1 by 1996 in the classic Amazon (northern region) 62 % of the population was urban. The same table shows how states such as Roraima and Tocantis have 70 % of the total population living in cities. Between 1960 and 1980 the urban population in the region grew by 10.9 % per year, while the rural population grew by only 3.9 % (IBGE, 1986). The other important phenomenon is the increasing urbanization of the Amazon. In 1980 already half of the population was urban, in the census that reached 59 % and in the latest census (1996) reaches 62%. Towns have appeared along roadsides at major intersections, near colonisation projects and near centres of ranching, wood extractions and mining.

Finally, it is important to mention that population growth in the Amazon is basically related to migration. That will be explained in more detail in the next section, however it should be stated that migration to the Amazon has been historically important. For instance, in 1980 according to Martine & Turchi 48,3% of the total population living in urban areas of the Amazon were migrants. Another important element is that even though there were regional differences in term of the total number of migrants, in all the Amazonian states the percentage of migrants were more than 25%.

*Table 1: Northern Region: Evolution of population growth by federal states - 1970/1996.*

| **State** | **Population (1.000)** | | | | **Population growth (%)** | | | **Urbani-zation (1996)** |
|---|---|---|---|---|---|---|---|---|
| | 1970 | 1980 | 1991 | 1996 | 1970/80 | 1980/91 | 1991/96 | |
| Rondônia | 116,6 | 491,1 | 1.132,7 | 1.231,0 | 15,5 | 7,9 | 1,7 | 62,0 |
| Acre | 218,0 | 301,3 | 417,7 | 483,7 | 3,3 | 3,0 | 3,0 | 65,2 |
| Amazonas | 960,9 | 1.430,1 | 2.103,2 | 2.389,3 | 4,1 | 3,6 | 2,6 | 73,9 |
| Roraima | 41,6 | 79,2 | 217,6 | 247,1 | 6,7 | 9,6 | 2,6 | 70,5 |
| Pará | 2.197,1 | 3.403,4 | 4.950,1 | 5.510,8 | 4,5 | 3,5 | 2,2 | 53,5 |
| Amapá | 116,5 | 175,3 | 289,4 | 379,5 | 4,2 | 4,7 | 5,6 | 87,1 |
| Tocantins | 537,6 | 738,9 | 919,9 | 1.048,6 | 3,2 | 2,0 | 2,7 | 70,6 |
| Região Norte | 4.188,3 | 6.619,2 | 10.030,6 | 11.290,1 | 4,7 | 3,9 | 2,4 | 62,3 |

Source: IBGE (1996). Demographic census

# 6. TRANSFORMATIONAL MIGRATION IN THE AMAZON

Although all the Amazonian states have experienced an important increase in population, the percentage of increase among them differs (see Table 1). Although Pará and Amazonas are the most populated states, Roraima and Rondônia as is shown in Figure 3, have experienced the most population growth in the 1980s. Roraima has a population growth of 9.5 % and Rondônia of 7.9 % the period 1980-1991.[7] What is interesting to note is that in both cases such growth cannot be attributed to high birth rates. In the case of Roraima, the most likely factor precipitating such growth was the mining economy, and in the case of Rondônia it was the colonisation schemes under *Polonoroeste*. Taking a look at the recent 1996 demographic census one realises that the above explanation is quite well grounded. From 1991 to 1996 the total population increase in those states has not been significant, coinciding with a decline in the mining economy in Roraima and the end of the *Polonoroeste* program in Rondônia. Thus, Rondônia went from a total population of 1.130.000 in 1991 to 1.229.306 in 1996 and Roraima from 216.000 in 1991 to 247.131 in 1996. The total population for the classic Amazon in 1996 was 11.288.259. This situation reinforces the criteria that migration in the Amazon can be called transformational.

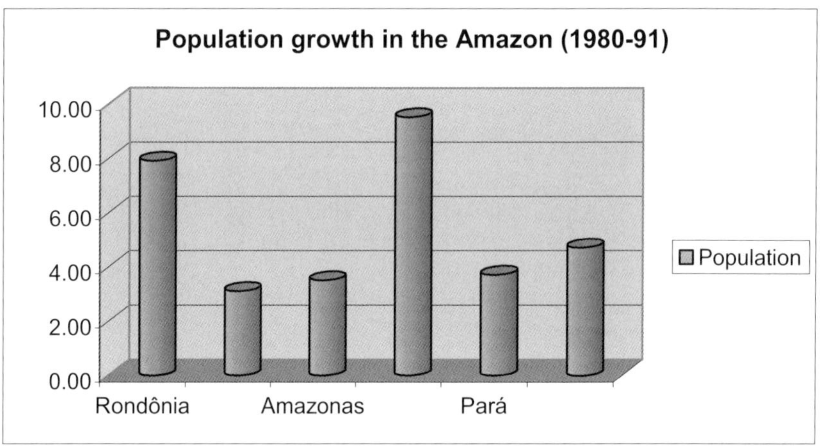


*Figure 3: Percentage of population growth in the classic Amazon (1980-1991). Source: IBGE, 1991*

As I said the population growth in the Amazon basin is basically the product of a migratory process, a product of the opening up of the Amazon due to governmental activities. This is partially confirmed by the fact that the places with the largest populations are located around colonisation schemes and in areas connected by regional roads. Examples of the population growth in the 1980s are the cases of Cacoal (49.6 %), Ji, Paraná (29.9 %), Pimenta Bueno (27.6 %), and Vilhena (27.03 %) all of them in Rondônia (Jackson, 1990: 82).[8] As a result of the previous situation most of the migration in the Amazon can be qualified as Transformational migration. This is linked to economic and social changes in a region or a country, and as the name implies, it tends to radically transform both the distribution of people and traditional livelihood.[9] It also implies a fundamental change from one way of life to another, for both the actors and the system (Lisansky, 1990: 106).

# 7. RORAIMA AS A CASE STUDY OF POPULATION MOVEMENTS: WHY DID RORAIMA ATTRACT MORE "JOB SEEKERS" THAN PEASANT PIONEERS?

## 7.1 Sources of Population Growth

An important element for understanding Roraima ′current situation is to look at population growth, and how this pattern has influenced the social composition of Roraima's population. The impact of population movements on the social structure and in the configuration of social conflicts can be seen in the fact that the traditional power of the *fazendeiros* has been challenged basically by the arrival of many *garimpeiros* as well as the influx of many new settlers. Another example is the social stress created by the thousands of *garimpeiros* working on Indian lands.

Let me start by giving a short overview of population growth. Roraima's population grew ninefold between 1940 and 1985, from 12,200 to 104,000. By 1980 the density was 0.35 per square kilometre, the lowest density of any state or territory, compared to 14.07 for Brazil. In the latest demographic census (1996) Roraima had a total population of 247.131, which means a density of 1.07. Despite this sparseness, Roraima's population has grown proportionally faster than the rest of the entire country. The rapid growth has come not only from a high birth rate, but also as has been said from the mining boom experienced in the 1980s. Moreover, the opening of the BR-174 road increased the immigration in the 1980s. It is important to note how this population growth did not have a significant impact in terms of increasing in the rate of deforestation. In that sense it can be assumed that although population growth is part of the problem of environmental change, it is not one of the main causes in Roraima.

Although population growth in the Amazon accelerated sharply after the 1960s, Roraima did not experience a dramatic change in terms of demographic density during the 1960s and first part of the 1970s. In fact, during these years Roraima remained as a remote and isolated place. However, from the 1970s an important increase can be noted. Thus, in the decade 1960-70 the rate of population growth was 3.26 %, but in the decade 1970-80 it rose dramatically to 7.04 %, and even more in the period 1980-90 when it reached 9 % (IBGE Censos Demográficos 1960, 1970, 1980, 1991). It is clear that this population explosion since the 1980s is not related to an enormous growth in birth rate, but to the enormous amount of people coming to the state as a result of the boom of some economic activities, notably gold mining.

The importance of population growth in Roraima in relation to the social structure is twofold: First, there is a clear connection between population growth and the dynamic of social conflict in the state. This point will be taken in the next section where I will analyse the conflict dynamics in the Yanomany land. The second aspect has to do with the fact that the majority of migrants, at least in Boa Vista, come from other cities. Many have migrated repeatedly from city to city, often within the Amazon region itself. The above situation leads one to think that the classical image of the Amazonian migrant as a peasant pioneer has to be reconsidered in Roraima.

The tendency in the social literature on Brazil is to associate urbanisation with rural-urban migration. However, for Boa Vista at least the rural-urban migration paradigm does not explain the city's demographic boom. As Neaera (1992) argues, migration to Roraima is not an expression of proletarization, but of a shift in an already existing urban labour market, in which a highly mobile urban population flows to wherever employment opportunities can be found. In the case of Roraima, mining was the motor of this dynamic. As a result it can be argued that most of the population coming to Roraima during the 1980s were mostly *job seekers*, rather than *peasant pioneers,* in search of land for living and working.

## 7.2 The Yanomani Area: The Richness of Maciso das Guianas as a Source of Conflict

The Yanomani are primitive hunter-gatherers in the state of Roraima, occupying the border area between Brazil and Venezuela. The current population is estimated between 7,000 and 9,000 Indians. They occupy an area of forest that, although inadequate for intensive agriculture, the rivers and hills were found to be rich in gold and tin. The region is known as *Maciso das Guianas* and is one of the oldest geological formations of South America.

The conflictive situation in the Yanomani area started with the discovery of mineral wealth, which encouraged *garimpeiros* to invade the area. The *RADAMBRASIL* project discovered an important granite formation around the Serra de Surucucús, at the same time *RADAMBRAZIL* showed that an important amount of gold and diamonds could be found in the river sediment that drained the Roraima conglomerates. The problem was that both formations lay in the heart of Yanomani land.

*Garimpo* activities started in this region from 1975 with the establishment of the *garimpo* on the Serra de Surucucús. In 1985 there were approximately 600 *garimpeiros* working in the Yanomani area. However, after 1987 Yanomani land was invaded by approximately 40,000 *garimpeiros* who took of all its main rivers at the headwaters (Albert, 1992: 43). The main reason for this was the discovery of sizeable gold deposits at the *garimpo* of Cambalacho at the headwaters of the Apiaú river. Following the invasion of Cambalacho, the intensity of mining at the watershed of the Apiaú River increased as smallholders walked the two-week long trail from the Apiaú colonisation project to seek out new deposits.

## 7.3 Warfare between the Yanomani and Garimpeiros

By July 1987, the emerging wave of conflicts had its first manifestation when Manuel Luiz, a farmer from Alto Alegre crossed the watershed and discovered a gold-mining operation belonging to the Yanomani along the middle reaches of the Mucajaí basin. Its owners were absent at the time, and so Manuel claimed the site naming it the *garimpo* of *Novo Cruzado*. Accounts differ over what happened next. It has been suggested that the Yanomani who returned to the site a few days later came into conflict with the intruder. Yet some *garimpeiros* argue that a working relationship was established with the Yanomani owners and violence broke out when this agreement subsequently collapsed. Whatever the case, a clash ensued which left four Yanomani and one *garimpeiro* dead (Folha de Boa Vista, August 12, 1987 and cited by MacMillan, 1995).

The Yanomani had reportedly expelled miners from that area on three previous occasions, and relations between the two parties were increasingly violent. Although this was not the first incident of this kind, it did prompt both FUNAI and the state governor Getúlio Cruz to threaten closure of these clandestine mines. The first plan to remove *garimpeiros* (known as operation Roraima) was launched in September 1987, but its inefficacy could be understood as a lack of political will to tackle the problem with conviction. It should be remembered that some governors of Roraima have been fierce defenders of the *garimpo* activities, for instance Romero Juca.

The number of casualties among the Yanomani due to conflicts with *garimpeiros* has been accounted for by the *Conselho Indígena de Roraima* (CIDR). Thirty-four murders were recorded between 1987 and 1996, however these numbers are certainly conservative, due to the fact that many of the violent incidents between these two groups took place in remote places which are very difficult to register.[10]

In conclusion, the Yanomani area has clearly been an area of open conflict due to the invasion of *garimpeiros*. This has resulted in the dispute over physical space, in particular for the mineral wealth of the subsoil due to the existence of gold. In this area, accommodating the interests of the *garimpeiros* and Yanomani has been extremely difficult, because first the demands involve the satisfaction of basic needs for both social groups, and second due to the existence of a quite different pattern of resource use between *garimpeiros* and Yanomani. In short, this area shows how the population movements (garimpeiros) is clearly related to the process of environmental change (river pollution), that in turns results in social conflicts (garimpeiros versus Yanomani).

## 8. SECOND CASE STUDY: ENVIRONMENTAL CHANGE BY MINING OPERATIONS IN PARÁ AND ITS IMPACTS ON POPULATION MOVEMENTS

The environmental side effects of mining operations are largely localised in Pará around the Carajás corridor and Serra Pelada. In principle, the consequences of mining *per se* are not dramatic, because of the limited area affected. However, the utilisation of charcoal has had repercussions in terms of forest cover in Pará. So, natural forest and *cerrados* have been used for the production of charcoal since then. It was estimated that by 1987, 372,000 hectares were being deforested annually in a 300 km corridor along the Carajás railroad (Silveira, 1993: 186). A series of over thirty pig-iron smelter and other mineral processing plants along the railway corridor are the most serious threats to the environment. According to Hall (1991), these schemes approved by the Carajás council in 1987 required the removal of over 5,000 hectares of forest annually to supply the one million tones of charcoal required to fire their burners.

The PGC was created by the CVRD, by then a state-owned enterprise, but privatised in 1996, which is the largest world exporter of iron.[11] Even though the rich deposits of iron-ore were discovered in 1967, the program was formally established in November 1980 via Executive act No.1,183. The work started in 1982, and the exportation began to take place in 1985

(Becker, 1991: 71). The project was totally supported by the government, and the decree Law 1,825 of December 1980 instituted the fundamental taxes and financial incentives for investors in the PGC.

When it comes to the social side effects one could see the formation of towns with the consequent population growth and the impact in terms of disturbance of some of the native people's way of living. Despite an official 1982 accord signed by CVRD, the World Bank, and FUNAI, the 13,000 Indians belonging to thirty-four tribal groups within the direct sphere of influence of the Carajás railway have apparently suffered numerous invasions of their territory by mining companies, *garimpeiros*, cattle ranchers, and loggers. The most important consequences of mining are derived from the activities with which it is associated. In the Carajás case this is quite evident, due to the fact that the infrastructure produced by this complex is responsible for the formation and growth of towns and population.

The most visible consequence of the PGC is the complete transformation of the social dynamics in the area and the increase in deforestation. The development of infrastructure including roads, railroad, and a harbour has had a major influence in the transformation of the social dynamics by allowing rapid occupation of the region. In addition, it is clear that this infrastructure contributes to the frontier dynamic by assimilating new regions into the new socio-economic dynamic. The size of this project is such that its impacts go beyond the project's immediate area, fostering huge movement of people and provoking large environmental damage.

The program has been also a magnet for job-seekers, a situation which has transformed small villages into real towns, as in the cases of Marabá and Açailandia. The Carajás project and Serra Pelada had a strong impact on Marabá, leading to its subdivision in 1988, when the municipalities of Parauapebas and Curionópolis were created. By 1991 Marabá had become the third largest city in Pará, only after Belém and Santarém. Parauapebas is the clear example of a town created to support the mining activities of Carajás and another case is Curionópolis which originated basically from the informal gold exploitation in Serra Pelada. Both towns today are the focus of intense social conflicts, most of them caused by land dispute.

*Table 2: Population growth in some municipalities along the Carajás corridor (1980-1991).*

| Municipalities | 1980 | 1991 | Annual growth % |
|---|---|---|---|
| São J. do Araguaia | 12,163 | 19,768 | 4.5 |
| Marabá | 37,713 | 1,218,114 | 11.2 |
| Cirionópolis | 13,625 | 38,458 | 9.9 |
| Parauapebas | 8,577 | 53,016 | 18.0 |

Source: IBGE Censos demográficos 1980, 1991. Rio de Janeiro.

## 8.1 Population Growth: The Relationship between Environmental Change, Population Movements and Social Conflicts in Pará

Population growth is particularly important in Pará because it shows that there is a geographical relationship between the areas of most intense frontier development, the areas with the highest population growth and the main scenarios of social conflicts.

According to Silveira (1994), mining activities and specially the Carajás Iron-Ore Project has a strong effect on the spatial configuration of the region and the dynamic of conflict. Thus, with the project in the region land values went up considerably, which soon closed the agrarian frontier for many landless migrants, and implied losses for large segments of the local population, especially farmers and groups dedicated to extraction activities. Competition between Indian groups, small farmers, nut collectors and large enterprises increased and has been the cause of many conflicts.

The development of the Carajás programme has generated great social unrest. Some of these elements are the attraction to an area that has a new set of social actors and the development of a totally new infrastructure. According to Cleary (1991), most concern has centred around the fate of the estimated 40,000 people displaced by separate projects, the integrity of Indian reserves within the PGC area, and the environmental consequences. In addition, the agricultural programs and the poverty of the soil also contribute to social unrest.

A large project such as PGC has the tendency to pressure peasant agriculture, which erodes the ability of small farmers to provide their own subsistence, as well as eliminates the possibility of additional income that generates more insecurity. This situation is complemented by the expulsion of small farmers from frontier to frontier, therefore producing settlement instability and the chances of conflicts consequently grow.

# 9. CONCLUDING REMARKS: INCOMPATIBILITY OF SOCIAL STRUCTURES AND GROUP IDENTITY CLASHES

The first conclusion from the case studies is that an important characteristic of the population in these states (Roraima and Pará), as well as in the entire Amazon, is their high level of mobility. Part of the explanation for this could be found in the fact that many of the activities that employ people in these states are predatory in the sense that they destroy resources in the production process, for instance through cattle ranching, logging, and mining. Hence, the rapid exhaustion of these resources forces the activity to relocate regularly over time as booms turn to bust. These boom and bust cycles in the Amazon could explain why migrants move repeatedly within the region. In Roraima the informal mining sector has been an important factor of attraction for this volatile population.

Both cases (Roraima and Pará) show how population movements have generated an acute incompatibility of social structures, and therefore group identity clashes. Group identity clashes are an important feature in Roraima and Pará due to the fact that an important number of groups with different social backgrounds have come together to share the same physical space. It could be claimed that in many cases the group identity conflicts arise from the incompatibility generated by the different social structures within the confines of the same physical space. Hence, the incompatibility of social structures in Roraima and Pará has been basically caused by a large movement of population coming into these states. The process of environmental change has displaced these large sectors of population in many cases. Thus, the combination of thousands of people coming to these states and the displacement suffered by them due to environmental change has provoked a condition of deprivation and unequal access to natural resources, thereby giving rise to inter-group hostility in which a given group will reinforce its own identity while attacking outsiders.

Roraima is a good example of the above complexity. In this state the incompatibility of the social structure has resulted in group identity conflicts. As showed in Roraima there is one set of values, beliefs and norms held by the native population such as the Yanomani and another, very different set used by the *garimpeiros* and *fazendeiros*. Thus, one can understand why the large arrival of *garimpeiros* to Roraima has resulted in violent conflicts between Indians and small-gold prospectors. In Pará, a similar pattern has been experienced with the difference that the web is far more complex due to the larger number of actors. In fact, Pará has not only experienced the arrival of thousands of *garimpeiros* as Roraima, but in addition the arrival of thousands of landless competing for physical space with large and medium landowners, Indians, nut-gatherers, miners, mineral companies, and logging companies.

## NOTES

[1] According to Pearce, Barbier, Markandya (1990) tax incentives tend to increase the demand for land in the Amazon, speed up conversion of land for agricultural uses, raise the price of

land, increase inequality in land ownership holdings, and finally increases the pace of migration of poor people to the frontier areas in search of land.

[2] For instance, Fearnside made some exponential projections based on the data published by INPE. Fearnside started a series of publications using mathematical projections. His first work formulated the problem in an exponential way. However, by 1989 he concluded a new projection on deforestation, now introducing a linear mathematical projection instead of an exponential one.

[3] The data presented by Tucker and Skole show that in 1988 the isolated forest in the legal Amazon was 16.228 $km^2$ and the edge effect was calculated in 341.052 $km^2$. Thus, if one adds the 230.324 of deforestation one will have 587.604 $km^2$ of total forest affected in 1988.

[4] The total area deforested annually ($km^2$) in the legal Amazon is as follows: 1989 (401.400); 1990 (415.200); 1991 (426.400); 1992 (440.186); 1994 (469.978); 1995 (497.055); 1996 (517.069).

[5] The annual rate in percent is as follows: 1989-90 (0.37); 1990-91 (0.30); 1991-92 (0.37); 1992-94 (0.40); 1994-95 (0.81); 1995-1996 (0.51).

[6] As cited by George Foy and Herman Daly in the 1970s, wood-processing industries in the Amazon received tax credits on their investment in approved projects of up to 50 % of their total income tax liability, and 75 % of the total costs. Between 1965 and 1983, about 500 million dollars of such funds were invested in wood-processing industries, which is 35 % of all tax funds committed to Amazonian investment.

[7] The percentage of population growth for the period 1980-1991 is as follows: Rondônia (7.9 %); Acre (3.0 %); Amazonas (3.5 %); Roraima (9.5 %); Pará (3.7 %); Amapá (4.7 %). The total for the classic Amazon was 3.9 %.

[8] This is data from SUDAM. Divisão Estatística. Amazonia Legal. Area População Residente Censitaria in 1970, 1980 e Projectada para Periódo 1981-1990. Belém 1984.

[9] Transformational migration is a category used by Stanley H. Brandes, to distinguish between institutional migration and transformational migration.

[10] The distribution is as follows: 1987 (4); 1988 (1); 1989 (5); 1990 (4); 1992 (2); 1993 (16); 1994 (1), 1995 (1).

[11] Brazil is the second largest producer of iron in the world only behind the former Soviet Union, and it has 11 % of the world's iron reserves. Half of its production is exported. Iron accounted for 8 % of the total Brazilian export.

# REFERENCES

Albert, B. 1993. Indian lands, Environmental Policy and Military Geopolitics in the Development of the Brazilian Amazon: The Case of the Yanomani". *Development and Change* 23:35-70.

Becker, B. 1991. *Amazonia*. Brasil: Editora Ática.

Binswanger, H. 1989. Brazilian Policies that Encourage Deforestation in the Amazon. Environ-ment Department Working Paper, no.16; Washington, D.C.: World Bank.

CIDR. 1992. *Relatório das Violencias Contra os Indios en Roraima*. Assesoria Jurídica da diocese de Roraima, e Conselho Indígena de Roraima.

Cleary, D. 1991. *Amazonia and the Environment Question in Brazil*. London: The Economist Intelligence Unit, Special Report no. 2100.

Fearnside, P. 1986. *Human Carrying Capacity of the Brazilian Rainforest*. New York: Columbia University Press.

Foy, G. and H. Daly 1992. "The Economic Causes of Ecological Disorder in Brazilian Amazonia: Allocation, Distribution, and Scale." In *Ecological Disorder in Amazonia: Social Aspects,* L. Kosinski, ed. Paris: UNESCO/ISCS/EDUCAM,.

Hall, A. 1991. *Developing Amazonia: Deforestation and Social Conflict in Brazil's Carajás Programme*. New York: Manchester University Press.

IBGE. 1984. *Censo Agropecuário*. Rio de Janeiro: IBGE IX Recennseamento Geral do Brasil. Rio de Janeiro.

IBGE. 1993. *Anuários Estadístico*. Rio de Janeiro, IBGE, 1986-1992.

INPE. 1998. *Instituto Nacional de Pesquisas Espaciales Actualiza os Dados do Desflorestamento na Amazonia*. INPE Noticias No. 13, Jan-Feb.

Jackson, M.J. 1990. *Demografía e Mano de Obra na Amazonia*. Igarape. NAEA, Belém, Pará: Igarape, NAEA.

Lisansky, J. 1990. *Migrants to Amazonia: Spontaneous Colonization in the Brazilian Frontier*. Boulder, Colorado: Westview Press.

MacMillan, G. 1995. *At the End of the Rainbow? Gold, Land and People in the Brazilian Amazon*. Earthscan Publications, London,.

May P. and E. Reis 1993. *The User Structure in Brazil's Tropical Rain Forest*. Kiel: Kiel Institute of World Economics.

Neaera, R. 1992. *Urbanization and City-Ward Migration on a Resource Frontier: The Amazon Gold Rush and the Case of Boa Vista, Roraima*. Thesis submitted in partial satisfaction of the requirements for the degree Master of Arts in Urban Planning. Los Angeles: University of California, Los Angeles.

Repetto, R. 1987. Creating Incentives for Sustainable Forest Development. *Ambio* 16: 2-3: 94-9.

Schmink, M. and C. Wood 1992. *Contested Frontier in Amazonia*. New York: Columbia University Press.

Silveira, S. 1994. *Transformation in Amazonia. The Spatial Reconfiguration of System*. Kugl., Stockolm: Kutgl. Tecknisca Högskolan.

# MIGRATION PATTERNS, LAND USE AND CLIMATE CHANGE

Urs Luterbacher
*Graduate Institute of International Studies, Geneva, Switzerland*

## 1. INTRODUCTION

Land-use and then land cover is one of the key links between human activities and climate change. Clearly land use is influenced by demographic evolutions that can also be constrained by climate change. Whereas there are some physical representations of land use and cover very few socio-economic models of land use have been proposed. This is surprising since land use is obviously the result of human decision-making processes. As usual in cases involving such decisions, the incentive structure and preference orders underlying human choices has to be explored. An important variable to be considered here is the relative price of different uses of the land. As one knows, land can have different types of use depending on the relative value associated to it. Obviously, a plot of land can be either left to natural cover in which case it still might have a specific value as forested or recreation area or transformed into agricultural land of various kinds or finally used as dwelling or industrial site. According to this perspective land use will then be determined by the relative prices associated to these specific uses. The land price will be established, provided no administrative restriction is introduced (such as zoning for instance), by the competing uses one can make of it. If dwellings bring in more than agricultural use, housing areas and urbanization will expand. However, if agriculture or lumber production or tourism is of high value, land will be used for these purposes and not for housing and so on.

In this paper I want to work toward the establishment of an analytical construction in the form of model of the relative values of land areas under the influence of changes. Changes to be considered should include: 1) Ecological parameters such as climate change 2) Socio-economic forces with special attention paid to the additional market forces unleashed by market liberalization and globalization. From a theoretical point of view I want to invoke the theories of exhaustible and slowly renewable resources to attempt to explain the relations between the categories of land areas that are being considered (for this see the pioneering work of Dasgupta and Heal 1978,



*J.D. Unruh, M.S. Krol and N. Kliot (eds.),*
*Environmental Change and Its Implications for Population Migration,* 165–175.

Chichilnisky 1994). If these areas are considered limited assets, they can be substituted by other productive assets. An increase or decrease of the productive value or the desirability of an asset will be reflected in its price. Under pure market competition without state intervention, landed assets will be exchanged, until equilibrium is reached. A greater demand for a particular asset, i.e. land for tourism and recreation will lower the price of agricultural land and wilderness or forests and lead to a transformation of the least demanded land into the most demanded land. State intervention will slow such a process through zoning and protection but will have trouble taking the demand pressure completely off especially if the transformation is perceived to be linked with economic growth. Natural processes also alter the quality of productive assets and contribute thus also to shifts in demand. The model thus should ultimately try to reflect these various demand pressures that will alter the landscape together with existing state policies. This model should be able to represent deforestation as well as reforestation (in terms of a hike or a lowering of the value of forested areas with respect to agriculture and dwellings).

Very few explicit models so far deal with the question of land use especially under conditions of demographic pressure. The model developed by Guillermo Owen (1987) is one of the few that come to mind. My purpose here is to describe the main characteristics of this model, to develop it further and then to draw some elementary conclusions from it under specific assumptions of climate change and demographic pressure in order to open the path for more thorough empirical investigations.

## 2. THE OWEN MODEL

The land use model developed by Owen assumes only two types of land use, agriculture and dwelling and examines the special case of areas around urban centers. Owen assumes further that all inhabitants in an area are to some extent dependent on local food production and that each of them receives subsidies equivalent in total to the number of households. These assumptions are fairly plausible for an urban area in a developing country. Since, I will use some of the detailed results established by Owen, I will give here a rather full account of its mathematical developments. Subsidies will be given in some numeraire and thus if there are $N$ households and a total agricultural production of $P$, $N$ units of numeraire will be distributed as subsidies and $P$ units will result from the sale of agricultural goods. The total income of the region is then $P + N$. The retail price of one unit of food will then be $(P + N) / P$ so that one unit of numeraire will buy $P / P + N$ units of agricultural goods. A farmer producing b units of goods will receive b amount of numeraire and buy goods equivalent to $bP / (P + N)$. Each household receives independently of any productive activity, $P / (P + N)$ either in goods or in the equivalent numeraire. The quantity b, which Owen maintains as fixed, can be thought of as the average production per farm at a given time. He assumes further that a plot can contain $\lambda\, b$ urban lots, in other words, that the number of urban lots is proportional to the average productivity of the farm. He also puts the value of $\lambda$ as much greater than 1, which means that many urban lots can be made out of a single farm plot. He also envisages a maximum number (or production) that would support households in the region. This maximum number is defined as $H$. Lets now consider the decision making process of a single landowner. This landowner is willing to sell his land at the following conditions:

He gets currently: $$(b + 1)\, P / (P + N) \qquad \text{(eq. 1)}$$

income out of the land (his agricultural production plus his subsidy). If he sells, he will allow $\lambda\, b$ households to get $P^* / (P^* + N^*)$ where $P^* = P - b$ and $N^* = N + \lambda\, b$. In the end, the stream of revenues generated by transforming the land into dwellings will be:

$$\lambda\, b\, (P - b) / (P + N + \lambda\, b - b) \qquad \text{(eq. 2)}$$

The problem presents itself thus in the short run as:

If $$(b + 1)\, P / (P + N) < \lambda\, b\, (P - b) / (P + N + \lambda\, b - b) \qquad \text{(eq. 3)}$$

then the landowner will sell. This particular condition reduces then to:

$$(\lambda\, b - b - 1)\, (P^2 + PN) + (b^2 + b - 2\, \lambda\, b^2 - \lambda\, b)\, P - \lambda\, b^2\, N > 0 \qquad \text{(eq. 4)}$$

Conversion of agricultural land to urban lots will continue as long as the above condition holds. For Owen, however, $N$ and $P$ are not independent since when $N$ increases, $P$ decreases. This relationship is expressed by:

$$N = \lambda\, (H - P) \qquad \text{(eq. 5)}$$

Substituting (eq. 5) into (eq. 4) yields:

$$[(\lambda - 1)^2\, b - (\lambda - 1)]\, P^2 - \\ [\lambda\, (\lambda - 1)\, b\, H - (\lambda - 1)(b^2 + b) - \lambda\, H] + \lambda^2 b^2 H < 0 \qquad \text{(eq. 6)}$$

The right hand side of (eq. 6) can be written as a quadratic function in $P$. One can now compute the two roots of this function which will give two critical values for $P$, $P_1$ and $P_2$ , the agricultural production with the following conclusion: if $P_1 < P < P_2$, selling will occur otherwise for $P < P_1$ or $P > P_2$ selling will not take place. Under considerable simplifying assumptions such as that $H$ is considerably larger than $b$ and that $(\lambda - 1)\, b > 1$, Owen finally gets the following approximate values for the two roots:

$$P_1 \cong \frac{\lambda b^2 H}{\lambda(\lambda - 1)bH - (\lambda - 1)(b^2 + b) - \lambda H} \qquad \text{(eq. 7)}$$

$$P_2 \cong \frac{\lambda(\lambda - 1)bH - (\lambda - 1)(b^2 + b) - \lambda H}{(\lambda - 1)^2 b \ - (\lambda - 1)} - P_1 \qquad \text{(eq. 8)}$$

Owen shows that $P_2$ is under his assumptions greater than $H$. This limits therefore the analysis to $P_1$. The question is now: Under which conditions will $P$ the agricultural production stay above $P_1$ , defined in expression (eq. 5)? Clearly, if $P_1$ is low, the urbanization process will continue. Thus the level of $P_1$ will determine what makes a given $P$ more or less likely to stay above it or fall below it. If $P_1$ is very small or close to 0, then urbanization will continue even though agricultural production is by then very small. If $P_1 < 0$, no level of agricultural production will stop urbanization. This is probably the case by now in many industrialized countries where the transportation costs for agricultural products from far away are so low that urbanization continues until distance from the urban center is too high. Therefore one has to look at the way $P_1$ varies as a function of the parameters involved in order to determine its level.

## 3. FURTHER INVESTIGATION OF THE MODEL

In order to enlarge the perspective introduced by Owen's model, several paths can be taken. I will here outline four such possibilities:

1. The stream of income coming from either agricultural activities or dwellings as outlined in expressions (eq. 1) and (eq. 2) should be discounted against the future in terms of present values. If a common interest rate $r$ is applied for discounting, the analysis of Owen's model is not different. The situation changes however if different discount rates $r_1$ and $r_2$ are applied:

$$1/r_1[\,(b+1)\,P\,/\,(P+N)]\ \text{and}\quad 1/r_2[\lambda\, b\,(P-b)\,/\,(P+N+\lambda\, b-b)] \qquad \text{(eq. 9)}$$

In this case, if as it seems plausible especially in a developing country, a higher interest rate is applied to agricultural activities because they are more subject to random fluctuations (a situation that could be aggravated by climate change) and thus incur a higher risk premium, the process of selling could be accelerated. One has then:

$$1/r_1[\,(b+1)\,P\,/\,(P+N)] < 1/r_2[\lambda\, b\,(P-b)\,/\,(P+N+\lambda\, b-b)] \qquad \text{(eq. 10)}$$

in part because

$$r_1 > r_2$$

2. The subsidies given to urban dwellers are not only dependent upon agricultural income but on some other income (which could be international aid). In this case:

$P^* = P - b + Ys$ where $Ys$ stands for other kinds of subsidies.

In this case, $Ys$ could more than offset the loss in agricultural production (the $-b$ term) and enhance inequality (eq. 10). Moreover, an increase in $Ys$ will attract more households if people know about the provision of international aid to them. This is a first inclusion of the migration factor. An increase in $N$ will as can easily be seen enhance inequality (eq. 10) even more:

$$1/r_1[\,(b+1)\,P\,/\,(P+N)] < 1/r_2[\lambda\, b\,(P-b+Ys)\,/\,(P+N+\lambda\, b-b+Ys)] \qquad \text{(eq. 11)}$$

3. Migration could be explicitly taken into account if one assumes that the maximum population absorption capacity $H$ is increasing. In the model, $N$ increases if $H$ increases as determined by relation (eq. 5). Clearly, the increase in $N$ can be the result of increased migration into the area. To analyze this occurrence, one should note that expression (eq. 7), the solution of the critical value can be rewritten in the following way:

$$P_1 \cong \frac{\lambda b^2}{(\lambda-1)b-1-\dfrac{b(\lambda-1)(b+1)}{\lambda H}} \qquad \text{(eq. 12)}$$

In this case, one can see that an increase in $H$ will render the fraction in the denominator of expression (eq. 12) smaller thus making $P_l$ marginally smaller but mostly, as $H \to \infty$, the fraction will tend toward zero, so that one is left with:

$$P_1 \cong \frac{\lambda b^2}{(\lambda - 1)b - 1} \qquad \text{(eq. 13)}$$

In other words, $P_l$ will be mostly a function of $b$, the average production of a farm, since $\lambda$, which is the ratio of urban lots to a particular farm surface should be close to constant. The investigation of the behavior of $b$ is important and deserves special attention.

4. Clearly, in reality, $b$ is a **net** profit (revenues minus costs), which is neither a constant in space nor in time. $b$ is in fact a distribution $b(., s)$ where $s$ is a surface measure and where other variables might be important such as for instance labor on the farm and ecological parameters (micro-climate, availability of water, exposure, soil, etc.). In an important study on the effects of climate change on US agriculture, Mendelsohn and Nordhaus (1994) claimed that from a Ricardian point of view (David Ricardo, British economist of the early 19$^{th}$ Century), such ecological variables should be reflected in the price of agricultural land at a given moment. Potential climate change will shift land prices (not necessarily decrease or increase them across the board), since some land will become more suitable to certain cultures and other land will diminish in productivity. Events such as climate change will then shift the shape of the distribution around and possibly modify overall agricultural production. $b$ is also a function of various inputs such as labor and capital (machinery, fertilizer, ability to adopt new crops and to take care of them) and evolving time trends in technology so that one has $b(.,s,t)$ and

$$\sum_i b_i(.,s,t) = P_t \qquad \text{(eq. 14)}$$

In other words, the sum of the production of all farms is the total agricultural production. The various $b_i$ could also be affected by the property right regime in the sense that ill-defined property rights could push the $b_i$ 's higher than they would be otherwise and thus first raise but then lower $P_l$, and thus further encourage transformation of agricultural land into urbanized areas. The analysis of this effect is complex and requires a special development.

# 4. LAND USE, MIGRATION AND PROPERTY RIGHTS

The impact of ill-defined property rights on environmental systems has by now been thoroughly analyzed. Of importance here is the classic work on this topic by Dasgupta and Heal (1979) and its extension to trade and special taxation policies by Chichilnisky (1994). Open access or ill-defined property rights systems usually lead to overuse and dissipation of any surplus (Ricardian) type rents[1]. Mathematically, this can be expressed by the production decision of producer i who wants to maximize his revenue, which is proportional to average product minus cost, which will be assumed fixed per unit of input.[2] Recall here that Owen's model only takes into account pure profits with no explicit inclusion of costs. In order to show the interactions between the inefficiencies generated by the Owen model and the ones due open access or ill-defined property rights, a relatively detailed presentation of the mathematical intricacies of the Dasgupta and Heal formulations will be undertaken. Assuming that the production function depends on a fixed factor with constant returns to scale that can be normalized to 1 (and the price is therefore 1) and a variable factor or input x with diminishing returns, so that one has $F(x)$ where $F''(x)<0$ and assuming that among $N$ producers each other producer will introduce an input $x^{\circ}$, producer i maximizes:

$$\frac{x_i F\left[(N-1)x^{\circ}+x_i\right]}{(N-1)x^{\circ}+x_i} - px_i \qquad \text{(eq. 15)}$$

In other words, producer i maximizes with respect to his input $x_i$, proportionately to the average product determined by all the other producers who constitute to him a negative externality.

Maximizing with respect to $x_i$ while assuming that here $x^{\circ} = x_i = x$, yields:

$$\frac{F(Nx)}{Nx} - \frac{1}{N}\left[\frac{F(Nx)}{Nx} - F'(Nx)\right] = p \qquad \text{(eq. 16)}$$

Alternatively this expression can also be written:

$$\frac{F(Nx)}{Nx} + \frac{1}{N}\left[F'(Nx) - \frac{F(Nx)}{Nx}\right] = p$$

Quite clearly the fraction $F(Nx)/N(x)$ is the average product. In the absence of jointness, i.e. if there is no open access, the optimal behavior of a producer[3] is simply to maximize total revenue given by the production function minus cost which is:

$$F(Nx) - pNx \qquad \text{(eq. 17)}$$

Then the maximum is:

$$F'(Nx) = p \qquad \text{(eq. 18)}$$

which simply states that the value of the marginal product has to be equal to marginal cost. A closer look at expression (eq. 2) shows that the common marginal product is higher than the private marginal product expressed in (eq. 4) which is precisely at the root of what Hardin has called "the tragedy of the commons". The use of the word common by Hardin is unfortunate since it gives the mistaken impression that what is called common property leads always to inefficiencies and overuse of resources. This is only the case with open access situations, in other words when no restrictive rules are placed on resource use. Ostrom (1990) (and others such as Bromley et al. 1992) have shown that non-private forms of property control but very different from open access can be relatively efficient. In the open access case however, inefficiencies will occur. To show how these interact with the previously discussed Owen model, one can see that:

$$\frac{F(Nx)}{Nx} + \frac{1}{N}\left[F'(Nx) - \frac{F(Nx)}{Nx}\right] - F'(Nx) = (1 - \frac{1}{N})\left[\frac{F(Nx)}{Nx} - F'(Nx)\right] > 0 \qquad (19)$$

which means that the open access marginal product is always higher than the private property marginal product if the production function $F(x)$ is assumed to be concave at least in the limit. The opposite obtains if $F(Nx)/N(x) < F'(Nx)$ when $F(x)$ is not concave.

Expression (eq. 16) tells us that if $N$ becomes very large, net profits are driven toward 0:

$$\frac{F(Nx)}{Nx} + \frac{1}{N}\left[F'(Nx) - \frac{F(Nx)}{Nx}\right] = p$$

if $N \rightarrow \infty$, one has:

$$\frac{F(Nx)}{Nx} = p \qquad \text{(eq. 20)}$$

In this case, which would be common when one has mass migration, since then $N$ increases considerably and in addition private property rights are more and more difficult to enforce[4]. Then, profits denominated b in Owen's approach tend to 0 ($b \downarrow 0$) and then $P_l$ would be, in Owen's model also be driven to 0, making urbanization inevitable. Initially, however, given the fact that as shown in (eq. 19), the marginal product under open access is higher than the one under a system of private property, $b$ under open access would be higher and urbanization would be slowed. The scenario to be expected under open access would thus be an initial slowing of urbanization, followed by an acceleration, as the number of households migrating to urban areas increases.

## 5. CONCLUSIONS

The Owen model under the assumptions of mass migration, aid resulting from non-agricultural income from the region and open access type property regime with respect to agricultural land can be extended easily from a mathematical point of view and leads to a series of interesting conclusions that complete its scope:

The base Owen model states that even if my additional points are not considered, as long as there is an attraction to moving into an urban area such as a subsidy or the hope of a job, farm land will be urbanized down to a critical value which can be very close to zero. Introducing the new assumptions enumerated earlier about interest rates, climate change associated or not with mass migration, foreign aid, and property regimes leads to interesting transformations of the Owen base model and to a more focused analysis and some time more balanced views of Owen's conclusions. The more general approach taken in this paper emphasizes the role of accelerating factors that can lead more quickly to urbanization. Some of these forces are quite obvious because they make agricultural production and investments more difficult to carry out: So evidently, higher interest rate for agricultural investments as opposed to investments for urban dwellings will reduce agricultural profits and thus speed up the construction of urban dwellings. The same can be said for climate change if it decreases farm profits because of more difficult growing conditions. In some regions, however, profits could increase as a result of climate change, thus slowing urbanization. However, if mass migration occurs, which can result from climate change and associated difficulties in farming as well as from natural disasters, the process of urbanization will obviously accelerate both as a result of the increased number of households in the area and diminishing agricultural profits. What is to some extend surprising is the fact that foreign aid and relief going directly to the incoming population speeds up the formation of big urban concentrations. Here the form of foreign aid appears to be important. An aid to agricultural production instead of a direct aid to households would of course slow down urbanization by making farming again more attractive. It is finally important to point out the importance of the property rights question: Paradoxically, a weakening of property rights arrangements will initially slow down urban concentration since it will produce temporarily higher yields and make agricultural production more attractive. This should however soon be followed by lower prices due to increased agricultural production and thus falling agricultural revenues and again a bigger push for urbanization. For all these reasons, increased urbanization is difficult to stop. However, a thorough investigation of the Owen model points to the factors that may slow the process or some times even reverse it: A more efficient and productive organization of agriculture in developing countries. Here again perhaps, a change in the agricultural import policies of advanced industrialized countries in Western Europe and North America might have a beneficial impact not only on economic relations but on environmental conditions as well.

## NOTES

[1] For an illustration of this problem and how it leads to unsustainability refer to Brander and Taylor (1998).
[2] I am just reproducing here the results given by Dasgupta and Heal (1979).
[3] In other words, if the whole open access area belonged to a particular owner, he would behave optimally by equating marginal product to marginal cost.
[4] The California gold rush provides an example of how difficult it is to maintain property rights under mass migration. Large tracts of land belonging to John Sutter were just seized by incoming immigrants.

## REFERENCES

Brander. J. and M. Taylor 1998. "The Simple Economics of Easter Island: A Ricardo-Malthus Model of Renewable Resource Use" *The American Economic Review* 88, 1, 119-138.

Bromley, D. ed. 1992. *Making the Commons Work: Theory, Practice, and Policy.* San Francisco: Institute for Contemporary Studies Press.

Chichilnisky, G. 1994. "North-South Trade and the Global Environment" *The American Economic Review* 84 (4): 851-874.

Coase, R. 1960. "The Problem of social Cost" *The Journal of Law and Economics* 3:1-44.
Dasgupta, P. and G. Heal 1979. *Economic Theory and Exhaustible Resources.* Cambridge: Cambridge University Press.
Hardin, G. 1968. "The Tragedy of the Commons" *Science* 162: 1243-1248.
Helpman, E. and P. Krugman 1985. *Market Structure and Foreign Trade*. Cambridge, Mass. : MIT Press.
Isaacs, R. 1967. *Differential Games*. New York. John Wiley
Luterbacher, U. 1994. "International Cooperation: The Problem of the Commons and the Special Case of the Antarctic Region" *Synthese*, 100: 413-440.
Mendelsohn, R., W. Nordhaus and D. Shaw 1994. "The Impact of Global Warming on Agriculture: A Ricardian Analysis" *The American Economic Review* 84 (4): 753-771.
Ostrom, E. (1990) *Governing the Commons: The Evolution of Institutions for Collective Action.* New York: Cambridge University Press.
Owen, G. 1987. "Application of Game Theoretic Analysis to a Problem in Demography" *Management Science* 33: 253-257.

# POPULATION GROWTH, MIGRATION AND URBANISATION

*Environmental Consequences in Kathmandu Valley, Nepal*

Pushkar K Pradhan
*Central Department of Geography, Tribhuvan University, Kathmandu, Nepal*

## 1. INTRODUCTION

Population growth is an important factor for local environment change. More importantly, the consumption level and the technology assume to play a vital role in the overall environmental change. Ehrlich and Ehrlich (1990) argue that population growth causes a disproportionate negative impact on the environment. Commoner et al (1971) in their study conclude that though population plays role in environmental deterioration, it is by no means only a major factor of environmental crisis. The technology factor plays a much more significant role in the ecological crisis.

Rapid population growth is one of the main reasons for increasing the number of people on the move for livelihood, and migration has been a major factor of rapid population growth in urban areas in less developed countries (LDCs). Lee (1969), Todaro (1976), and Goldscheider (1983) emphasize the pull of the urban sector for urban bound migration. However for the urban pull to be operative, the population in the hinterland must be conducive to dislocation. While others argue that the stimulus for migration in LDCs lies in the rural condition. In other words, the accretion of unemployed migrants in urban areas in LDCs is indicative less of urban dynamism than of strong repulsive forces operating in rural areas (Gurung 1989). Zelinsky et al (1970) and Ng (1975) argue that out-migration from rural areas is primarily due to the growth of the local population beyond the carrying capacity of the land. Excessive pressure on land resource and increasing poverty generate a large volume of migration (Kosinski and Prothero 1975; Zelinsky 1983). All poor people do not often move out despite intensive resource pressure. Goldscheider (1983) concludes that the poorer the people, the stronger the impediments to moving. This observation may be true for rural-urban movement in Nepal (Gurung 1989).

With nearly 15 percent urban population, Nepal is one of the least urbanized countries in the world. However, the growth of population in the country's large urban areas has been rapid since the last few years. The recent



*J.D. Unruh, M.S. Krol and N. Kliot (eds.),*
*Environmental Change and Its Implications for Population Migration*, 177–199.

annual growth rate of urban population is 6 percent as against to slightly over 2 percent national average growth rate of population. High growth rate of urban population may be due to natural growth, migration, designation of additional municipal towns, and expansion of municipal towns amalgamating rural areas. Recently, migration of population is said to be one of the important factors for rapid growth of large urban areas. Declining environmental quality in the large urban areas is a recent phenomenon, as well as a major concern among the urban environmental planners.

This paper analyses population growth and migration, urbanisation and their impacts on environmental condition of the urban areas of Kathmandu Valley, Nepal. Based on the available information, the magnitude of environmental impacts has been examined. In the end, policy measures so far undertaken on the part of the government of Nepal have also been presented.

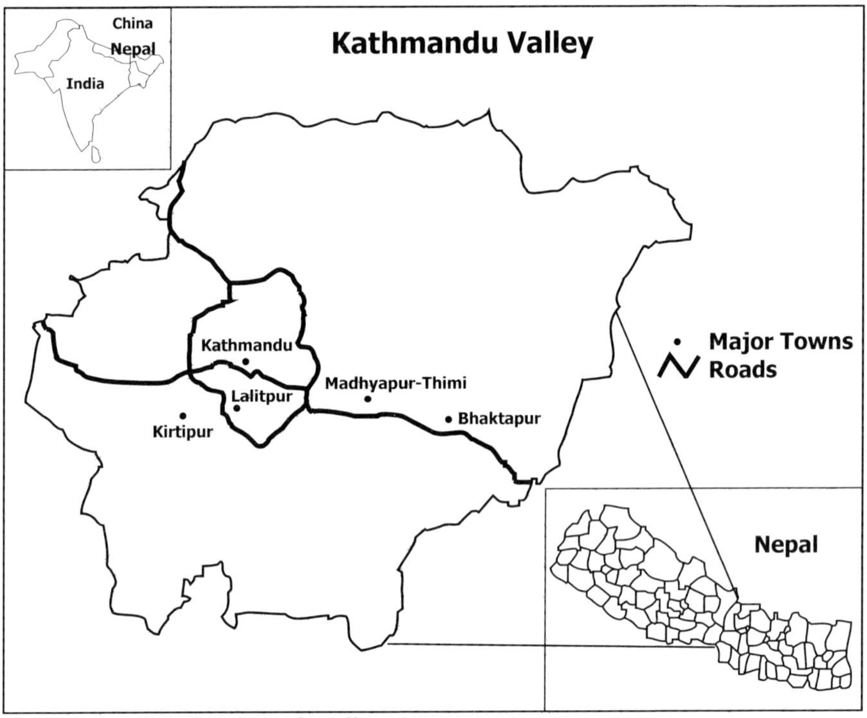


*Figure 1: Location of Kathmandu Valley*

## 2. BACKGROUND

The Kathmandu Valley[1] of Nepal lying in the central middle hill region measures about 19 by 25 kilometers (Fig. 1). The valley floor lies at an average elevation of 1,250 metres from the mean sea level. This tectonic valley is a tertiary structural basin covered by fluvial and lacustrine sediments, and encircled by mountains on all sides with above 1,800 metres high. The valley lies in the temperate climatic regime with a relatively well-balanced climate.

This valley now has five municipal towns[2] viz. Bhaktapur, Kathmandu, Kirtipur, Lalitpur, and Madhyapur-Thimi and over a dozen of small market settlements. Kathmandu is capital as well as the only metropolitan city, and Lalitpur is one of the four sub-metropolitan cities of the country. In Nepal, the Kathmandu Valley Urban Region (KVUR) comprising its five municipal towns is the largest urban region in terms of population size. The valley's urban population is nearly 1 million that constitutes almost 89% of its total district population, and makes up slightly less than one-third of the country's total urban population[3]. These cities and towns have evolved eminently over the last few decades. As the valley became more and more important as the cultural, economic and administrative heart of the country, it naturally began to attract people from different parts of the country. This has brought about several environmental changes in the valley's urban landscape.

## 3. HISTORICAL EXPOSÉ OF URBAN GROWTH

The Kathmandu Valley has a long history of urban development. The valley is among the oldest settled areas in the central Himalaya (Depuis 1962). The early development of the towns in the valley is believed to have taken place probably during the period ca 300-800 AD and the most of the prominent settlements appear to have enlarged and consolidated by the beginning of the thirteenth century (HMG/UN/UNESCO 1975). Two fundamental factors appear to have played role for the sustenance of urban growth in the valley. One was the rich agricultural resource base with productive soil and water supply[4], and the other was long distance trading activity, also known as entrepôt trade[5]. The former made it possible to provide sufficient food supply to its inhabitants while the latter provided revenue for a strong economic base (Regmi 1978). These two activities together led to flourish small-scale production of textiles and metalwork, which entered as major export items in the foreign trade. By developing these, the valley towns also flourished in the field of art, culture and civilisation, and hence grew up as ceremonial complexes consisting of temples, shrines and palaces.

After 1769, Kathmandu as one of three major principalities (the other two were Lalitpur and Bhaktapur) of the valley became the capital of unified

Nepal[6] and therefore the seat of political, economic and social power. This indeed marks the beginning of movement of people from several parts of the kingdom to the valley. By that time, according to Father Giuseppe (Wolfgang 1993), the size of the three cities-cum-principalities – Kathmandu, Lalitpur and Bhaktapur - was pretty small with the number of houses of about 18,000, 24,000 and 12,000 respectively. They together with other small towns were initially developed on the upland plains and terraces, fairly above the flood level of major rivers such as the Bagmati, the Bishnumati, the Manohara and the Dhobikhola. A great transformation in the pattern of physical expansion of the cities took place in the beginning of the twentieth century. This involved the construction of extravagant palaces and other buildings particularly along the lines of newer roads. The outsiders were being attracted to the roadside locations for taking an advantage of the road accessibility. Other factors that exhilarated for the expansion of the cities in the 1950s included the change in political system[7], and therefore the increased needs of government infrastructure and functions in administration, health, education, banking and so on. Since the 1960s, increase in the number of government offices and their relocation, growth in foreign aid, tourism and tourist related activities, and expansion of road networks led to further influx of migrants into the valley (Pradhan 1988).

*Table 1: Growth of Population of the Kathmandu Valley Urban Region by Census Year*

| Cities | 1952-1954 | 1961 | 1971 | 1981 | 1991 | 2001* | 1952-2001 |
|---|---|---|---|---|---|---|---|
| Bhaktapur | 32,320 | 33,877 | 40,112 | 48,472 | 61,405 | 77,764 | 140.6 % |
| Kathmandu | 106,579 | 121,019 | 150,402 | 235,160 | 421,258 | 754,409 | 607.8 % |
| Kirtipur | - | - | - | - | 31,338 | 49,606 | - |
| Lalitpur | 42,183 | 47,713 | 59,049 | 79,875 | 115,865 | 168,077 | 298.4 % |
| Madhyapur-Thimi | - | - | - | - | 31,970 | 34,587 | - |
| **Valley total** | **181,082** | **202,609** | **249,563** | **363,507** | **661,836** | **1084,443** | **499%** |
| **Growth** | **-** | **11.9%** | **23.2%** | **45.7%** | **82.1%** | **63.9%** | |

Source: Pradhan (2000), UNEP et al (2001). *Projected population.

## 4. POPULATION GROWTH AND MIGRATION

The growth trend of urban population in the valley is rapid. In 1952-54[8], the valley had urban population of 181,082 (Table 1). By 2001, the valley's urban population is estimated to be slightly over 1 million. This means an increase of almost 500 percent within five decades. Among the five centres, Kathmandu city has shown the highest growth percentile. Between 1991 and 2001, the urban population of Kathmandu has increased with 7.9% per annum. Next to it is Kirtipur with 5.8 growth percentile per annum.

Madhyapur-Thimi has the lowest growth rate of urban population with 0.8% per annum, while Lalitpur and Bhaktapur have shown 4.5 and 2.7 percentiles of urban population growth during the last decennial census.

The valley urban region has surpassed over both the nation and the valley districts in cumulative growth percentile of population in all successive decennial censuses since 1952 (Table 2, Fig. 2). Table 2 also shows that the migration has contributed to total population growth of the Kathmandu urban region at varying magnitudes; the most remarkable being the years 1981-1991 during which 59% of the total population growth was estimated to be contributed by migration. The growth trend of population measured in terms of cumulative percentile as shown in Figure 2 appeared much more rapid in Kathmandu urban region than in the country and Kathmandu districts since the decennial year 1981. This was mainly due to the improvement of health conditions that caused to substantial decline in mortality, while birth rates have continued to remain stable (HMG 1983).

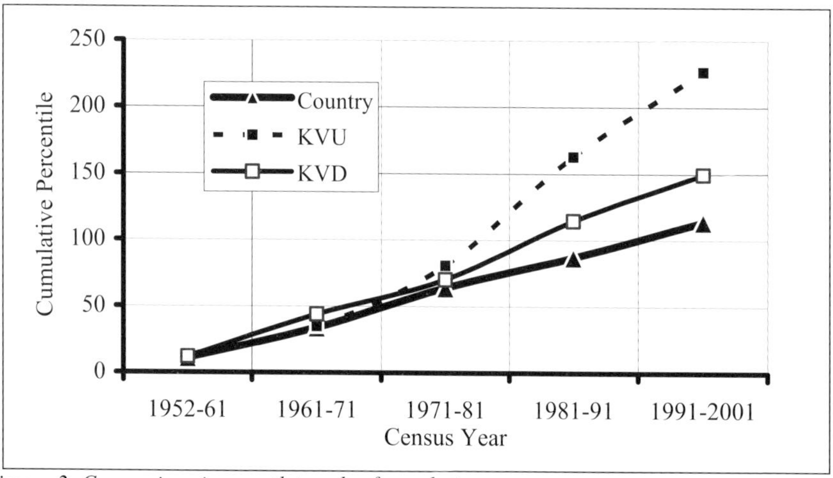


*Figure. 2: Comparison in growth trends of population.*
Note that the growth percentile of population for each of the country, Kathmandu valley districts (KVD), and Kathmandu valley urban region (KVUR) has been obtained for the decennial census years by using the following formula: $R = 100*(P_n - P_o)/P_o$, where R = growth percentile, $P_n$ = population of the last year, $P_o$ = population of earlier year. Total cumulative growth percentile for each spatial unit (country, KVD, KVUR) has been obtained from adding up percentile of the beginning decennial census year to percentile of each successive upper decennial census year, and then the values have been plotted in graph.

Migration is assumed to be one of the major factors for rapid growth of population in the valley. Between 1981and 91 the valley's urban population increased by over 82%, to which migration alone was assumed to contribute 59%[9] (Table 2). A recent study of demographic characteristics of labour forces indulged in informal sector activities in Kathmandu city found that slightly over 72% of the total sample households were migrants with predominant proportion (87%) of productive age group of 20-50 years (Joshi 2002). Larger proportion of productive age in population, designation of new municipal towns, expansion of municipal boundaries over contiguous rural region[10] etc. were other factors. In 1981, the young productive age constituted 39% in the total population (Pradhan and Joshi 1983). In 2001, the annual growth rate of population of Kathmandu city for instance had 4.82% as against the national growth rate of population of 2.27%[11]. By 1991, two additional towns viz. Kirtipur and Madhyapur-Thimi were designated as municipalities. The municipal areas of three principal cities -Bhaktapur, Kathmandu and Lalitpur- were increased tremendously as described below.

*Table 2: Population growth percentiles and contribution of migration to population growth in Kathmandu Valley Urban Region (KVUR).*

| Year | Population growth percentile | | | Estimated migrants* of KVUR's population | |
|---|---|---|---|---|---|
| | Country | KVD | KVUR | Population | % |
| 1952-1961 | 11.1 | 11.9 | 11.9 | 1,449 | 0.8 |
| 1961-1971 | 22.8 | 34.5 | 23.2 | 810 | 0.4 |
| 1971-1981 | 29.9 | 23.8 | 45.7 | 39,431 | 15.8 |
| 1981-1991 | 23.1 | 44.2 | 82.1 | 214,469 | 59.0 |
| 1991-2001 | 26.8 | 49.9 | 63.9 | 245,541 | 37.1 |

Note that the migration estimation is calculated based on the following formulae: $ME_R = \frac{R_P - N_P}{100} \times P_{OR}$, where, $R_P = \frac{P_{nR} - P_{oR}}{P_{oR}} \times 100$, $N_P = \frac{P_{nN} - P_{oN}}{P_{oN}} \times 100$, MER = migration estimation at region 'R'; RP = population growth percentile of a region 'R' (or Kathmandu Valley); NP = population growth percentile of nation; Pn = population of last census year; and Po = population of earlier census year.

## 5. URBAN GROWTH AND EXPANSION

As a result of rapid growth of population, the urban areas of the Kathmandu valley have expanded in uncontrolled and haphazard manner. The present valley urban area comprising five municipal towns is 8,253 ha, which signifies an increase of 278.6% over the year 1955 or a mean increase rate of 6.2% per annum. There has constantly increased in the urban area with varying ratios. Between 1984-1991, the urban area grew with the highest mean annual rate of 10% (Table 3). This was due mainly to designation of two additional municipal towns. This considerable growing size of the urban areas can be compared with the alarmingly growing proportion of the population, though the years for which the available data on urban area expansion did not match with the census years.

*Table 3: Expansion of Kathmandu Valley Urban Areas*

| Year | Area (ha) | Growth (ha) | Mean annual growth rate (%) |
|---|---|---|---|
| 1955 | 2180 | - | - |
| 1978 | 2600 | 420 | 0.84 |
| 1984 | 3096 | 490 | 3.13 |
| 1991 | 5270 | 2174 | 10.02 |
| 1996 | 6800 | 1530 | 5.80 |
| 2000 | 8253 | 1453 | 5.35 |

Source: Tuladhar (1998); UNEP et al (2001).

The density of population for the valley urban area shows 131 persons per hectare, with the highest density of 156 for Kathmandu and the lowest density of 62 for Madhyapur-Thimi (Table 4). Both newly designated municipalities –Kirtipur and Madhyapur-Thimi- show much low population densities. There is a great variation in housing density between urban core and periphery areas. The housing density in the core areas ranges 500-1000 persons per hectare. The built-up area has slow growth rate with lower than 3% per year (Pradhan 1998). This can be evidenced by the fact that there is virtually no physical space being left in the core areas (Plate 1). The higher densification of houses in the core areas is also due to building height limitation, which is 80 feet. The historical factor as noted above also resulted into the densely built houses in the urban cores. Thus, the built-up area has expanded towards outward to accommodate additional populations.

*Plate 1: A part of the cityscape of Kathmandu*

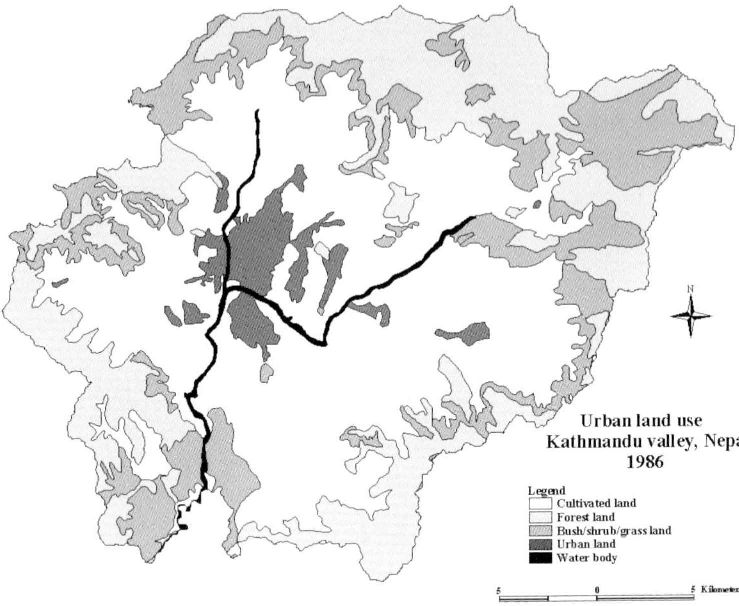


*Figure 3: Land-use of Kathmandu Valley, Nepal (1986).*

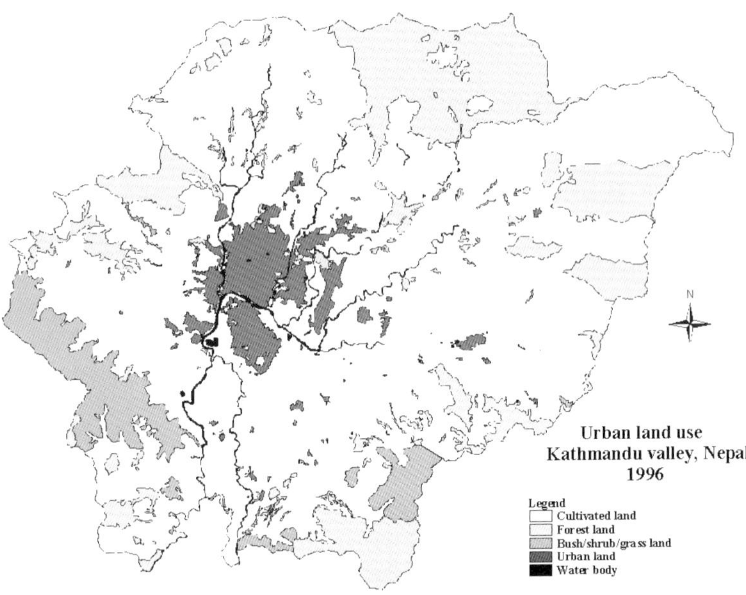


*Figure 4: Land-use of Kathmandu Valley, Nepal (1996).*

It means there is more of horizontal than vertical expansion of the buildings over the urban space. The housing study of Joshi (2002) estimated that 27,087 housing units were added during 1991-2001 in Kathmandu city alone. On the whole, the periphery growth rate in the valley urban area is over 7.5% per year (MOPE 1999).

*Table 4: Urban Population Density, 2001*

| Municipality | Population | Area (ha) | Density/ha |
|---|---|---|---|
| Bhaktapur | 77,764 | 688 | 113 |
| Kathmandu | 754,409 | 4829 | 156 |
| Kirtipur | 49,606 | 570 | 87 |
| Lalitpur | 168,077 | 1610 | 104 |
| Madhyapur-Thimi | 34,587 | 556 | 62 |
| Valley Total | 1,084,443 | 8253 | 131 |

Available documents indicate that increased need of the government infrastructure and functions in administration, health, education, banking, etc is found among several factors to contribute to the accelerated pace of influx of people into the valley and hence the rapidity in the urban expansion.

Other factors including growth in foreign aid, tourism, industries, hotels and vehicular traffic, expansion of external transport lines linking the valley with the country's other parts, increase in government expenditure on urban infrastructure development such as water pipelines, electricity, sewerage system, telephone and street network, land and housing, and construction of ring road enclosing Kathmandu and Lalitpur cities are also responsible for the growth of urban areas (Pradhan 1988; 2000).

The valley has owned over 2,500 institutions of education, government and private companies. In the health sector alone, 173 institutions with a bed capacity of around 4,000 exist in the valley. Out of about 300 standard hotels and resorts of the country, 87% are in the valley. The valley has a total of 943 km of road on which over hundreds of thousand vehicles move every day. All these have undoubtedly generated ample opportunities of ancillary employments for job seekers, attracting large mass of migrants. It is argued that when the supply of labour exceeds the demand, excess labour forces that are mostly unskilled, illiterate and poor will not go back to their homes but rather tend to indulge in low paid informal activities in the cities. The urbanization in the valley has turned into haphazard, out of the control of planning agencies and beyond the capacity of limited resources. It is predicted that if the present trend of urbanization continues and if all the factors said above continue to act, the valley's fertile agricultural floor will completely turn into urban use by the year 2020 (Pradhan 2000) and the valley of ancient civilization will turn into a concrete slum (Poudel 1998). While prepared planning of the Kathmandu Valley for the year 2020, DHUD (1991) estimated that about 3,614 hectares of land would be required to accommodate the additional populations and half a million new employment will be required.

## 6. ENVIRONMENTAL IMPLICATIONS OF THE VALLEY'S URBAN GROWTH

Rapid growth of population has occurred in the valley's urban areas due to relatively better opportunities available in economic, social, administration and other dimensions than in the country's other urban areas.

It implies various environmental implications. Indeed, urbanization in itself is not necessarily a problem, as it promotes economic activities and alleviates pressure on land resources in the hinterland areas. But haphazard and uncontrolled growth of urban areas generally invites many environmental problems. When they become more complex, they incur high cost, usually beyond the capacity of municipal, as well as central government expenditure. This has been the case of the Kathmandu valley urban areas.

### 6.1 Change in Land Use Pattern

One of the consequences of rapid urbanization is the change in urban land use and land cover. The valley has witnessed a remarkable change in land use pattern over the last few decades (Figs. 3, 4)[12]. The urban built-up area increased by 46 $km^2$ in 1996 or expanded with almost 212% as compared with that in 1955, where as the rural built-up area increased by 24 $km^2$ during the same interval of years (Table 5).

*Table 5: Change in land uses, Kathmandu Valley (1955 –1996)*

| Land use | 1955 | | 1978 | | 1991 | | 1996 | | 1955-1996 | |
|---|---|---|---|---|---|---|---|---|---|---|
| Categories | Area | % | Area | % | Area | % | Area | % | Area | % |
| Urban built-up | 21.8 | 3.7 | 26.0 | 4.4 | 52.7 | 8.9 | 68.0 | 11.6 | 46.2 | 211.9 |
| Rural built-up | 1.6 | 0.3 | 4.6 | 0.8 | 5.3 | 0.9 | 2.1 | 0.4 | 0.5 | 31.3 |
| Rural built-up with farm area | 9.6 | 1.6 | 30.6 | 5.2 | 31.8 | 5.4 | 33.0 | 5.6 | 23.4 | 243.7 |
| Vegetation cover | 163.6 | 27.7 | 149.5 | 25.3 | 146.7 | 24.9 | 116.0 | 19.6 | -47.6 | -29.1 |
| Agricultural land | 368.3 | 62.4 | 354.8 | 60.1 | 335.1 | 56.8 | 319.9 | 54.2 | -48.4 | -13.1 |
| Others | 25.1 | 4.3 | 24.5 | 4.2 | 18.4 | 3.1 | 51.0 | 8.6 | 25.9 | 103.2 |
| Total ($km^2$) | 590 | 100 | 590 | 100 | 590 | 100 | 590 | 100 | - | - |

Source: Tuladhar (1998).

The housing units found to be increased with 27,087 in Kathmandu city alone during 1991-2001 (Joshi 2002). These built-up areas seem to have expanded encroaching upon agriculture land, forest area, riverbank and open spaces. Both forest and agricultural areas have decreased by 29 and 13% respectively between 1955 and 1996 (Table 5, Fig. 5). The decline in the forest area has affected not only the water recharge capacity of groundwater sources, but also caused frequent landslides and soil erosion on the surrounding hill slopes and flash floods in the valley floor (UNEP et al 2001). 'Others' category including ponds, bare land, rock cliff, landslide, etc. has increased remarkably from 18.4 $km^2$ in 1991 to 51 $km^2$ in 1996, which is mainly due to variation in base map scale[13].

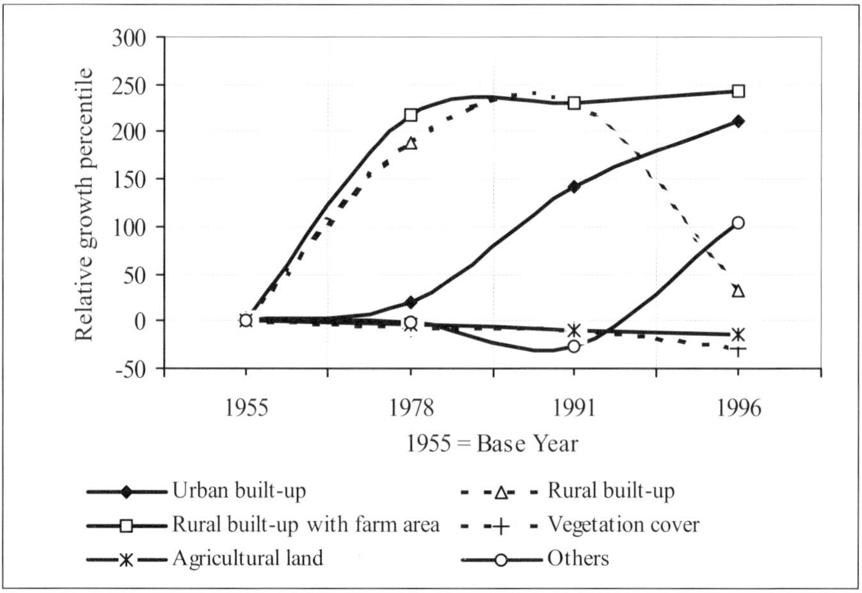


*Figure 5 Relative Change in Land Use Categories in Kathmandu Valley.*

The history of urbanization planning in Nepal is nearly four decades old. In 1962, the first urbanization plan was formulated and after that other 6 plans were also drafted, but never been implemented them due mainly to too ambitious plans with respect to limited financial resource (Poudel 1998). Despite the regulations and code of conduct for construction, new house have been built in newly developed residential localities. Absence of soft loans from the government and lack of official organized investment have contributed to the growth of unplanned development of housing. There are not regulations on development rights, sub-divisions and land uses (Joshi 2002). One can build a house wherever he/she likes. What the general practice is that usually the medium income city-dwellers think housing as a basic requirement especially in old age and they are often not able to afford planning for other facilities during the time of construction. Further, the exploitation of low land and the use of relatively cheaper land with little or no facilities are some of the reasons for severe pollution of the urban environment. It is believed that the traditional urbanization was more scientific and hygienic than the present one. As stated above in earlier time, the high land terraces were allocated for the settlement and the low lands for farming.

Until 1981, 75% of the valley's population was counted on agriculture, while in 1991 only one-third was engaged in any agro-profession. This change reflected the diversifying of economic activities of the residents.

## 6.2 Environmental Consequences

Studies show that the accelerated and haphazard growth of the urban areas in Kathmandu valley has resulted into squatting of open-public space, mismanagement of solid wastes and sewers, increasing in slums and level of pollution of water and air, depleting water sources, and so on. The urbanization is characterised mostly by unplanned, haphazard and irregular manner. Buildings and their access paths or roads have been built without following the urban development code and building norms; this being due to lack of comprehensive urban policy and a weak implementation of legal measures (Joshi 2002). The quality of basic utility services has not only declined rapidly but become grossly inadequate as well. Heavy funds are badly needed to develop additional infrastructure services, as well as to maintain them. The pronounced environmental consequences due to uncontrolled growth of urban areas have been discussed on the following components.

### 6.2.1 Squatters and Slums

Squatting public open space is a recent urban phenomenon in the valley. High level of unemployment in the rural areas is considered as principal factor for the people to leave their villages and move towards the cities. Large stream of rural migration has contributed to the proliferation of slums and squatters in the urban areas. In the absence of adequate income to own or rent houses, people choose the most neglected urban land for building shelter. The river banks, public land, and open space are the most favourable locations for squatters. Till early 1970s, there were almost no squatters and slums in the valley. In 1985, there were 17 squatter settlements in the valley with a population of 3,000. By 1990, their number increased to 33 with population of about 15,000. Of the total, about 60% of the squatting was on public land and the remaining 40% in public buildings, temple sites and rest houses (Rabenau 1990). In most cases, each squatter settlement has at least 130 houses. Another survey showed 39 squatter settlements in the valley (Gajabsiddhi 1988). Of these, 64% were on the banks of major river including the Bagmati, the Bishnumati, the Tukucha, the Dhobikhola, and the Samakhusi and the rest in the urban areas (Plate 2). Yet, the squatting activity in the recent years appears to be increasing in the valley due to tremendous influx of labour force from different parts of country, as well as adjoining states of India. The year 2000 saw 61 squatter settlements with 2,031 households and population size of 11,826 (CDS 2000 as quoted in Joshi 2002).

*Plate 2: Squatter houses on the local riverbank*

The squatter settlements are usually not entitled to get urban utility facilities. In a way they have managed to get electricity, other facilities such as piped water and private toilet are virtually lacking (CDG 1994). As a result, they used to go to riverside for defecation to pollute the river water and the surroundings. The main economic activity of these squatter families is to scavenge the waste and sell it to the waste dealers. If the former activity is responsible for deteriorating river environment, the latter is considered to be prudent use of wastes, mitigating environmental problem. ILO estimates that 60% of the present total population of Kathmandu city are indulged in informal sector activities such as petty street vendors, repairing, personal service, crafts and manufacturing, construction, etc. (Joshi 2002)[14].

Apart from squatter settlements, the urban environment is found to be poor quality due to slums in the valley urban areas. Particularly the inner city areas have gradually been turned into slum condition where there is inadequacy of basic utility facilities. Studies show that there is high occupancy rate with 4.8 persons per rented single room with an average area of 120 square feet. Out of 24 cases, 87% had piped drinking water, and on average 36 persons had shared a toilet (Lumanti 1997 as quoted in Joshi 2002). It is estimated that Kathmandu city alone has about 50 slums with population of 40,000.

There are not any agencies that can provide low cost housing and residential quarters particularly to the urban poor. In the absence of cheaper accommodation, the poor either go for cheap renting house or room or squat public space or buy a small piece of land wherever they can find and start constructing houses on piecemeal basis, as and when they earn some money. The government has already approved National Housing Policy, 1996 with a view to ensure private sector participation in the land development and housing constructions. Local government units and some non-government organisations are being involved to improve the environmental conditions of squatter settlements.

### 6.2.2 Unmanaged Solid Waste

The volume of solid waste generation in the valley is increased tremendously; as it is directly related to rapidly growing population, change in consumption pattern and social behaviour. Increasing density in housing and petty enterprises, and diversifying of economic and social activities including hotels, industries, commercial establishments, hospital/nursing homes, etc. are responsible for generating solid wastes. It is estimated that the valley cities and towns have generated 480-500 tonnes of solid wastes every day (UNEP et al 2001). Migration to the large urban areas in the valley in search of jobs, education, and other opportunities, is the main cause of rapid growth in the urban population. A survey revealed that 41% of the urban population consisted of lifetime migrants, of which 45.6% were farmers prior to migration, while 9.6% were students and 15% were service holders (CBS 1997). This signifies that the main cause of rapid urbanization in the country is the inability of villagers in the rural areas to sustain their livelihoods with agriculture only.

As a result, heaping of solid waste seems to be the most visible problem, which is now out of management capability of the municipal government. What practices are being developed by the city dwellers are that they throw household waste and discharge domestic, as well as industrial effluents directly into local streams without treatment? Most of the cities' corners, roadsides and riverbanks in the valley are being the places for household garbage (Plate 3). Until a few decades ago, almost all of wastes in the valley was organic and was recycled to produce compost. In the traditional system, waste generated in households was either sold directly to farmers or placed in pits that were located in between the houses. Nowadays, the traditional system has vanished and the composition of urban waste has also changed with the introduction of new materials such as plastics, paper and glass. Studies carried out in Kathmandu city show that there is not much varied in the proportion of organic materials in the total composition of solid waste between past and present years (RESTUC 1999). In 1976, the proportion of organic materials was 67.8% and it decreased to 67.5% in 1999. The contribution of paper to the total waste has increased gradually from 6.5% in 1976 to 8.8% in 1999, while that of plastics has increased tremendously with 0.3% in 1976 to 11.4% in 1999. Other wastes comprise dust and construction debris, textile, glass, metal scraper, wood and rubber and leather.

*Plate 3: Pilling of unmanaged household garbage.*

A few studies have estimated the per capita waste generation in Kathmandu. For instance, Lohani and Thanh (1978) estimated that the per capita waste generation rate was 0.25 kg/day; GTZ (1988) estimated that it was 0.4 kg/day, and a recent study estimated the average amount to be 0.48 kg/day (RESTUC 1999). At present, about 93% of the pollution loads are from domestic sewage and the remaining 7% from industrial effluents. The valley's household wastes constitute about 40% of the country's total waste generation of 406,830 tonnes (UNEP et al 2001). Most of these wastes go to rivers, which clearly indicate that the valley's rivers are gradually being turned into open sewers (Devkota and Neupane 1994).

As a result of direct dumping of domestic waste and other discharge, much negative impact appears to be on aquatic macro-invertebrates in the rivers. The core urban rivers have shown a dominant of few types with small number of aquatic macro-invertebrates as compared to their rich variety in the headwater region in the valley. This attests the deteriorating condition of the aquatic environment (UNEP et al 2001). It has also affected the quality of river water and the aesthetic value of temples, monuments and the rivers (Pradhan 1998).

Majority of the waste generation is composed of biodegradable (79%) and the rest are non-degradable. Metal scraps, plastics, paper and glass accounted for 30% of the waste generated (UNEP et al 2001). Over the years, composition of solid waste has changed from organic to non-organic matters and chemicals. The wastes from industries and hospitals or nursing homes contain toxic chemical materials, hazardous to public health.

Organised solid waste management at institutional level was initiated in 1980 and the organic wastes collected from the garbage containers were processed into compost manure. This system was terminated in 1991. Likewise, a sanitary landfill site that developed for the final waste disposal was also closed down in 1994. All these were due to strong opposition of local residents who had got ill-health impacts of the waste deposition or location. At present, all municipal wastes are dumped at the bank of the rivers without treatment. They are compacted and covered by earth, sand and gravel to make future roads along the rivers as temporary management until long-term solution is available for solid waste management.

### 6.2.3 Depletion of Water Resources

One of the visible impacts of rapid urbanization in the valley is an inadequacy of drinking water supply to increasing demand of population. To meet this, almost all the valley's major rivers have been tapped and groundwater source has been utilized for drinking water. As a result, the water level of most of the valley's rivers has been lowered down and the pumping of the groundwater has been dropped from 9 meters to as below as 68 meters in the valley (Metcalf and Eddy 2000). The current withdrawal of groundwater is 58.6 million litres per day, which is far above the sustainable level of aquifer of 26.3 million litres per day (NWSC 1999). Yet, only 79% of the total demand of water of the urban population has been met. However, due to lack of regular monitoring program, groundwater depletion and recharge rate is uncertain. In order to meet the water demand, many households at private level have drawn water through water pump from shallow aquifer, though there is no information on it. Stanley (1994) estimated that the groundwater level has lowered down to about 4 meter per year particularly in northern and eastern parts of the valley. This practice is a serious concern in terms of health hazard, as the shallow aquifer is likely to be contaminated with organic and non-organic pollutants. The valley's water sources drawn from the shallow aquifer were polluted (non-coliform free) because of seepage from run off, farmland, road, as well as leakage from sewers (Pradhan 2000).

Quality deterioration of the river water is another consequence of the rapid urbanization. The quality of drinking water is below the general acceptable standard of health. The river water gets much polluted when they enter the urban built-up areas (Plate 4). The main cause of the river water pollution is direct disposal of solid waste and sewerage, as well as discharge of industrial effluents into them, as noted above. Contamination of drinking water has also occurred due to mixing up of sewer material with the piped water through leakage in the old pipeline. The pipelines in the valley are laid parallel to sewage lines. The water pipelines are sometimes broken for irrigation. The city dwellers dig the ditch to lower the tap for water collection. Intermittent water supply is another reason causing the mixing of piped water with sewerage. About 15% of the urban houses in the valley have accessed to a sewerage facility (NWSC 1999). These sewers are directly discharged into the rivers. During the dry season, the flow in the lower ranches of the rivers is virtually raw sewage. This constitutes a serious health hazard, as the people have the tradition to use river water as holy water. Construction of buildings on flood plains or the river banks has caused drainage problem in the valley. Over a quarter of the total urban houses was often flooded (DHUD 1991). Inadequate drainage induces a rise in the groundwater level, compounding the problem of contamination of drinking water.

*Plate 4. Polluted local river by household waste*

Disposal of urban solid waste and domestic and industrial effluents in the rivers is responsible for polluting the water and causing water-borne diseases, such as diarrhoea and typhoid. One hospital in Kathmandu shows that 16.5% of all deaths were due to water-borne diseases (Metcalf and Eddy 2000).

As the river water is a major source of drinking water in the valley, conservation of surrounding watershed is very much essential. But the fact is that due to excessive use of both surface and groundwater sources, its all water sources have been depleted and there is little support from the surrounding watershed areas to replenish the water sources in the valley. As noted above, the valley has already lost nearly one-third vegetation cover for urban and other uses. Few industries have wastewater treatment plants but they are not at working condition. Recently an NGO has operated wastewater treatment plants in some schools, hospitals and private houses in the form of trial based on constructed wetland system before discharging sludge into the rivers (UNEP et al 2001). It is yet to be made for widespread use in the valley, be it feasible for cases.

*Plate5. Industrial smoke in Kathmandu city.*

### 6.2.4 Air Pollution

Kathmandu valley has attracted not only a large mass of population but also a variety of activities including industries, and public and private vehicles. The valley now hosts over 72% of the total water-polluting industries and nearly two-thirds of the total vehicles of the country. The industries are responsible for releasing 104 tonnes of dust particles and big columns of smoke (Plate 5). Lead particles found in the air around the localities of the urban core are responsible for growing cases of respiratory illnesses (UNEP et al 2001). The transport sector relies mostly on petroleum products; the largest usage is for diesel that accounts for 66% of total fuel consumption.

In 1999, electric vehicles and unleaded petrol for automobiles were introduced. The level of pollution related to the transport sector is expected to decrease in coming years. The government has also proposed a new concept for taking the industrial and academic institutions out of the valley with the objective of resolving the problem of environmental degradation.

At last, the aesthetic value of the cultural heritage including temples and shrines of the valley is not left untouched to deteriorate due to haphazard urbanization and cultural pollution. The invaluable cultural heritage sites renowned for their archeology and historical monuments, which have been inscribed in the World Heritage List, are now increasingly being threatened. As a result, most of them now are proposed to be included in the heritage danger list.

# 7. CONCLUSIONS AND POLICY RECOMMENDATIONS

The urban settlements of the Kathmandu valley are one of the oldest human settlements in Asia. The valley is a legendary for its arts and culture, as well as for its agricultural land. The art and culture however have been threatened because of losing of their original traditional arts and culture.

Within five decades, the urban areas in the valley grew by nearly 279% encroaching upon prime agricultural land, public land, riverbank, and forestland. During these years, the forest and the agriculture lands decreased by 29% and 13% respectively. The urban population grew by almost 500% between 1952 and 2001, in which migration was assumed to contribute 113%. The main cause of rapid urbanization in Kathmandu valley is the inability of villagers in the rural areas to sustain their livelihoods with agriculture only and therefore move out to urban areas for seeking jobs. Further, the increased needs of the government infrastructure and functions in administration, health, education, banking, tourism and industry, as well as improvements in transport network linkage both within and outside the valley are responsible for the rapid growth of the urban areas.

As a result, the urban expansion in the valley has taken place in a haphazard and uncontrolled manner, which has caused serious environmental consequences. The valley that was once self-sufficient in food production, fuel wood supply, drinking water, irrigation water as well as rich in traditional arts and crafts is now facing environmental pollution and deterioration in land, water and air due to rapid urbanization. At present, planning of the urban areas in the valley has innumerable challenges. One basic problem of the urban developing planning in Nepal is the lack of seriousness or weak enforcement in the implementation of the measures or strategies so far developed. The fact that many efforts were made in the past and to introduce urban planning but no real progress has been achieved so far. But rather the towns and cities continue to expand at an alarming rate and associated problems get worse.

Based on the above discussions, some of the policy recommendations may be provided to mitigate urban environmental problems and correct haphazard urbanization in the valley.

- Efforts should be made to educate and create awareness among the urban, as well as rural dwellers towards conserving the environment. Awareness can be raised through involving local inhabitants in planned urban development and expansion activities.
- With the lack of modest urban land use zone policy, urban planning is taking place in a piecemeal basis. The valley requires a comprehensive mapping to delineate land use components such as urban built-up, historical, cultural and religious sites, forest areas, agriculture areas, archeological sites, industrial zones, road network, rivers and ponds, and others. This is to be made in view of preserving urban environments and to maintain desirable level of quality of water, air and noise, as well as to regulate urban growth in the valley. Direct discharge of raw domestic sludge and industrial effluents into the rivers should be stopped immediately. This requires to developing legislative measures.
- Private sector should be encouraged to invest on urban land development and on low cost housing development to supply housing plots to poor urban communities, such as slums and squatters. Private sector together with NGOs need to be encouraged in household solid waste collection and management.
- Several agencies are indulging in urban development activities in the valley. It appears that there is no agency that acts for linkage and coordination among the concerned agencies. There is absence of monitoring mechanism. There should be an agency in the valley that coordinates among different sectors and development agencies related to urban activities.
- Small market towns at the outskirts of the valley's cities should be developed so that they can also offer suitable living environment for both urban dwellers and migrants.
- Urban planning is entirely a technical issue but its implementation depends very much on the will of the political leadership. Without a strong leadership, no implementation of the plans is possible. The support of the international communities appears important for putting pressure on the Government of Nepal to enforce the policies effectively.

## NOTES

[1] The Kathmandu Valley has been defined by considering its surrounding mountain ridge or watershed, which covers 475 $km^2$. The Valley also consists of three districts viz. Bhaktapur,

Kathmandu and Lalitpur that extend beyond the valley ridge. Juxtaposing these three districts, the Valley has an area of 899 $km^2$. The country has altogether 75 districts that consist of both municipal towns and village areas or village areas only.

[2] The municipal towns of Nepal are defined as incorporated urban centres that have population size of >10,000.

[3] In 2000, there were 58 designated municipal towns, which had population of 3,172,094, making up about 15% of the total population of the country. This shows that Nepal is one of the least urbanised countries in the world.

[4] The valley had an intensive cultivation system, which became clear from the great attention paid to irrigation canal network system called *king canals* during the period of 1200-1769 AD. Elaborate arrangements existed for the maintenance and repairing of such irrigation canals and rights to utilise irrigation facilities were carefully regulated (Regmi 1978). The historical documents also evidenced the fact that the valley people during that time appear to have motivated by a strong concern for preserving the agricultural land in order to avoid its inefficient use. Because, towns and villages were built on upland plains with terrace fields on either side of the rivers. Houses were built in compact form proximity to cultivated fields but some distance from the flood. These all made it possible to provide sufficient food supply to the people of the valley.

[5] The valley had played as a centre of entrepôt trade by virtue of its geographical position on one of the major trade routes between India in the south and Tibet in north, which facilitated to control the trade route between the two countries. In the valley's north were there opened two most accessible passes, namely *Kuti* and *Kerung* over the Himalayas to Tibet, which gave the rulers of the valley benefits of controlling and organising the trade between Tibet and India.

[6] Until 1769, Nepal was divided into several small principalities, known as *Baise-Chaubise Rajyes* (22-24 principalities).

[7] The Rana autocracy that had remained for 104 years in the country ended in 1950.

[8] The first population census of Nepal was taken in 1911 and then the census continued at about every 10 years interval. However, the census of 1952-54 is considered more reliable than the previous efforts and the information on designated incorporated urban centres is available since this census year.

[9] No data on migration or birth statistics at settlement localities is available in population census of Nepal. Therefore, migration estimation for the town has been made based on differential growth percentiles of population between nation and municipal town (Edler et al 1976).

[10] For instance, a total of 2,073 ha of contagious rural area was annexed to Kathmandu municipal boundary, increasing from 29 wards in 1971 to 33 wards in 1981(PADCO 1986). It now has 35 wards with 50.77 $km^2$.

[11] The average household size of the five municipal towns was 5.8 as against to 6.5 of the country as a whole.

[12] The base map scales for land use analysis were 1:50,000 for the year 1996 and 1:1 mile, air-photograph and TM (satellite imagery) for other years and therefore information of ground resolution size of the components varied greatly. However, the data do not consistent with the information given in Table 5.

[13] Information of area coverage of the land use components on 1:50,000 base map has successfully detected only those with ≥10 ha (Tuladhar 1998), while in other base maps greater than this size was being able to be detected.

[14] There is no information available on informal sector activities and squatter and slum settlements for other municipal towns of Nepal.

# REFERENCES


CBS. 1995. *Statistical Year Book of Nepal*. Kathmandu: Central Bureau of Statistics, National Planning Secretariat, His Majesty of Government, Nepal.

CBS. 1997. *Urban Population Survey-1996*, Kathmandu: Central Bureau of Statistics, National Planning Secretariat, His Majesty of Government, Nepal.

CDG. 1994. *A Study on Population and Environment in Nepal, A Project Report*, Central Department of Geography (CDG), Tribhuvan University, Kathmandu, Nepal.

Commoner, B., M. Corr and P. Stamler. 1971. "The causes of pollution." *Environment* 13:2-19.

Depuis, J. 1962. „Les basins interiéurs du Kashmir et du Nepal." *Annals de Geographie* 71(384):156-66.

Devkota, S.R. and C.P. Neupane. 1994. *Industrial Pollution Inventory of the Kathmandu Valley and Nepal*. Kathmandu: Industrial Pollution Control Management Project.

DHUD. 1991. *Kathmandu Valley Urban Development Plans and Programmes*, Kathmandu: Department of Housing and Urban Development, Ministry of Housing and Physical Planning.

Edler, J. et al., 1976. *Planned Resettlement in Nepal's Tarai: A Social Analysis of the Khajuri-Bardia Punarvas Project*, Kathmandu: INAS, Tribhuvan University.

Ehrlich, P. and A. Ehrlich. 1990. *The Population Explosion*, New York: Simon and Schuster.

Gajabsiddhi, H.P. 1988. *Squatter Settlements in Kathmandu City*, Unpublished Thesis, Central Department of Geography, Tribhuvan University, Kirtipur.

Golden, H.H. 1981. Ur*banisation and Cities: Historical and Comparative Perspectives on Our Urbanising World*. Lexington: D.C. Heath.

Goldscheider, C. 1983. "Modernization, migration, and urbanisation." Pp. 47-66 in *Population Movements: Their Forms and Functions in Urbanisation and Development*, editor P.A. Morrison. Liege: Ordina Editions.

Gurung, H. 1989. "Regional Patterns of Migration in Nepal". *Papers of the East-West Population Institute* 113. Honolulu (Hawaii): East-West Center.

GTZ. 1988. *Solid Waste management with People's Participation-An Example in Nepal*, No. 231, Kathmandu: GTZ.

HMG. 1983. *National Population Strategy*, Kathmandu: National Commission on Population.

HMG/UN/UNESCO. 1975. *Kathmandu Valley: The Preservation of Physical Environment and Cultural Heritage Protective Inventory. Vol. 1. Kathmandu*: His Majesty of Government, Nepal.

Joshi, A.M. 2002. *Housing for Urban Informal Sector Workers: A Case Study of Kathmandu City*, Nepal, An Unpublished M.Sc. Thesis, Institute of Engineering, Tribhuvan University, Nepal.

Kosinski, L.A. and R. Mansell Prothero. (eds.) 1975. *People on the Move: Studies on Internal Migration*, London: Methuen.

Lee, E.S.., 1969. "A theory of migration." In *Migration*, editor J.A. Jackson. London: Cambridge University Press.

Lohani, B. and N.C. Thanh. 1978. *Problems and Practices in Solid Waste Management in Asia*. Bangkok: Asian Institute of Technology.

Malla, U.M. 1978. "Settlement Geography of Kathmandu Valley." *Geographical Journal of Nepal* 1.

Metcalf and Eddy Inc. 2000. *Groundwater and Wastewater*. A Paper presented at the Seminar on Groundwater and Wastewater, Organised by the Melamchi Water Supply Development Board, Kathmandu.

MOPE. 1999. *Environmental Planning and Management of the Kathmandu Valley*, Kathmandu: Ministry of Population and Environment (MoPE), His Majesty's Government of Nepal.

Ng, R.C.Y. 1975. "International migration in Southeast Asian countries." Pp 181-192 in *People on the Move: Studies on Internal Migration*, editors L.A. Kosinski and R. Mansell Prothero. London: Methuen.

NWSC. 1999. *Water Supply and Coverage in Urban Area of Nepal*. Kathmandu: Nepal Water Supply Corporation.

PADCO. 1986. *Kathmandu Valley: Urban Land Policy Study*, Kathmandu: Kathmandu Valley town Planning Office/Planning & Development Collaborative International.

Pradhan, B. 1998. *Water Quality Assessment of the Bagmati River and its Tributaries*, Kathmandu Valley, Nepal. Ph.D. Dissertation, Vienna, Austria: Institute of Water and Wastewater Engineering, BOKU.

Pradhan, B. 2000. "Assessing Drinking Water Quality in Kathmandu Valley, Nepal." *Health Prospects* 2(2):7-9.

Pradhan, P.K. 2000. "Urban Growth Pattern and Food Security in Kathmandu Valley." International Workshop on Integrative Strategies for Food Security Improvements in the Kathmandu Valley, April 30 – May 1, 2000, Kathmandu, Organised by the University for Agricultural Sciences, Vienna (Austria) in collaboration with ENPHO and ICA Nepal.

Pradhan, P.K. 1998. "Urban Growth Pattern in Kathmandu Valley, Nepal: A GIS/RS Analysis." Unpublished Paper. Nepal: Geography Department, Tribhuvan University.

Pradhan, P.K. and M.P. Joshi. 1983. *Implications of Population Growth on Basic Need Services in Kathmandu Valley*. Project Report, Kathmandu: Research Division, TU.

Pradhan, S.M. 1988. "Kathmandu: metropolitan development and conservation issues." *Regional Development Dialogue* 9(3):203-21.

Poudel, K. 1998. "Kathmandu Valley: City of Slums." *Spotlight*, The National Newsmagazine (September 11-17):16-21.

Rabenau, B.V. 1990. *Issues in Housing Development*. Kathmandu: Urban Development through Local Efforts.

RESTUC. 1999. *A Study of Solid Waste and Its Management in Kathmandu*. Kathmandu: WHO.

Regmi, M.C. 1978. *A Study of Nepali Economic History (1768-1846)*. New Delhi, Manjushri.

Stanley. 1994. *The Bagmati Basin Water Management Strategy and Investment Program-Final Report*. Kathmandu: Stanley International Ltd.

TKP. 2001. *The Kathmandu Post* (English Daily), April, Kathmandu.

Todaro, M.P. 1976. *Internal Migration in Developing Countries: A Review of Theory, Evidence, Methodology and Research Priorities*, Geneva: International Labour Office.

Tuladhar, T.M. 1998. *Urban Growth, Land Use Change and Impact Environment: A Case Study of Kathmandu Valley*. Post-Graduate Thesis, CSSTE-AP, Dehradun, India.

UNEP, MoPE/HMGN, SACEP, ICIMOD and NORAD. 2001. *Nepal: State of the Environment 2001*. Bangkok: United Nations Environment Programme (UNEP).

Wolfgang, K. 1993. *The Traditional Architecture of the Kathmandu Valley*. Kathmandu: Ratna Pustak Bhandar.

Zelinsky, W. 1983. "The impasse in migration theory: A sketch map for potential escapees." Pp. 19-46 in *Population Movements: Their Forms and Functions in Urbanisation and Development*, editor P.A. Morrison. Liège: Ordina Edition.

Zelinsky, W., L.A. Kosinski and R. Mansell Prothero. (eds.) 1970. *Geography and a Crowding World*. New York: Oxford University Press.

# CONSIDERING MIGRATION AND ITS EFFECTS ON COASTAL ECOSYSTEMS

Sara Curran and Tundi Agardy.
*Department of Sociology and Office of Population Research, Princeton University, Princeton, NJ*

## 1. INTRODUCTION

For decades people have tried to understand and define the relationships between population growth and migration, consumption, and environmental condition. Early simplistic formulas claiming a linear relationship between population size and environmental degradation have been shown to be deficient, especially when nuances of demographics such as migration and factors driving and organizing human behavior are not taken into account. Discarding outdated paradigms has not immediately led to the development of new ones that can easily explain the relationship between human population and environment, however. Yet much recent research has moved us towards elaborating the elements that will compose the foundation for a new paradigm. We describe some of these foundational elements from the perspective of migration scholars and with some attention to the research and theory of common property resource management scholars.

While we are far from a conceptual framework that adequately incorporates all the variables in such complex relationships between humans and nature, there are common threads in the way researchers have addressed these questions. We believe that there is a timely convergence of ideas and demand for empirical evidence for understanding the relationship between population size, migration, consumption and the health and productivity of ecosystems. Much of this convergence has focused on coastal ecosystems: some of the most complicated, and increasingly most stressed, global environments. In this paper we explore a limited set of theoretical pathways from which hypotheses might be derived about the positive and negative impacts of population growth and migration upon the environment. We focus upon migration, in particular, since it is an understudied phenomenon in coastal areas, but the largest contributor to population growth in coastal areas (Cohen and Small 1998). We suggest three critical questions that need to be answered in order to link migration processes to coastal ecosystem health: 1) Who migrates? 2) How are they received in the place of destination? 3) And, do they maintain ties to their place of origin? In the



*J.D. Unruh, M.S. Krol and N. Kliot (eds.),*
*Environmental Change and Its Implications for Population Migration,* 201–229.

parlance of migration scholars these concerns are migrant selectivity and migrant social networks. Recent theorizing and empirical evidence from migration scholars implicates embedded relations as critical for understanding both selectivity and the role of social networks. Embedded relations also emerge as a key concept in the literature about coastal ecosystems, marine resources, and fisheries management. We present a synthesis of the migration literature and the coastal and marine management literature and use recent research from coastal ecosystems to propose measurement, methods, and modeling approaches for refining our understanding of the possible impact of population, particularly migration, upon coastal ecosystems.

Among the world's many biomes, coastal zones are perhaps the areas most subject to rapidly changing demographics and environments. The broad band of continental coastal areas, and small islands in their entirety, provide the best alluvial soils for agriculture, good opportunities for transport and waste disposal, access to fisheries (and increasingly, access to lands and near-shore areas suitable for aquaculture), and space deemed valuable for recreational use and housing. Population doubling rates are almost uniformly higher in the world's coastal zones than in inland areas. At the same time, environmental change in coastal areas is dramatic – sometimes happening rapidly under ambitious development schemes and at other times happening more slowly as impacts on the coastal environment cumulate. With many anthropogenic impacts and natural changes occurring simultaneously in coastal zones, and with inadequately integrated coastal management being the norm throughout the world, the need to assess the complex human and natural environment of the coasts is great. Thus coastal areas and their inhabitants present the greatest challenge for investigators examining the population/environment nexus – and are the regions where learning about the relationship between humans and environment is most urgently needed.

One of the weakest points in empirical studies tying together demographic analysis and ecological assessment has been the omission of how social networks and embedded relations evolve and relate to the physical environment (Curran 2002). A new paradigm to replace Malthusian ideas of population will not arise until we develop better understanding of human behavior and the drivers behind such behavior, particularly how human behavior is organized and predicted. We must assume that the set of social relations defining natural resource use by people (consumption, exploitation, management, investment) is a critical intervening variable between population growth, migration and environmental outcomes. The set of social relations defining natural resource use are described in the environmental literature generally as property relations, which can vary from open access, to common property (with varying degrees of local and state participation), and to private or market-based relations(Ostrom 1998). We discuss these social relations, and what we have begun to learn about them in coastal areas, in the following pages.

## 2. PARAMETERS OF HUMAN POPULATION DYNAMICS AND ENVIRONMENTAL CHANGE IN COASTAL ECOSYSTEMS

Myriad studies have addressed one aspect or another of human populations and behavior in coastal communities, from small rural villages to large urban areas. Similarly, the last three decades have witnessed a proliferation of studies that assess the current condition and trends in environmental health and productivity of coastal ecosystems, both terrestrial and nearshore marine. Relatively little research has attempted to uncover correlations between population changes and environmental conditions in coastal ecosystems, and even fewer have explored the causality behind correlations that are revealed (Dietz and Rosa 1994). Although a large literature exists concerning coastal zone management issues (e.g. Clark, 1996; Ruddle et al., 1988), quantitative analysis of coastal environmental changes in this literature tends to focus almost exclusively on environmental degradation of the terrestrial portions of the coastal zone, not the marine. Furthermore, policy prescriptions such as those suggested in the Noordwijk Guidelines (World Bank, 1993) have been based on general overviews of human impacts in the coastal zone, and not on site-specific empirical studies, nor on an understanding of the demographic and other drivers behind observed or expected impacts. Nonetheless, a critical mass of studies now exists to allow us to begin to develop a conceptual framework for revealing the relationship between demographics and environment in the coastal zone.

Parameters that sociologists and anthropologists quantify in these important interdisciplinary studies are far too numerous to list comprehensively here. Some of the more important data upon which empirical studies are based include population size and doubling rate, fertility, immigration and emigration, population age structure, and information about education, health, employment, etc. Social scientists also assess consumption, and both household- and market-driven demand (Princen 1999). At the gross level much of this data can be obtained through government census and other records, though the best studies attempt to ground truth census data with more nuanced interpretations of aggregate census figures. Many population researchers also attempt to assess the collective behavior of individuals, through finer scale household studies, and rapid rural appraisal. At the same time, social scientists analyze issues of management and governance that influence human behavior and community development (Bavinck 2001). Comprehensive investigations of population dynamics then attempt to bridge these various scales, and to uncover drivers behind population changes that are, or have been, taking place (Jolly and Torrey 1993; World Resources Institute 2001).

On the environmental side of the equation, the kinds of parameters that are the focus of interdisciplinary population/environment research are those that give indication of environmental quality (water quality, erosion rates, presence of algal blooms, etc.) or levels of productivity (biomass estimates, levels of resource, phytoplankton production, mangrove forest coverage and density, etc.). Sophisticated new Geographic Information Systems (GIS) models are also being employed to display historical trends in land and/or coastal ocean use and environmental condition, and even to allow predictive modeling of future conditions (Bryant, Burke, McManus, and Spaulding 1998; Australian Institute of Marine Science 2001). Since marine environments are notoriously difficult to study, scientists often use proxies for productivity such as catch per unit effort rather than direct stock assessment of a target fishery. Bio-indicators are also increasingly being used to show changes in ecosystem condition, including situations when ecosystems change at dramatic rates to alternative states (known as phase shifts). Human health in coastal communities is also sometimes used as a proxy for changes in coastal environments, but only in cases where causality between pathology and environmental degradation can be confirmed (e.g. incidence of *E. coli* related sickness in bathers).

Correlating population and ecological data is not meaningful unless two conditions are met: 1) the parameters being investigated are the correct choices for accurately assessing the human and environmental dynamics, and 2) the causal factors driving changes in human behavior and ecosystem function are revealed. For instance, not all environmental changes are indicative of a decline in environmental health or decreased productivity (from the perspective of human beings). Investigators are now spending significant effort in satisfying these two primary conditions, although their approaches show a wide diversity in the tools used for assessment, scales of study, and methods of analysis. Most are also careful to avoid basing empirical analysis on variables with inherent endogeneity. The diversity of approaches and the care taken to identify meaningful dependent and independent variables confers robustness to this interdisciplinary field of study, and suggests that the development of an entirely new paradigm to replace the Malthusian model is now within reach.

As we have come to shed the old Malthusian paradigm that population growth leads to environmental decline (neatly summarized by the IPAT equation, a shorthand for a multiplicative model, Environmental Impact = Population*Affluence*Technology), we have begun to learn that the interplay between humans and environment is a complex, non-linear relationship characterized by positive and negative feedback loops. Based on studies of developing and developed country demographics, it is clear that accelerated growth of human populations in coastal areas does not in and of itself explain the trend towards degraded coastal environments and decreased ecosystem health. More important are shifts in collective human behavior brought about by erosion of traditional management practices and tenure or other property rights, unchecked competition for increasingly scarce resources, and creation of subsidies or other, usually central government-led incentives to increase consumption of coastal resources. Patterns of in-migration into coastal areas are far more critical a factor in assessing population/ environment links than absolute population size, primarily because of the selectivity of migrant flows and because existing social networks and embedded social relations in coastal communities are often upset by influxes of non-indigenous peoples with differing customs, technologies, and levels of investment in resource management. For this reason, we focus this paper on migration, its influence upon social networks and embedded social relations as they each relate to resulting environmental changes.

Migration has been described as "an extremely varied and complex manifestation and component of equally complex economic, social, cultural, demographic, and political processes operating at the local, regional, national, and international levels…" (Castles and Miller 1993). As complex as migration is, the environment is equally so. And it is similarly problematic to remove environmental processes from the social, economic, political and institutional structures of which they are a part (Bilsborrow and Carr 2000). Therefore, drawing a linear, deterministic relationship between environmental degradation and migration is inappropriate and impossible (Zaba and Clarke 1994; Lonergan 1998). The current preference in the literature is to conceptualize the relationship as complex system meets complex system demanding longitudinal and multilevel approaches (Marquette and Bilsborrow 1999; Marquette and Bilsborrow 1997; Zaba and Clarke 1994). And several empirical case studies make evident support for this perspective (Sunderlin and Resosudarmo 1999; Burns, Kick, and Davis 1998; Silliman and King 1999).

## 3. UNRAVELING MIGRATION AND ITS EFFECTS ON THE COASTAL ENVIRONMENT

The current theoretical paradigm dominating the migration and environment literature modifies a long-held demographic theory, multi-phasic response (Davis 1963). The modification of multi-phasic response theory specifies that social relations and behavioral responses modify how population growth impacts environmental outcomes, specifically land-use practices (Bilsborrow and Ogendo 1992). The theory postulates that population growth affects land-use change through four stages, which can be consecutive, concurrent, or cumulative. These four stages are tenure regime change, appropriation of land (extensification), technological innovation, and demographic (out-migration) (Bilsborrow and Ogendo 1992).

This multi-phasic explanation for understanding the effect of population growth upon land-use change made important theoretical advances for many scholars in the field. However, it only considers two migration-environment relations – migration to places where there is "available" land (presumably organized under open-access or common property relations) and out-migration in response to limited environmental resources. Although more dynamic than previous theoretical models, it does not consider the varying forms of migration – return, repeat, circular, permanent, temporary – nor the selectivity of migration, nor how social networks and embedded relations may be important intervening variables for understanding migration impacts upon the environment. Despite these limitations, the empirical research that builds on this original theory has begun to complicate the explanation through the recognition of four dimensions of the migration process: selectivity, origin/destination differences, remittances, and social institutions (Curran 2002).

Summarizing the findings from the empirical literature shows that the selective nature of migration has an impact on environmental outcomes, including variability in the type of migration. Also, environmental considerations at both places of origin and destination can serve as push or pull factors respectively (highlighting the reciprocal character of the relationship). Further, remittances back to places of origin may play an important role in redirecting consumption in either positive or negative ways for environmental outcomes. Finally, migration affects the environment through social and economic institutions, such as land tenure and poverty(Curran 2002).

The preceding summary of findings in the migration and environment literature suggests that two questions must be answered simultaneously to refine our understanding of the impact of migration upon the environment: *Which migrants have access to which resources? And, how are these migrants embedded in the set of social relations defining ecosystem use in a place of destination?* The first question addresses a longstanding concern of migration scholars, that of selectivity. To understand both the causes and consequences of migration – whether they are environmental or not – researchers must examine the selection processes that include or exclude particular people from becoming migrants. The second question addresses recent theoretical concepts of social networks and embedded relationships, which can be used to generate hypotheses that predict positive or negative environmental outcomes in a given context, or system of property relations.

In this section of the paper we elaborate upon these two questions and use examples from recent studies of coastal and marine ecosystems to suggest how selectivity and social networks might mediate the impact of migration upon ecosystem health. First, we explain what is meant by migrant selectivity and use examples from the Galapagos, Ecuador and Goa, India to demonstrate how selectivity might affect ecosystem health. Second, we elaborate upon the second question by discussing three literatures: migration and social networks, common property resource management, and migrant assimilation. In all of these literatures social networks and embedded social relations are critical conceptual tools for understanding the distribution of people and services. And, third, we use four case studies to demonstrate the importance of social networks for understanding migration effects for ecosystem health.

## 3.1 Migrant Selectivity and Environmental Outcomes

One way in which the literature on migrant selectivity has attempted to show the variability of migrant impacts upon the environment is to compare spontaneous versus forced migration. The findings show varying results. In some cases, spontaneous migrants appeared to cause more destruction of the environment than forced migrants. In Indonesia, spontaneous migrants were associated with rates of deforestation twice those of the transmigrants (Bilsborrow and Ogendo 1992). In other cases, forced migration is worse for environmental outcomes in places of destination than is spontaneous migration. In Mexico, resettlement schemes of the 1960s and 1970s led to extensive deforestation (Alcorn and Toledo 1998; DeWalt and Rees 1994; Ewell and Poleman 1980). In the case of the Mexican tropical rainforest, in fact, spontaneous migrants adopted local management practices (Alcorn and Toledo 1998). In Zimbabwe, migrants resettled through government programs were not interested in farming and did not invest in sustainable land use practices (McIntosh 1993).

Another way in which the literature on migrant selectivity has demonstrated differential impacts is to examine the impact of return migrants upon environmental consumption and valuation in places of origin. For example, return migrants to rural places in Ecuador have a lower impact on deforestation than do new migrants, although Bilsborrow (1992) does not specify a reason. In other instances, return migrants, especially if they are professionals or gained professional training while migrating, return to places of origin with a different valuation of ecosystem services and commitment to preserving the environment (Conway and Lorah 1995). In the case of the Caribbean islands, Conway and Lorah (1995) find that return migrants invest in the establishment of local NGOs for the protection of the environment. In another study of the Caribbean, return migrants invested in secure land holdings and proceeded to invest in long-term, sustainable agro-forestry projects (Thomas-Hope 1999).

Most of the research on migration and the environment does not consider measurement of migrant selectivity in the traditional ways in which migrant selectivity is considered. Variation in age, life course stage, sex, or human capital of migrants has not been considered in regards to how variability in migrant stream composition might imply different environmental outcomes. Further, these selectivity issues can also be considered in relation to both origin and destination environmental outcomes, especially as these factors will differentially interact with the social institutions governing ecosystem management.

Two recent case studies illustrate the importance of understanding migrant selectivity issues in relation to environmental resource exploitation in marine and coastal ecosystems. In the case of the Galapagos, the selectivity must be inferred, but the description of the case reveals questions about who migrates and what type of migration they employ. Although the Galapagos Islands are world renowned for their unique flora, fauna and world heritage status, they also represent economic opportunity to many Ecuadorians, particularly poor fishers living along the South American coast (Bremner, Perez, and Borja forthcoming). The most recent marine resource to come under extraction pressure is the sea cucumber (Isosthichopus fuscus). Resource extraction has been largely undertaken by migrants, and individual migration has led to family migration and more permanent residence.

The young men migrating to harvest sea cucumbers came from particular communities along the Ecuadorian mainland coast where they became experienced in sea cucumber harvesting methods. Through their connection to Asian trading networks (the market for sea cucumbers is primarily in Japan and China), which gave them access to financing, technology, and knowledge, they were able to completely diminish the sea cucumber population along the mainland coast. Following their Asian beneficiaries they were then able to locate a new population of sea cucumbers on the Galapagos Islands. These new fishers to the Galapagos Islands had important demonstration effects upon the local fishers. Soon most fishers (part time or full time) were participating in the extraction of sea cucumbers until harvesting was closed by the Ecuadorian government between 1992 and 1994. Since then the harvesting season has only periodically opened for very limited time periods (Bremner et al. forthcoming), but each time more and more fishers participate in the harvest and the catch per unit effort has declined dramatically from the beginning to the end of the season and across years.

This example shows how migrant selectivity is important for the way the resource base is exploited in a destination. The particular environmental resource draws a particular type of person to a locale. In the case of the Galapagos Islands, the sea cucumbers attracted young male migrants. More specifically, it is young male migrants with particular human (fishing skills), financial (Asian financial backing), and physical (boats and technologies) capital resources from one location on the mainland coast. Migrant selectivity is also associated with the type of migration and the age of the migrant network. During the initial stages of the establishment of the migrant stream, selectivity is strongest and migration is temporary. The longer the migration origin-destination path is established the less selective the composition of the stream and the more permanent the migration.

Exploitation of sea cucumbers initially drew particular migrants to the destination, however given their limited residence in the destination the effects on the overall ecosystem well-being may have been relatively limited. Certainly, their dramatic harvesting of sea cucumbers led to the closing of the harvest by the government. Unfortunately, it is not clear whether this diminished migration, or whether the migrants consequently settled on the island and brought their families with them. The qualitative interviews from the case study, suggest this is so. Although the case raises more questions than it answers, there is strong evidence about how initial resource exploitation was spurred by a particular type of migrant and migration. From a migration scholar's more needs to asked about the selectivity and type of migration in order to more completely understand the character of migration and the composition of the migrant stream.

In another example from Goa, India, selective out migration and selective in migration changed social relations concerning ecosystem management in the coastal tidal plain (Noronha et al. forthcoming). Goa is known to the world as a place of beautiful beaches and wonderful sunsets, but this characterization is relatively recent. Prior to the 1970s (before Goa was "discovered") the Goan coastal tidal plain was comprised of a relatively complex set of land use relations making the most of land resources, fresh water and brackish tidal water to grow rice and coconuts, farm fish, and pan salt. Goa is also known for significant historical variability of migration patterns. In the early 20$^{th}$ century Goa was characterized by out migration to British India and Portuguese colonies in Africa. After colonial independence from Portugal in 1961, Goa experienced a surge of return migration from other Portuguese colonies and British India. Then, in the 1970's Goa experienced selective out migration of young men (both single and married) to the gulf state nations. As a result, the male labor force was significantly diminished and it became very difficult to maintain the complex irrigation system. Further, male migrants' remittance income was used to supplement livelihoods lost as rice, fish, and salt could not be harvested as the irrigation system deteriorated. The gradual shift in land use resulting from selective out migration set the stage for the consequent migration of tourists and tourist related service industries.

The timing of the migrant exodus to the Gulf states slightly preceded the "discovery" of Goa's beaches. Incoming tourists and associated migration of tourist industry service sector labor and capital also shifted relative use values of land. The result was a rapid decline in paddy land, fish ponds, and salt flats and an increase in housing construction. The combination of both selective out migration and selective in migration redefined land use along the coastal tidal plains. What this means for ecosystem health, particularly pollution and effluent management is not evaluated. Nevertheless, selective patterns of migration are critical for our understanding of the relationship between migration and land use change.

The two examples illustrated above highlight the importance of migrant selectivity for understanding ecological impacts. Specifically, from these two examples we can see that who migrates (selection based on age and sex, as well as choosing to move independently or with family) and what form the migration takes (permanent, temporary, return) are critical pieces of evidence for understanding the population-environment relationship. Further, the examples offer perspectives from both an origin and destination, reminding us that selectivity has impacts in both places. Finally, it is clear that there is a reciprocally caused relationship between migration and the environment. Environmental characteristics can selectively pull migrants, but subsequent resource exploitation may change the form and composition of future migrant streams, which can further affect environmental outcomes in both origin and destination.

## 3.2 Social Network Explanations in Migration and Environment Research

Explanations for variation in the selection of migrants to a destination community or from an origin community began with relatively simple metaphors of pushing and pulling people out of, and towards, places. Recent evidence suggests that this powerful metaphor may not be as strong an explanation nor help predict the consequences of migration(Curran 2002; Massey 1990). Consequently migration scholars have attempted to better theorize intervening explanations. In this section we briefly trace the development of this literature and show how it has influenced the migration / environment literature to some extent, but not completely. Most importantly, we show how the migration / environment literature could benefit from a more systematic incorporation of the concepts of social networks and embedded relations, which have become so important to migration scholars' research agendas. We then describe current evidence and theory in the human ecology literature on common property resource regimes to show that here, too, important conceptual developments point to the importance of embedded social relations for understanding successful resource management schemes. We conclude this section with a brief discussion of the concept of embedded relations as it is understood by scholars of migrant assimilation processes.

### 3.2.1. Current Explanations for Migration

The push-pull paradigm holds considerable sway in the migration-environment literature – however, it begs the question of why and how people move to particular destinations and not others of equally high ecosystem quality, or why and how people move from particular places of origin and not others of equally poor environmental quality. Much of the explanation lies in understanding the intervening set of social relations organizing people's lives and their relationship to the environment. Some of the most important, new concepts in migration research are social networks and embedded relations. These two concepts have yet to be deployed in analyses of migration and the environment. Lutz and Scherbov argue that they may be very useful concepts, because where people move depends in large part upon social networks (Lutz and Scherbov 2000) and this factor may be equally as important for explaining migrant impacts upon the environment than sheer numbers of migrants, since social networks imply a degree of integration in both places of origin and destination affecting access to resources and resource valuation in both the short- and long-term.

Migration theory now conceptualizes an individual's migration decision as conditional upon their social relationships, rather than merely atomistic. These social relationships can be observed within families, households, communities, markets, and nation-states (Stark 1991). It is from this line of research inquiry that social networks emerged as a critical conceptual and measurement tool for understanding the decision to move (Massey 1990). Social networks in relation to migration are commonly understood as the links between residents in a community of origin and individuals living in another place, or with individuals who have migrated before regardless of their current residence (Massey 1990; Hugo 1998). Social networks increase the propensity of an individual to migrate to a specific destination through three mechanisms: (a) demonstrating feasibility (This includes informing the individual about the possibility of migrating to alternative destinations. The contact with former migrants makes individuals realize that they may be better off in a place other than their current residence (Hugo 1991)); (b) reducing the expected costs and risks. Social networks may reduce "assimilation shock" if immigrants arrive in an environment where others speak their language (Choldin 1973) and where living among other foreigners can easily prevent deportation (Massey 1990)); and, (c) increasing the expected benefits (This happens when contact with previous migrants helps in the job search process, by both reducing the "opportunity costs" of movement and increasing the long-term benefits (Massey and García-España 1987) (Stark 1991; Taylor 1986). Social networks can also help to save and reduce living expenses and provide financial assistance upon arrival.)

One of the most important insights from this research has been that social networks are cumulatively caused. In other words, as migrant experiences multiply the marginal risks of migration decrease and the marginal benefits of migration increase. Cumulative causation theory predicts, therefore, that moves by individuals who previously would have been unlikely to assume the risks of migration and migrate are more likely to migrate when they know more about migration through multiple experiences of people within their social networks. This means that older migrant streams will be composed of a greater diversity of individuals with much more variable human and financial capital at their disposal. Again, the variability in the history of migrant streams as it relates to the characteristics of migrants has not been systematically evaluated in relation to environmental impacts.

The variable impacts of migrant social networks upon the environment are the least theorized or conceptually evolved in the migration-environment literature. The general finding with regard to the relationship between migration and environmental outcomes, especially in developing countries, implicates deteriorating social institutions in both place of origin or destination. In the case of migration impacting deforestation, most findings include important intervening characteristics of poverty, land tenure, export cropping systems and global demand, and agricultural mechanization (Bilsborrow and DeLargy 1990; Bilsborrow and Carr 2000; Adger 1999). The migration literature would suggest that social networks play an important role in migrant abilities to avoid poverty, gain tenure access, participate in export cropping systems, and have financial capital for agricultural mechanization. Further, evidence regarding the impact of return migration suggests the importance of stable social relations diminishing the negative impact of migration upon the environment. By implication return migrants are embedded in a set of social relations which diminishes the impact of migration upon the environment in the destination (Bilsborrow and DeLargy 1990; Brechin, Surapaty, Heydir, and Roflin 1993).

### 3.2.2. Common Property Resource Regimes

The human ecology literature on common property resource regimes, particularly the resilience of such institutions, also points to the importance of understanding social networks and embedded relations. The human relationship to coastal and marine ecosystems has increasingly recognized some form of community property resource regimes predominating (Ostrom 1998; Ostrom 1987; Ostrom, Burger, Field, Norgaard, and Policansky 1999; McCay and Acheson 1987; Pretty and Ward 2001; Begossi 1998; Berkes 1995; Palsson 1998; Hanna 1998). Open access conditions used to be assumed to prevail in marine systems, although there is a growing chorus of dissidents on this point (McCay and Jentoft 1998; McCay and Acheson 1987; Begossi 1998; Berkes 1995; Berkes and Folke 1998; Folke and Berkes 1995; Palsson 1998). In fact, fishing in many locales is often regulated to more or less success by, at the very least, norms (Palsson 1998; National Research Council 1997). And, Pretty and Ward demonstrate the predominance of common property institutions in a wide variety of ecological or environmentally important contexts, including marine and water systems (Pretty and Ward 2001). The growing recognition of the prevalence of common property regimes has generated concern about the institutions' resilience in the face of social change, their dynamics, and the varying role of local and national governance.

Some have argued that common property regimes are necessarily diminished by population growth (either through natural increase or migration) and thereby population growth contributes to environmental deterioration (Panayatou 2000). Others counter that common property resource institutions respond resiliently to the pressures of population, economy, and politics if the tools of management are in the hands of local communities with the support of national governments for enforcement (McCay and Acheson 1987; Ostrom et al. 1999). However, there has to date been little empirical research or theoretical motivation to examine the ways in which migration patterns may differentially impact common property resource institutions across a wide array of ecosystems, including coastal ecosystems.

The human ecology literature finds that there is rarely a condition of open-access and therefore, by implication, Malthusian predictions of population size overwhelming environment resource quality unlikely (Berkes and Folke 1998; Folke and Berkes 1995). The findings in this literature also argue against a solution that involves privatization or socialization (Ostrom 1998; Ostrom 1987; Ostrom, Gardner, and Walker 1994; Ostrom 1990; Ostrom and Walker 1997; Ostrom et al. 1999; Ostrom 1998). More solutions exist than Hardin's (1965) two extremes (privatization or socialization), but predicting environmental outcomes remains inherently complex. To understand and model this complexity we must assess individuals' (or communities') embedded relationships within social networks of relations (McCay and Jentoft 1998; Pretty and Ward 2001). (Portes and Sensenbrenner 1993; Portes 1996)

Scholars of common property resource management make explicit the distinction between the resource and the institution governing the social relations among people that affect the exploitation of the resource. For the former, common-pool resources (Berkes and Folke 1998; Ostrom et al. 1999) are defined by two problems: *exclusion* and *subtractibility*. Or, in other words, *exclusive* use of common-pool resources is very costly and individual exploitation *reduces* resource availability for other users. The latter characteristic is typically described as: people's short-term interests in using the resources may not be in theirs or others' long-term interests. Common-pool resources are found in both marine and terrestrial ecosystems, are both renewable and depletable, and can be either man-made or earth-made (Ostrom et al. 1999; Pretty and Ward 2001).

Common property resource institutions are the formal or informal set of social relations governing people's relationships within a particular ecosystem as they relate to resource exploitation (Ostrom et al. 1999; McCay and Jentoft 1998; Berkes and Folke 1998). This convoluted definition has emerged as a result of a decade of debate about what is a common-pool resource and to what extent and under what conditions local stakeholders manage resources for both current and future sustainable consumption (McCay and Jentoft 1998). Common property relations govern human interaction and resource exploitation through solving the exclusion and subtractibility problems – by restricting access and creating incentives for investment in the resource base. The past decade of research reveals a wide variety of institutions (formal and informal) with varying participation of local and national stakeholders, and varying success (Alcorn and Toledo 1998; Begossi 1998; Joseph 1995; McCay and Jentoft 1998). Property rights are complex because they involve the management of a bundle of citizenship rights – the right to use a resource, the right to exclude others, the rights to manage resource, and the right to sell a resource.

Thus, the "tragedy of the commons" results, not from an inherent failure associated with a particular common pool resource, but from institutional failure to control access to the resource, and to make and enforce internal decisions for collective, long-term use (Berkes and Folke 1998; McCay and Jentoft 1998; Ostrom et al. 1999). Until recently, most policy approaches failed to recognize the formal and informal collective action of people within ecosystems, instead focusing upon individual behavior and attempting to modify individual motives (Pretty and Ward 2001). Despite recent recognition of this failing, policymakers are still faced with the dilemma of accomplishing a balance between individual, community, and national interests of cooperation and conflict (McCay and Jentoft 1998).

Recent research has demonstrated two approaches – one that draws upon the concept of social bonds among people reinforcing normative behavior and expectations (Ostrom et al. 1999), the other focusing upon the "embeddedness" of human action (Giddens 1994; Granovetter and Swedberg 1992). Using the first approach, Pretty and Ward demonstrate that the social bonds among people enhances collective capacity to manage watershed/catchment areas, agricultural irrigation, forests, integrated pest applications, and farmers' research. Katz (2000) demonstrates that communities where social bonds have been disrupted through migration into the community by "outsiders" are less likely to have the capacity to take collective action for long-term natural capital enhancement. Implicit in much of this research is that social bonds are diminished through migration because migration disrupts the social bonds of reciprocity and trust which are required for collective action.

The other approach to understanding common property resource regimes, which is not exclusive of the first, is to emphasize how human action is embedded within social relations. In contrast to the preceding approach, however, embeddedness emphasizes location within historically contingent social, cultural, economic, and political relations, as well as environmental conditions. Varying degrees of embeddedness (disembeddedness being the antithesis) can lead to either positive or negative outcomes for individuals, groups, or the environment. The key difference between the two approaches is not the existence of social bonds (which both approaches highlight as important), or that more social bonds are better (as in the first approach), but that each extractive action carried out by an individual or group has variable meaning to the individual or community. This meaning emerges because resource users are embedded in a variety of social institutions, which are themselves embedded, these institutions can include family, community, market or nation-state. McCay and Jentoft (1998) suggest a working hypothesis oriented toward explaining individual behavior, as opposed to group behavior.

Typically, migration into an area is presumed to weaken the social bonds in a place of destination. This appears to be the case in Guatemala (Katz 2000), Ecuador (Bilsborrow 1992), Mexico (Howard and Homer-Dixon 1996; Izazola, Martinez, and Marquette 1998), the Himilayas (Jodha 1998), and Brazil (Martine 1993) (McIntosh 1993)where movement into a community not only puts added pressure on resource extraction, but diminishes trust, reciprocity, exchange and social bonds (Ostrom et al. 1999). But, this is not always the case. In Indonesia, transmigrants clear half as much forest as spontaneous migrants because, by implication, they have greater collective action capacities through greater embeddedness in political and social institutions at all levels (Bilsborrow 1992). In Ethiopia particular property systems are deployed to attract migrants to communities (Bauer 1987). Further, migration out of a community may serve to embed an origin community more effectively, enhancing capacity for long term resource management (Begossi 1998; Bauer 1987).

There is much variability in the success of common property management regimes (Ostrom et al. 1999). Evaluating success or failure of marine resource management is hampered by the fact that measurement of coastal ecosystem viability or sustainability requires intensive longitudinal observation over wide ranging spatial areas on a scale much larger than that of terrestrially-based ecosystems (Agardy 1997). Furthermore, there has been no systematic analysis of how migration affects common property resource regimes (or vice versa) through embedding processes. To understand how migration might affect embedding processes we briefly review the migrant assimilation literature.

### 3.2.3. Migrant Assimilation and Embedded Relations

The literature on migrate assimilation mirrors that of the human ecology literature about common property resource relations, insofar as embedded relations are implicated in the process and are considered critical intervening conditions for understanding migrant success in places of destination and migrants' continued ties to places of origin. Accounts of migrant assimilation describe variability in social networks not only in terms of the characteristic of the members (as was done in the earlier discussion of migrant selectivity), but in terms of the character of the relationships between a migrant and other members of the migrant community and in relation to external linkages to other social institutions and non-migrants in the place of destination. The measurement of these phenomena occurs at the level of the individual, but can be usefully aggregated to characterize groups or communities.

The literature about migrant assimilation provides greater conceptual clarity about the embedding processes described in the human ecology literature. Firstly, it emphasizes understanding the social ties and resources (social and economic) available to immigrants in places of destination, depending on their relative position within a community – through their normative obligations and normatively induced behaviors within communities in places of destination and through the structure of their instrumental relationships. Secondly, it emphasizes the importance of understanding the type, flow and distribution of resources (social and economic) within and between social groups and social institutions. Thirdly, it emphasizes the importance of the length of temporal vision for affecting social network formation, growth and its impacts, given the way migration occurs, who migrates, and the context of reception in the place of destination. These three aspects suggest a way for understanding the differential impact of migration upon the environment as mediated by the migrants' relationship to common property resource relations in places of destination.

In the first sense, Portes and Sensenbrenner (1993) identify how embedding processes can work to create opportunities and constraints for individual immigrants within immigrant communities. Immigrants who experience discrimination from the native born community, based on phenotypic or cultural differences, and/or whose options for exit out of an immigrant community are blocked (because of limited legal, political, or economic resources), and whose immigrant community in the place of destination has maintained an autonomous cultural repertoire, are likely to experience bounded solidarity. In this case, the social context of arrival in the place of destination enhances normative obligations towards the immigrant community and is expressed in a variety of behaviors (e.g. voluntarism, charity, and preference for co-ethnics in economic transactions). Embedded relations of this type are consummatory in nature, i.e. those making the claims rely on the normatively invoked generosity of others, whose generosity is unlikely to be repaid directly.

A second way that embedded relations may affect immigrant assimilation is through enforceable trust, which relies on instrumental motives and the sanctioning capacity of the community. Limited social and economic opportunities outside of the immigrant enclave, available in-group economic resources, and sanctioning capacity of communities to enforce reciprocity arrangements yield flexibility in economic transactions (fewer formal contracts), privileged access to economic resources (like employment or start-up funds), and reliable expectations that malfeasance will be addressed. In the short-run this can lead to the emergence of economically important ethnic enclaves of entrepreneurial vibrancy (Portes and Sensenbrenner 1993). These types of embedded relations are instrumental in nature, i.e. those making claims are expected to repay those claims to those who agree to the demands. The formation of the social bonds among individuals in this case relies on transfers of assets among group members.

For both expressions of embedded relations, the longer an immigrant community experiences blockage and discrimination the greater the likelihood that cultural and linguistic resources are undermined, depriving the collectivity of the resources necessary to reward or punish members independently. Instead the collectivity has to rely on externally derived sanctions, primarily discrimination, which has the effect of downwardly leveling expectations and aspirations and limiting individual opportunity. Community ostracism by those left behind towards those who leave and abandonment of the community of origin by those who escape is a likely outcome (see examples in Portes and Sensenbrenner 1993). The result is a disembedding of the immigrant enclave from the larger social and economic community and subsequently limited access to opportunities for all members (Stepick 1992)

How does this discussion contribute towards a more complete understanding of the relationship of migration, social networks, embedded relations and the environment? In the next section we focus on four more case studies as illustration of the importance of social networks and embedded relations for understanding the variable impact of migration upon coastal and marine ecosystems.

## 3.3 Migrant Networks, Embedded Relations and Coastal Environmental Outcomes

Migrant networks can serve to increase the heterogeneity of migrant stream composition – through the provision of information and resources. Migrant networks can also affect social relations in places of destination through these processes. Two examples from Macarthur Foundation-funded research are used to illustrate these processes. One shows how social relations of communal property management are maintained among migrant fishers (Ghana) and one shows how variability in embeddedness results in variability in resource exploitation (Guatemala).

Ghanaian canoe fishers are among the most mobile along the West African coast. Documentation shows they have migrated as far north as Mauritania and as far south as the Congo throughout the 20$^{th}$ century (Overaa 2000). Historical accounts point to both push and pull factors contributing to Ghanaian fishers' high degree of mobility. Push factors include population pressure and land shortages. Pull factors have also been compelling explanations and include relatively better fishing grounds, lower input prices, and currency differentials. But in recent decades political conflict and turmoil have limited Ghanaian's access to other nations' fishing grounds. Nevertheless, the striking feature of Ghanaian migrant fishers is the replication of social institutions in places of destination - replete with recognized tribal authority (recognized in places of origin and destination). Migrant fishers must register with local tribal authorities in places of destination before fishing and conform to the same fishing regulations as those in their place of origin. The social networks linking places of origin and destination ensure strong normative and instrumental embedding relations (Overaa 2000). Working through these already established social networks, some interventions, such as co-management of fisheries along the West African coast appear to have been successfully established and may forbode sustainable fisheries in the future.

Quite different embedding processes are evident on the Guatemalan coast (Ross and Mendez 2001). A study of the growing commercialization of fishing in Livingston, Guatemala shows how varying degrees of embeddedness in larger economic and social institutions has marginalized some groups and benefited others, resulting in disparate impacts upon the marine and coastal ecosystem. Q'eqchi, Garifuna, and Ladino migrants moved into the coastal region during the last half of the twentieth century, but their insertion in the local economy is very different with different consequences for resource extraction. Nevertheless they are all involved in fishing activities to greater or lesser extent, an activity that has grown in economic significance for the region in the last five years. Ladinos' greater degree of embeddedness within regional and national socio-economic institutions and networks affords them access to financial and legal resources for investment in large scale fishing or employment opportunities in such operations as wage laborers. The Garifuna have much lower access to these networks of social support and consequently exploit fisheries resources of lower commercial value, but high nutritional value, supplementing their diets. The Q'echi exploit commercial species as well, but in much smaller quantities and only to supplement their income through sales in local markets or to commercial buyers. The impact of these diversified approaches to livelihood strategies is not fully analyzed in the preceding study, but the implication is that diversification results in overexploitation of the fisheries resources to the detriment of the ecosystem (Ross and Mendez 2001).

In both of the preceding cases social networks and their relationship to embedding processes are implicated in the way migration impacts coastal ecosystems. However, the two cases present differing outcomes. In the first, the embedding processes link origin and destination communities and integrate both normative and instrumental social motives to affect individual behavior. There is some indication that these embedding processes will ensure better management and greater resilience for both humans and the ecosystem. In the second, instrumental constraints are most at work and to the benefit of some more than others. More importantly, the lack of normative embedding processes results in significant over exploitation of the fisheries resource. Again, neither of these studies offers definitive results, rather they are suggestive of the importance of concepts already well developed in the migration and human ecology literatures, but which have not been systematically applied within the field of migration and the environment.

A key element for understanding social networks, embedded relations, and migration is to understand the role of reciprocity in this process. In particular, who does the reciprocating and with whom does reciprocation occurs or is expected to occur? Are there reciprocal exchanges of resources (either natural, financial or social) among migrants at a place of destination, between migrants and non-migrants in a place of destination, or between migrants and their origin communities? Does the variability in reciprocal systems of exchanges affect the embeddedness of migrants and non-migrants within their social and ecological community of origin or destination? Asking and answering these questions, which naturally emerge from migration studies, may yield important insights for why migration has both positive and negative outcomes for common property regimes and subsequently inconclusive impacts upon the environment.

The emphasis in the preceding discussion of migration is upon migrant relations in a place of destination. The role of remittances in the migration process mostly addresses migrant relations to places of origin. The literature on the impact of migrant remittances upon development outcomes in places of destination has focused on investment flows. First, remittances are found to increase income and wealth inequality in places of origin (Massey 1988; Massey, Alarcon, Durand, and Gonzalez 1987; Stark, Taylor, and Shlomo Yitzhaki 1986; Stark, Taylor, and Yitzhaki 1986). Second, remittances are found to increase consumption, but not greater investments in productivity (Taylor et al. 1996; Taylor 1999). Third, who remits and their relationship to places of origin affects the character of the investment (Lucas and Stark 1985; Portes, Guarnizo, and Landolt 1999; Curran and Saguy 2001). Finally, varying degrees of embeddedness in places of destination and relative to ties to place of origin affects the level and flow of remittances (Curran and Saguy 2001; Lauby and Stark 1988; Lucas 1997; Durand, Parrado, and Massey 1996).

All of these aspects can be systematically studied in relation to migration impacts upon the environment. Few studies have systematically considered what is known about migration, remittances, and development in relation to environmental outcomes in places of origin and destination. In the next section, evidence from recent studies in the Asia-Pacific region, suggest the importance of migrant remittances for altering the social relations governing coastal environmental resource use and management in places of origin.

An important, but rarely discussed aspect, is the impact of migration upon the environment in places of origin. One possibility is the alleviation of population pressure upon the environmental resource base through out migration. To our knowledge there is no empirical research addressing this possibility. Beyond noting the limited theoretical and nonexistent empirical attention to this possibility, this remainder of this section of the essay turns instead to the impact of migrant remittances upon environmental outcomes in the place of origin.

Measuring resource flows within migrant networks is a critical component for an improved understanding about the impact of migration upon the environment. These are implied within the preceding section, but not explicit. In the next two examples, these resource flows are especially implicated in regard migrant origin communities. Specifically, these two examples highlight the importance of understanding the stock, flow, and meaning of remittances exchanged between migrants and their origin communities. Asking questions about the level and frequency of remittances, as well as who remits and who receives the remittances, adds a layer of complexity to the previous discussions about migrant selectivity and social networks. An example from Vietnam illustrates how the combination of both selectivity and remittances has an important impact upon ecosystem management. An example from Micronesia illustrates how both selectivity and social networks combined with remittance patterns have a detrimental effect upon crab populations and mangrove forest stands.

In a small-scale, longitudinal study of Vietnamese households located in the Red River Delta on the northern coast, Adger et al. (2001) analyze the role of migration and remittance income for affecting livelihood outcomes between 1995 and 2000. With regards to environmental outcomes, they examine household investments in aquaculture. Aquaculture has been shown to have significant negative impacts upon mangrove forests and other fisheries (Naylor et al. 2000; Goldburg and Triplett 1997). Deforestation of mangroves also increases risks of floods and undermines the availability of marine and fish resources for the entire community. They find that over the five-year period of study the households in the study site have significantly increased their reliance on remittance income, expanded aquaculture production, and reduced agricultural production. They conclude that it is the combination of the loss of labor through migration and the remittance income which has shifted local production activities away from labor intensive paddy cultivation to less, labor intensive aquaculture investment, especially for wealthier households (Adger, Kelly, and Locke 2001). They note that the combination of out migration, remittances and shifts in agricultural production have increased income inequality in their study site. Implicit in this explanation is the importance of migrant selectivity particularly how it affects remittances and consequent investments.

In another study of mangrove resources, Naylor et al. (2001), examine migration and remittance patterns in Micronesia. Micronesia is characterized by extensive, but temporary, out migration of the working age population (25-34 years old) to the United States through the terms of agreement of the compact association which gives Micronesians open access to living and working in the U.S. (mostly in Guam and Hawaii) (Naylor, Bonine, Ewel, and Waguk 2001). Several processes seem to be at work in relation to migration, embeddedness, and ecological resources use. Households with migrants in the U.S. are more likely to have a household economy based on subsistence.

These households, in turn, were more likely to use mangrove wood for fuel (twice as much fuel as households linked to the formal economy).

Besides mangrove fuelwood extraction, crab harvesting also takes place. Crabs occupy an important ecological niche within mangrove forest systems, as well as an important economic niche for Micronesians. Crab consumption also appears linked to migration in two ways, yet to be completely explored by Naylor et al. First, until recently the most important reason for crab harvesting was gift export to Micronesians abroad. Importantly they find that households with greater levels of migration and remittance income are more likely engaged in crab harvesting. One way to consider the gift exports is to see them as a way of increasing ties to migrants in order to ensure steady, remittance flows. Secondly, commercial exports of crab have increased eight-fold between 1996 and 2000, matching gift exports. Crabs are being sent to seafood restaurants in Guam, presumably because of prior migration networks. Despite the increases in crab harvest, Naylor et al. also show that crab abundance is declining as the per unit effort has dramatically increased with time. Thus, in two ways migration and remittances may be driving particular types of resource exploitation behaviors on the part of Micronesians. First social networks have increased origin village contacts and opportunities to trade and market ecological resources to places like Guam and, in the future Hawaii. Second, remittance income is so important to the maintenance of Micronesian households that crab gift exports are used to ensure a steady resource flow. These last statements are conjectures and not tested directly by Naylor et al. However, given the literature on migration, social networks and remittances these are not unlikely suppositions.

The preceding two examples demonstrate the importance of considering the role of migrant remittances in relation to ecological outcomes in places of origin. Admittedly, the two examples are relatively sketchy in their detail, but from a migration scholar's perspective they point to further, more systematic inquiries. In the first example the combination of migrant selectivity and the flow of remittances back to a place of origin appear to have had a significant impact on reorganizing agricultural production or resource exploitation – from paddy rice farming to aquaculture. In the second example, migration, social networks and remittances appear to have increased mangrove deforestation and crab harvests, significantly affecting the abundance of crab.

## 4. CONCLUSIONS

This paper shows that there is a timely convergence of ideas and demand for empirical evidence for understanding the relationship between population and coastal ecosystems. Coastal ecosystems are under increasing pressure from population growth as a result of migration, industrial development and ecological resource exploitation. Through a review of the migration and environment literature several themes are identified as being extremely important explanations for particular environmental outcomes. These themes include migrant selectivity, social networks, and remittances. However, the review highlights how these three themes have not been as systematically studied as one might expect from a migration scholar's perspective. Migrant selectivity has not been systematically explored (especially regarding sex, age, and human and financial capital). Neither have migrant social networks been systematically linked to resource use or property relations in either place of origin or destination. Further, very little research has focused upon the relationship between migration and coastal ecosystems.

We are undeniably a long way from fully understanding the connections between human population growth and aspects of consumption such as subsistence demand, commercial production, and international markets. We are even farther from a clear understanding of how the resource use and the indirect impacts on the environment related to these changes in population and consumption affect environmental health and the ability of ecosystems to provide goods and services. On the population side, we have been confounded by how to bridge differences in scale from the household to the national level, and remain constrained in understanding the drivers behind demographic growth, migration, consumer behavior, and other dynamics of human populations. On the environment side, we have yet to understand which measures or environmental parameters serve as the best indicators of environmental health and production potential. And determining causal links between population change and environmental change has been notoriously difficult, requiring decades of ecological study.

Interdisciplinary studies on the relationship between population change, consumption, and environment constitute an important new specialized field. The growth of this research is fueled by more than intellectual curiosity; as resource management agencies, urban and land use planners, government jurisdictions dealing with demographics at the local, provincial and national levels, and the private sector are all demanding better information about how population growth and resource use affect environmental quality and corresponding delivery of goods and services. Though this paper draws evidence from cases in less developed nations, the generalizations may not be limited to such contexts. Demands for better understanding of the population / environment relationship are rising in developed and developing countries simultaneously.

A review of the human ecology literature with regard to coastal and marine ecosystems reveals the prevailing attention to common property resource management. Several important theoretical concepts emerge from this review, dovetailing nicely with recent concepts in migration theory. Particularly important to understanding the success of common property resource management institutions is describing variability in the function of social ties and embedding processes. A similarly focused review of recent developments in the migration literature also highlights the importance of social networks and embedded relations. We put forward six examples from very recent research examining migration and coastal ecosystems that highlight the importance of migrant selectivity, social networks, and remittances for understanding environmental outcomes in places of origin and destination.

These examples stress the importance and variable impact of migrant selectivity upon environmental outcomes (Ecuador and India) and the interaction of migrant social networks through instrumental and normative embedding processes with environmental resource valuation and use (Ghana and Guatemala). And, finally, the role of remittances is shown to affect resource extraction in places of origin through differences in social networks and migrant selectivity (Vietnam and Micronesia).

Measuring migration and its causes is a difficult and complex task, but rigorous attention to it is extremely important. Unlike other demographic events which have distinct beginnings or endings (pregnancy, birth, marriage, death), migration is a fuzzier concept and consequently more difficult to measure (Shryock and Siegel 1976; Smith 1992). This is particularly true in developing country contexts. Analysis of migration will prove increasingly important in understanding the population /environment nexus. Such studies in coastal ecosystems present particular challenges, including the common pool resource nature of many of the elements within a coastal ecosystem. Privatization is difficult to achieve in such common pool resources, except in the most traditional societies (Cordell, 1989) and full government control unlikely, especially where legal and enforcement institutions are weak. Pluralistic arrangements that allow the synergy between community-based management and government control are relatively rare (Bavinck, 2001). Another aspect is the difficulty of assessing environmental resource damage. Unlike deforestation, where the impact of logging is measurable and visible, fish stocks and most other marine resources (like people) are mobile. Finally, measurement of property relations across many elements within an ecosystem is necessary. Understanding the variable roles, responsibilities, and rights of individuals, local communities and nation-states is paramount for understanding the impact of migration upon these very institutions and subsequent environmental outcomes.

In conclusion, tantalizing insights can be drawn from numerous studies around the world that point to the need for a reformulation and reapplication of effort in collecting data and conducting research on the relationship between migration and the environment. Based on the evidence from the literature we suggest that there is a convergence around particularly important concepts, especially embedded relations (as a combination of both migrant selectivity and social networks), to explain the relationship between migration and the environment. The limited scientific attention among migration and environment scholars towards coastal ecosystems, the significance of coastal ecosystems to human livelihoods of all forms, and the growing population along the world's coasts, all necessitate an approach that incorporates the most recent conceptual and methodological approaches in the fields of migration, common property resource management, and coastal ecology.

## REFERENCES

Adger, N. 1999. "Social Vulnerability to Climate Change and Extremes in Coastal Vietnam." *World Development* 27(2):249-269.

Adger, W.N., P.M. Kelly and C. Locke. 2001. "Migration, Remittances, Livelihood Trajectories and Social Resilience." In *Cyber Seminar Series: Population and Environment Dynamics and Relationships in Coastal Areas*, www.populationenvironmentresearch.org.

Alcorn, J. and V. Toledo. 1998. "Resilient Resource Management in Mexico's Forest Ecosystems: The Contribution of Property Rights." Pp. 216-249 in *Linking Social and Ecological Systems*, editors F. Berkes and C. Folke. Cambridge, UK: Cambridge University Press.

Australian Institute of Marine Science. 2001. *Status of Coral Reefs of the World: 2000*. Cape Ferguson, Australia: Australian Institute of Marine Science.

Bauer, D. 1987. "The Dynamics of Communal and Hereditary Land Tenure Among the Tigray of Ethiopia". Pp. 217-230 in *The Question of the Commons*, editors B. McCay and J. Acheson. Tucson, AZ: The University of Arizona Press.

Bavinck, M. 2001. *Conflict and Regulation in the Fisheries of the Coromandel Coast*. New Delhi, India: Sage Publications.

Begossi, A. 1998. "Resilience and Neo-Traditional Populations: The Caicaras (Atlantic Forest) and Caboclos (Amazon, Brazil)." Pp. 129-157 in *Linking Social and Ecological Systems*, editors F. Berkes and C. Folke. Cambridge, UK: Cambridge University Press.

Berkes, F. 1995. "Indigenous Knowledge and Resource Management Systems: A Native Canadian Case Study From James Bay." Pp. 99-109 in *Property Rights in a Social and Ecological Context: Case Studies and Design Applications*, editors S. Hanna and M. Munasinghe. Washington, D.C.: The World Bank.

Berkes, F. and C. Folke. 1998. *Linking Social and Ecological Systems: Management Practices and Social Mechanisms for Building Resilience*. New York: Cambridge University Press.

Bilsborrow, R. and D. Carr. 2000. "Population, Agricultural Land Use, and the Environment in Latin America." Pp. 35-55 in *Tradeoffs or Synergy? Agricultural Intensification, Economic Development and the Environment*, editors D.R. Lee and C.R. Barrett. Wallingford, UK: CAB International.

Bilsborrow, R.E. 1992. *Rural Poverty, Migration and the Environment in Developing Countries: Three Case Studies*. Background Paper for World Development Report ed. Washington, D.C.: The World Bank.

Bilsborrow, R.E. and P.F. DeLargy. 1990. "Land Use, Migration, and Natural Resource Deterioration: The Experience of Guatemala and Sudan." *Population and Development Review (Resources, Environment, and Population: Present Knowledge, Future Options)*. A Supplement to Volume 16125-47.

Bilsborrow, R.E. and H.W.O.O. Ogendo. 1992. "Population-Driven Changes in Land Use in Developing Countries." *Ambio* 21(1):37-45.

Brechin, S., S.C. Surapaty, L. Heydir and E. Roflin. 1993. "Protected Area Deforestation in South Sumatra, Indonesia." Pp. 225-251 in *Population-Environment Dynamics: Ideas and Observations*, editors G. Ness, W. Drake and S. Brechin. Ann Arbor, Michigan: University of Michigan.

Bremner, J., J. Perez and R. Borja. forthcoming. "A Case Study of Migration, Fishing Practices, and the Decline of the Sea Cucumber in the Galapagos Islands." *Ambio*.

Bryant, D., L. Burke, J. McManus and M. Spaulding. 1998. *Reefs at Risk: A Map-Based Indicator of Potential Threats to the World's Coral Reefs*. Washington, D.C.: World Resources Institute.

Burns, T.J., E.L. Kick and B.L. Davis. 1998. *Theorizing and Rethinking Linkages Between Population and Environment*.

Castles, S. and M.J. Miller. 1993. *The Age of Migration: International Population Movements in the Modern World*. New York, NY: Basingstoke MacMillan.

Choldin, H.M. 1973. "Kinship Networks in the Migration Process." *International Migration Review* 10:163-175.

Clark, J. 1996. *Coastal Zone Management Handbook*. New York: J. Lewis Publishers.

Cohen, J. and C. Small. 1998. *Hypsographic Demography: The Distribution of Human Population by Altitude*. Washington, D.C.: National Academy of Sciences.

Conway, D. and P. Lorah. 1995. "Environmental Protection Policies in Caribbean Small Islands: Some St. Lucian Examples." *Caribbean Geography* 6(1):16-27.

Curran, S. and A.C. Saguy. 2001. "Migration and Cultural Change: A Role for Gender and Social Networks." *Journal of International Women's Studies* 2(3):54-77.

Curran, S.R. 2002. "Migration, Social Capital and the Environment: Considering Migrant Selectivity and Networks in Relation to Coastal Ecosystems." *Population and Development Review* (Special Issue: Population and Environment - Methods of Analysis).

Davis, K. 1963. "The Theory of Change and Response in Modern Demographic History." *Population Index* 29:345-366.

DeWalt, B.R. and M.W. Rees. 1994. *The End of Agrarian Reform in Mexico: Past Lessons, Future Prospects*. San Diego, CA: Center for US-Mexican Studies, University of California.

Dietz, T. and E.A. Rosa. 1994. "Rethinking the Environmental Impacts of Population, Affluence and Technology." *Human Ecology Review*.

Durand, J., E.A. Parrado and D.S. Massey. 1996. "Migradollars and Development: A Reconsideration of the Mexican Case." *International Migration Review* 30(2):423-444.

Ewell, P.T. and T.T. Poleman. 1980. *Uxpanapa: Agricultural Development in the Mexican Tropics*. New York: Pergamon Press.

Folke, C. and F. Berkes. 1995. "Mechanisms That Link Property Rights to Ecological Systems." Pp. 121-137 in *Property Rights and the Environment: Social and Ecological Issues*, editors S. Hanna and M. Munasinghe. Washington, D.C. The World Bank.

Giddens, A. 1994. *The Consequences of Modernity*. Stanford, CA: Polity Press.

Goldburg, R. and T. Triplett. 1997. *Murky Waters: Environmental Effects of Aquaculture in the US*. Washington, D.C.: Environmental Defense Fund.

Granovetter, M. and R. Swedberg. 1992. *The Sociology of Economic Life*. Boulder, CO: Westview Press.
Hanna, S.S. 1998. "Managing for Human and Ecological Context in the Maine Soft Shell Clam Fishery." Pp. 190-211 in *Linking Social and Ecological Systems*, editors F. Berkes and C. Folke. Cambridge, UK: Cambridge University Press.
Hardin, G. 1968. "The Tragedy of the Commons." *Science*. Vol. 162 (December 13):1243-1248.
Howard, P. and T. Homer-Dixon 1996. "Environmental scarcity and violent conflict: the case of Chiapas, Mexico." Web page www.library.utoronto.ca/pcs/eps/chiapas/chaipas1.html
Hugo, G. 1991. "Village-Community Ties, Village Norms, and Ethnic and Social Networks: A Review of the Evidence From the Third World." Pp. 186-225 in *Migration Decision Making*, editors G. DeJong and R.W. Gardner. New York: Pergamon Press.
Hugo, G. 1998. "The Demographic Underpinnings of Current and Future International Migration in Asia." *Asian and Pacific Migration Journal* 7(1):1-25.
Izazola, H., C. Martinez and C. Marquette. 1998. "Environmental Perceptions, Social Class and Demographic Change in Mexico City: A Comparative Approach." *Environment-and-Urbanization;* 10(1):107-118.
Jodha, N. 1998. "Reviving the Social System-Ecosystem Links in the Himalayas." Pp. 285-210 in *Linking Social and Ecological Systems*, editors F. Berkes and C. Folke. Cambridge, UK: Cambridge University Press.
Jolly, C. and B.B. Torrey (eds.) 1993. *Population and Land Use in Developing Countries*. Washington, D.C.: National Academy Press.
Joseph, S. 1995. "Coping With Seasonality: Implications for Fisheries Community Development." *Social-Action* 45(4):431-442.
Katz, E. 2000. "Social Capital and Natural Capital: A Comparative Analysis of Land Tenure and Natural Resource Management in Guatemala." *Land Economics* 76(1):114-132.
Lauby, J. and O. Stark. 1988. "Individual Migration As a Family Strategy: Young Women in the Philippines." *Population Studies* 42(3):476-486.
Lonergan, S. 1998. "The role of environmental degradation in population displacement." Web document ecsp.si.edu/pdf/Report4b.pdf.
Lucas, R. 1997. "Internal Migration in Developing Countries". Pp. 722-798 in *Handbook of Population and Family Economics*, vol. 1B, editors M.R. Rosenzweig and O. Stark. New York: Elsevier.
Lucas, R.E. and O. Stark. 1985. "Motivations to Remit: Evidence From Botswana." *Journal of Political Economy* 93(5):901-918.
Lutz, W. and S. Scherbov. 2000. *Quantifying Vicious Circle Dynamics: The PEDA Model for Population, Environment, Development and Agriculture in African Countries*.
Marquette, C. and R. Bilsborrow. 1997. "Population and Environment in Developing Countries: A Select Review of Approaches and Methods." Pp. 611-638 in *The Population, Environment, Security Equation*, vol. 13, editors B. Baudot and W. Moomaw. New York: MacMillan.
Marquette, C. and R. Bilsborrow. 1999. "Population and environment in developing countries: Literature survey and research bibliography," Web page www.cnie.org/pop/marquette/Marque2a.htm.
Martine, G. 1993. "The Phases of Agricultural Modernization in Brazil." Pp. 167-186 in *Population-Environment Dynamics: Ideas and Observations*, editors G. Ness, W. Drake and S. Brechin. Ann Arbor, Michigan: University of Michigan.
Massey, D. 1990. "Social Structure, Household Strategies, and the Cumulative Causation of Migration." *Population Index* 56(1):3-26.

Massey, D., R. Alarcon, J. Durand and H. Gonzalez. 1987. *Return to Azatlán: The Social Process of International Migration From Western Mexico*. Berkeley, CA: University of California.

Massey, D. and F. García-España. 1987. "The Social Process of Migration." *Science* 237:733-738.

Massey, D.S. 1988. "Economic Development and International Migration in Comparative Perspective." *Population and Development Review* 14(3):383-413.

Massey, D.S. 1990. "The Social and Economic Origins of Immigration." *ANNALS, AAPSS* 510:60-72.

McCay, B. and J.M. Acheson. 1987. *The Question of the Commons : the Culture and Ecology of Communal Resources* . Tucson: University of Arizona Press.

McCay, B. and S. Jentoft. 1998. "Market or Community Failure? Critical Perspectives on Common Property Research." *Human Organization* 57(1):21-29.

McIntosh, A. 1993. "Population Policy and Environmental Impacts in Rural Zimbabwe." Pp. 133-151 in *Population-Environment Dynamics: Ideas and Observations*, editors G. Ness, W. Drake and S. Brechin. Ann Arbor, Michigan: University of Michigan.

National Research Council. 1997. *Environmentally Significant Consumption: Research Directions*. Washington, D.C.: National Academy of Sciences.

Naylor, R., K. Bonine, K. Ewel and E. Waguk. 2001. "Migration, Markets and Mangrove Resource Use on Kosrae, Federated States of Micronesia." In *Cyber Seminar Series: Population and Environment Dynamics and Relationships in Coastal Areas*. www.populationenvironmentresearch.org/.

Naylor, R.L., R.J. Goldburg, J.H. Primavera, N. Kautsky, M.C.M. Beveridge, J. Clay, C. Folke, J. Lubchenco, H. Mooney and M. Troell. 2000. "Effect of Aquaculture on World Fish Supplies." *Nature* 405:1017-1024.

Noronha, L., S. Sreekesh, L. Qureshy, S. Kazi, S. Nairy and A. Siqueira. forthcoming. "Understanding Ecosystem Change: Reflections From a Study in a Tourist Destination." *Ambio*.

Ostrom, E. and J.M. Walker. 1997. Pp. 35-72 in *Perspectives on Public Choice: a Handbook*, editor D. Mueller. New York: Cambridge University Press.

Ostrom, E. 1987. "Institutional Arrangements for Resolving the Commons Dilemma: Some Contending Approaches." Pp. 250-265 in *The Question of the Commons*, editors B. McCay and J. Acheson. Tucson, AZ: The University of Arizona Press.

Ostrom, E. 1990. *Governing the Commons: the Evolution of Institutions for Collective Action*. Cambridge: Cambridge University Press.

Ostrom, E. 1998. "A Behavioral Approach to the Rational Choice Theory of Collective Action." *American Political Science Review* 92(1):1-22.

Ostrom, E., J. Burger, C. Field, R.B. Norgaard and D. Policansky. 1999. "Revisiting the Commons: Local Lessons, Global Challenges." *Science* 284 (April 9):278-282.

Ostrom, E. R. Gardner and J. Walker. 1994. *Rules, Games, and Common-Pool Resources* . Ann Arbor: University of Michigan Press.

Overaa, R. 2000. Institutions, Mobility and Resilience in the Fisheries of Moree and Its Diaspora. *Final Report Prepared for Project on Small Scale Fisheries of Moree and Its Diaspora*. Christian Michilsen Institute, Bergen, Norway and the John D. and Catherine T. MacArthur Foundation, Chicago, Illinois.

Palsson, G. 1998. "Learning by Fishing: Practical Engagement and Environmental Concerns." Pp. 48-66 in *Linking Social and Ecological Systems*, editors F. Berkes and C. Folke. Cambridge, UK: Cambridge University Press.

Panayatou, T. 2000. *Population and Environment*. Working Paper No. 54 ed. Cambridge, Massachusetts: Center for International Development.

Portes, A. 1996. "Transnational Communities: Their Emergence and Significance in the Contemporary World-System." Pp. 151-168 in *Latin America In The World-Economy*, editors R.P. Korzeniewicz and W.C. Smith. Westport, Connecticut, London: Greenwood Press.

Portes, A., L.E. Guarnizo and P. Landolt. 1999. "The Study of Transnationalism: Pitfalls and Promise of an Emergent Research Field." *Ethnic and Racial Studies* 22(2):217-237.

Portes, A. and J. Sensenbrenner. 1993. "Embeddedness and Immigration: Notes on the Social Determinants of Economic Action." *American Journal of Sociology* 98(6):1320-1350.

Pretty, J. and H. Ward. 2001. "Social Capital and the Environment." *World Development* 29(2):209-227.

Princen, T. 1999. "Consumption and Environment: Some Conceptual Issues." *Ecological Economics* 31:347-363.

Ross, E. and C.M. Mendez. 2001. "On the Shores of the World Economy: Commercial Fishing, Coastal Resource Scarcity and Livelihood Strategies in Livingston, Guatemala." In *Population, Consumption and Environment Initiative*, John D. and Catherine T. MacArthur Foundation.

Ruddle, K., W.B. Morgan and J.R. Pfafflin. 1988.Chur, Switzerland: Harwood Publishers.

Shryock, H. and J.S. Siegel. 1976. *The Methods and Materials of Demography*. New York: Academic Press.

Silliman, J. and Y. King. (eds.) 1999. *Dangerous Intersections*. Cambridge MA: South End Press.

Smith, D. 1992. "Migration in Population Analysis." Pp. 283-301 in *Formal Demography*, editor D. Smith. New York: Plenum Press.

Stark, O. 1991. *The Migration of Labor*. Cambridge, MA: Basil Blackwell.

Stark, O., E. Taylor and S. Yitzhaki. 1986. "Remittances and Inequality." *Economic Journal* 96:722-741.

Stark, O., J.E. Taylor and S. Yitzhaki. 1986. Migration, Remittances and Inequality: A Sensitivity Analysis Using the Extended Gini Index. *Discussion Paper No. 23*. Migration and Development Program, Harvard University, Cambridge, Massachusetts.

Stepick, A. 1992. "The Refugees Nobody Wants: Haitians in Miami." Pp. 57-82 in *Miami Now*, editors G.J. Grenier and A. Stepick. Gainesville, FL: University of Florida Press.

Sunderlin, W. and I.A.P. Resosudarmo. 1999. "The Effect of Population and Migration on Forest Cover in Indonesia." *Journal-of-Environment-and-Development* 8(2):152-169.

Taylor, E., J. Arango, G. Hugo, A. Kouaouci, D.S. Massey and A. Pellegrino. 1996. "Current Items: International Migration and National Development." *Population Index* 62(2):181-212.

Taylor, J.E. 1986. "Differential Migration, Networks, Information and Risk." *Research in Human Capital and Development: Migration, Human Capital and Development* 4:147-171.

Taylor, J.E. 1999. "The New Economics of Labour Migration and the Role of Remittances in the Migration Process." *International-Migration* 37(1):63-88.

Thomas-Hope, E. 1999. "Return Migration to Jamaica and Its Development Potential." *International-Migration* 37(1):183-207.

World Bank. 1993. *Noordwijk Guidelines for Integrated Coastal Management.* Washington, DC : The World Bank.

World Resources Institute. 2001. *World Resources 2000-2001: People and Ecosystems - The Fraying of the Web of Life*. Washington, D.C.: World Resources Institute.

Zaba, B. and J. Clarke. 1994. *Environment and Population Change*. Belgium: International Union for the Scientific Study of Population.

# DESERTIFICATION AND MIGRATION

*A Political Ecology of Environmental Migration in West Africa*

Thomas Hammer
*University of Fribourg, Switzerland*

## 1. ENVIRONMENTAL MIGRATION IN WEST AFRICA AND THE SAHEL REGION

*"While Africans constitute only 12 per cent of the global population, around 28 per cent (i.e. 3.2 million) of the world's 11.5 million refugees and just under 50 per cent (i.e. 9.5 million) of the world's internally displaced persons are to be found in Africa. The total number of displaced Africans thus stands in the region of 12.7 million. To this number can be added more than two million returnees, who, according to UNHCR, have not been able to reintegrate in their country of origin and who continue to need some form of international protection and assistance."* (Crisp 2000: 2)

These numbers do not include the so-called "economic refugees" and the "environmental refugees". Unfortunately, it can be assumed that Africa will also maintain its world records in these categories of refugees, at least concerning the population size (Richter 1998). For decades, Africa has been the "continent of refugees" (Nuscheler 1988). Africa's refugee drama conceals not only a wide set of conflicts, disasters and human suffering, but also a wider displacement problem: the migration phenomenon. The mostly short-term refugee phenomenon offers spectacular stuff for the media, but the normally mid- and long-term migration problem is much more important for the societies involved. In general, the short-term events constitute only the obvious tip of the iceberg: refugee movements are often an indication of larger population displacements. This is especially valid for the environmentally motivated displacements in West Africa and the Sahel region, to which the following reflections refer.

I will therefore demonstrate by the example of environmental migration in West Africa,

- that the refugee problem must be seen as part of a larger phenomenon – the migration and desertification syndromes – if an adequate and holistic solution is to be developed (Roithner 1998),
- that (human induced) environmental change, socio-economic migration and environmental migration are intensely related, and



*J.D. Unruh, M.S. Krol and N. Kliot (eds.),*
*Environmental Change and Its Implications for Population Migration,* 231–246.

- that political factors are – at least indirectly – of basic significance for economically and for ecologically motivated migration.

In reference to the above mentioned three-part distinction between politically, economically and environmentally motivated migration, there are three main flows of refugees in West Africa: migration motivated by politics or disorder, mainly in coastal states (e.g. Sierra Leone, Liberia, Guinea); economically motivated migration from within the country to large cities, coastal areas and coastal agglomerations; and ecologically motivated migration from the rural and especially from the Sahel region to cities and to coastal states (Hammer 1999a, 255-271). So generally, there is a large migration movement to the (coastal) agglomerations, in and to the coastal agglomerations and to the coastal states, especially from the rural Sahel regions southwards and to the coastal zones.

*Table 1: Environmental events in West Africa with more than one hundred thousand refugees.*

| Year | State | Refugees (number) | Environmental disaster |
|---|---|---|---|
| 1973 | Burkina Faso | 300,000 | Drought |
| 1973 | Mali | 300,000 | Drought |
| 1973 | Mauritania | 100,000 | Drought |
| 1973 | Niger | 110,000 | Drought |
| 1982 | Benin | 500,000 | Flood |
| 1985 | Burkina Faso | 222,000 | Drought |
| 1985 | Mali | 200,000 | Drought |
| 1985 | Niger | 1,000,000 | Drought |
| 1985 | Benin | 250,000 | Flood |
| 1986 | Mauritania | 220,000 | Drought |
| 1988 | Nigeria | 200,000 | Flood |
| 1988 | Benin | 100,000 | Flood |
| 1994 | Nigeria | 400,000 | Flood |
| 1995 | Ghana | 200,000 | Flood |
| 1996 | Sierra Leone | 200,000 | Flood |
| 1998 | Nigeria | 100,000 | Flood |
| 1999 | Ghana | 144,000 | Flood |

Sources: Richter 2000: 15-16, see also Anhuf 1991: 104-106, Klaus 1986.

After the recent repatriation of displaced Tuaregs, the Sahel region, as the economically and environmentally most marginal region, has – in comparison to other West African regions – no significant refugee population in the official sense (Crisp 2000). There is altogether a great economic migration and a distinct, environmentally motivated displacement. This makes the Sahel region ideal for research, to demonstrate the links between human-induced environmental change and economic and environmental migration.

While there are existing data and estimations for greater environmental disasters, it is relatively difficult to estimate environmentally motivated migration. Table 1 shows an overview of the number of refugees after greater environmental disasters with more than a hundred thousand refugees, based on the researches of Richter (1998, 2000). The table is incomplete and does not contain the relatively frequent events involving thousands and tens of thousands of refugees. In general, the coastal countries are increasingly affected being by floods, and the Sahel region was afflicted by the two extensive droughts, in 1972-73 and 1984-85.

The absence of an environmental disaster in the Sahel region since the mid-1980s might give the impression that no serious environmental problem exists. This is not the case. The problem is less obvious. The phenomenon of creeping desertification (UNEP 1997) has its counterpart: creeping migration.

*"Desertification ... has irreparably damaged millions of hectares of once productive land and made refugees out of millions of sub-Saharan African farmers. Migration is the signal that land degradation has reached its sorry end.*" (Jacobson 1988: 6 in Black 2001: 4)

It is, however, very difficult to quantify desertification-induced migration, because the relationship between desertification and migration is often *indirect* and in addition, the definition of desertification allows a wide range of interpretation:

*"Desertification is a man-made process of degradation of the biological potentials in the semi-arid tropics and sub-tropics tending to desert-like conditions".* (e.g. Definition of the UNCOD: United Nations Conference on Desertification 1977, UNEP 1991: 1, Hammer 2000a: 9, see also Aménagement et Nature 1998).

Altogether, though, there are many indications of *direct or indirect* relations between the environmental conditions, desertification and the decision to migrate for a large part of the rural migration in the Sahel region, at least since the mid 1970s (Hammer 1999a: 256-268).

- The relationship is *direct*, when decisions to migrate are the consequence of short- term disasters such as droughts or extensive man-made environmental degradations, as on the Yatenga Plateau in Burkina Faso, in the Senegal river valley, or in the Senegal peanut basin.

- The link is *indirect* in regions troubled with slight or gradual desertification, where other motivations, often economic reasons, have an effect (Pourtier 1992): *When the environmental situation deteriorates slowly, and agrarian techniques fail to adapt to changing conditions, few new jobs are created in non-agricultural sectors, and many people are forced to leave the rural areas.*

The great majority of the more than two million emigrants from Burkina Faso and Mali – more than a fifth of the current population – were motivated by direct or indirect environmental degradation (inland migration not included, Hammer 1999a: 259). It seems very likely that hundreds of thousands of people from rural Sahel regions are displaced every year as a consequence of environmental change and desertification (Club du Sahel 1995, Hammer 1999a: 153-170, 255-268).

## 2. THE POLITICAL ECOLOGY OF DESERTIFICATION AND ENVIRONMENTAL MIGRATION

What are the so-called "root" or "underlying" causes of the related phenomenon of desertification and environmental migration? – To answer this question, I choose the comparatively new approach of political ecology in geographical research on environment and development (Blaikie 1999, Bryant 1999, Bryant & Bailey 1997). This approach is based on the thesis that environmental problems are created by individuals, societies and groups of action, and thus by actions or political decisions in a wider sense, or by non-decisions, non-actions and omissions (Geist 1999, Krings 2000, Mayer-Tasch 1999).

This approach assumes that only a multitude of decisions, omissions and processes on several levels (e.g. individual, local, national, and international) can create greater environmental problems (Kohout 1999). The most difficult challenge is to understand and emphasise the central categories of relevant actions contributing to environmental problems (Krings 1994). Adequate and fundamental solutions can only be developed if the societal and, in particular, the political mechanisms of environmental degradation and their interdependence with geo-systemic relations are well known (Kreibich and Simonis 2000).

First, lets get back to history: in the pre-colonial systems, environmentally motivated migration was an important element of use of resources. After short and intensive use of resources, activities were moved to allow the regeneration of resources (e.g. nomadism, mobile cattle-breeding, shifting cultivation). The traditional systems were characterised by a "cyclic sustainability": short use of resources alternated with longer regeneration phases. This "cyclic sustainability" is not compatible with the current understanding of sustainability, in respect to the central European forest cultivation and management with the three elements a) steady in use of resources, b) continuous regeneration, and c) stability in the resource stock. Although in time the majority of the population settled down and the population grew steadily, "cyclic sustainability" was not adapted to the requirements of continuous sustainability. So the degradation of the systems of use of resources can be interpreted as a consequence of the deficient adaptation process. Interest now focuses on why adaptation was either imperfect or non-existent.

After twelve years of human research activity in desertification and migration in the Sahel region and on the basis of case studies (Hammer 1992 and 1997a) and fundamental research (Hammer 1999a and c), I have identified the following political and decisional fields of high importance (Hammer 2001, Vernet 1994).

## 2.1 The Colonial Policy.

The colonial policy promoted the introduction and extension of innovations, which heightened the pressure on resources. Among these innovations were rent production, such as peanuts, cotton, and sugarcane, mechanical treatment of the soil, the spreading of market- and rent-oriented values, methods, techniques, and forms of land utilisation, the displacement of mobile, cyclic forms of soil utilisation, and the settlement of the migrant population. Together, these innovations have contributed to the weakening of the environmental dimension in the local production systems. Subsistence production continued to be the central element for survival, but new, more intensive uses of resources were added with systems which were not adapted to the local environmental conditions. In view of these facts, the degradation of the environment must be seen as a logical consequence of colonial policy, which placed priority on the production of cheap resources for French industries.

## 2.2 The development policy since political independence in 1960.

After the political independence of the Sahel states in 1960, development policy took over the colonial policy with almost the same methods and instruments. As early as 1962, the agrarian expert René Dumont was motivated to write the classic L'Afrique noire est mal partie. The motivation to carry on the colonial policy had various aspects: the building-up and financing of the independent administrations and states from the profits obtained with agrarian resources, the lack of other financial resources, and the neo-colonial elite in the cities with a personal interest in maintaining their special relationships with European countries (Giri 1993).

## 2.3 The Agrarian and Rural Policies.

Within development policy, the agrarian and the rural policies are of basic significance. The agrarian policy is further oriented by the market, rent and export production, the cities, and the supply of the large, urban lower classes. A policy in favour of the rural areas or to ameliorate and promote subsistence production hardly exists. On the contrary, subsistence production often subsidises market production and is damaged by the interdependence with other sectors (Hammer 1992, 146-260). As subsistence production is the basis for survival for the greater part of the rural population, pressure on rural resources is increasing.

## 2.4 The Economic Policy in General.

The economic policy in general is oriented primarily by the cities, industries, and the services, so that rural areas, where the great majority of the population lives, are usually almost totally neglected. The industrial sector is weak and undeveloped, the industrialisation of import substitution has largely failed, and the export structures are dominated by raw materials and semi-finished products. Economic globalisation does not pay out/off for the Sahel region, because preconditions, such as developed communication systems and a relatively high educational level, do not exist. Transfer contributions flow from the rural to the urban areas and not the other way around. So it is not surprising that the agrarian sector – with agriculture and cattle-breeding as the most significant resource users in the rural areas – could not be developed.

## 2.5 The Policy of Land and Resource Use.

On the contrary, the policy in general, and more specifically the policy on land and use of resources, has intensified the pressure on natural resources. Until the last decades, there were widely accepted agreements, regulations, and rights of users, which integrated the ecological dimension in accordance with the traditional understanding of long-term strategies of survival. Now, however, there are frequently no accepted regulations for users at all. Correspondingly, resources like water, vegetation, pastures, forests, and woods are over-exploited, because of precisely the restricting regulation policies of local communities. Traditional regulations and agreements of users have been cancelled, so that widespread legal insecurities and user conflicts predominate, for example with the co-existence of local, Islamic, and public laws.

## 2.6 The Policies of Colonisation and Migration.

The policies of the opening up of land for development, colonisation and migration have led to the exploitation of new areas and to an intensification of the use of resources, as, for example, the great agrarian colonisation projects, such as the development of the Senegal river valley (Crousse et al. 1991), the so-called Senegal peanut basin, and the Niger inland delta in Mali. However, many small projects also have a direct or indirect impact on resources at the local level.

## 2.7 The Population, Health and Social Policies.

The health policy, in combination with the population and social policies, plays a very important part. The policy of health made great progresses – e.g. basic vaccinations, decrease in child mortality, and increase in life expectancy – but no population and social policies have followed, so that in the last decades populations have grown more than ever before. Children and the family system continue to represent a social precaution against emergency (e.g. illness, accident, pregnancy, age, or death). Social control of reproduction has decreased or been abandoned, and as a consequence of increasing individualisation, decreasing solidarity, and increasing economical insecurities, families must have more children to secure their survival in emergency situations. In view of this situation, the partial modernisation of health, population and social policy has led to a greater number of children per family and to increased population growth. The world-wide highest rate of population growth must be seen as a product of partial innovations.

## 2.8 The Environmental Policy.

The environmental policy also has undesired impacts. The delimitation of protected areas (forêts classées) has abolished the user agreements and regulations of the local societies. This has contributed to uncontrolled, arbitrary use of previously controlled resources. Meanwhile, many protected areas are becoming more and more degraded (Boubacar 1995). Even more important than the undesired impacts are the omissions in environmental policy, and the lack of integration of environmental considerations in the other policy fields.

This catalogue of policy fields, relevant for desertification and migration, is incomplete. For example, energy policy has not been able to find a substitute for primary energy (wood) or to promote its production sufficiently. All in all, between the different policy fields there exists a number of contradictions and undesired impacts. The partial modernisation, heterogeneous structures and incompatibilities between traditional, informal, and modern sectors and ways of production have created the ecological threats in the Sahel region (Hammer 1999c). In view of this fact, the large, creeping environmental migration must be interpreted as an expression of extensive crises, conflicts, and incoherent policies (Aumüller 1993).

# 3. PERSPECTIVES CONSIDERING GEO-HISTORICAL AND ENVIRONMENTAL CONSTRAINTS

In order to demonstrate the significance of and perspectives between environmental changes and migrations, longer geo-historical contexts are relevant. Four cycles can be distinguished (Afrique Contemporaine 1992).

## 3.1 Long-Term Cycles: Aridification, Desertisation, and Migration

The history of the human race can be seen as a reflection of climate evolution (Hsü 2000). The genesis of the old, very advanced civilisations, human settlements in the river valleys, and the great migrations are linked to general climate and environmental changes, with a general drying of the arid and semi-arid zones of the old world. Thus, the Sahara has been drying for millennia, which results in expansion of the Sahara desert. For example, the drying of the Sahara and the Arab peninsula led to immigration into the Nile valley, and intensified the pressure for the genesis of the old Egyptian civilisations. The same is valid for the genesis of agrarian techniques thousands of years ago in the Sahel region.

Historians of climate evolution show that the Sahel region is drying in episodes, but has been doing so continuously over the centuries. For example, the average, annual precipitation in the north of Burkina Faso (around 16$^{th}$ parallel and 0 degree of longitude) has decreased from 800 mm to half of this since the year 950 (Reichelt et al. 1992: 74). The drying episodes have provoked greater migration movements. The continued drying process and a long-term threat to the Sahel societies can be assumed.

In view of this fact, it is doubtful if the IPCC-Scenario is valid: "Parts of the Sahel could experience rainfall increases of as much as 15% over the 1961-90 average" (IPCC 2001, Box 2-3, Climate Scenarios). This scenario involves a marked increase in temperature and evapotranspiration, which might strengthen the aridification and desertisation. In each case, environmentally motivated migration will probably accelerate. The major problem will be the lack of unsettled or sparsely settled areas available.

## 3.2 Mid-Term Cycles: Desertification and Migration.

Historical references also demonstrate the alternation of phases with higher and lower precipitation, for example over decades. Consequently, good and bad phases can be expected in the next decades. Bad phases with little precipitation might lead to an increase in environmentally motivated migration, as described above: the relationship between lower precipitation, desertification and migration is a circular and cumulative process. The main problem is the absence of local adaptation strategies, and the impracticability of reintroducing traditional strategies.

## 3.3 Short-Term Cycles: Droughts and Floods, Environmentally Motivated Flight and Migrations.

Short-term events like droughts and floods generally have two impacts: first, the flight, and second, the greater migration. Even if the negative impacts can be limited by special programs, as they have been during the droughts of the last years, wider environmental migrations must be expected, for example as the consequence of reduced capacity in acting to secure the survival of the local population.

## 3.4 Seasonal Cycles: Seasonal Migration.

Seasonal migration represents another, very distinct type of migration. Arising parallel to colonisation, settlement and increase in population, seasonal migration involves many young, mainly masculine people, often up to half or three quarters of the young men of a region. The long dry season and the short rain season with a normal precipitation variability of 20 to 30% determine this form of migration to a great extent. If there is less precipitation, and consequently lower crop yields, the pressure to migrate and obtain supplementary income increases.

## 3.5 Fundamental Elements for a Solution.

Accordingly, a number of phenomena are concealed behind the notion of environmentally motivated migration. Only extreme short-time events produce environmental refugees. Creeping desertification remains the main reason for migration in the Sahel region. As a result of emigration, innovative potentials and know-how is exported from rural areas as well. Immigration into other regions, often into cities or agglomerations, causes the ecological stresses to increase in the host region. But there is no return to lost traditional uses of resources. Under current conditions, these are no longer ecologically compatible. Fundamentally new approaches, integrating the genesis of environmental migration and desertification, are indispensable if basic solutions are to be found.

## 4. POSSIBLE SOLUTIONS

Two findings are central for solutions. First, there are many indications that environmental migration will grow stronger and create more challenges in rural and urban areas. This conclusion results from the arguments proposed, and from what is known of the so-called "tendency scenario" from the studies by the OECD, the Club du Sahel, and Jacques Giri (Giri 1989, OCDE et al. 1988). Second, it seems to be indispensable to fight desertification in a fundamental way, if the decline of agriculture and a greater emptying of rural areas is to be prevented. In this respect, a sound policy requires a *double strategy*: first, a fight against the negative impacts of environmental migration in both the emigration *and* the immigration areas, and second, a fight against desertification and environmental degradation.

The first aspect, the fight against the negative impacts of environmental migration, requires, according to different studies, a wide spectrum of measures. For the emigration areas, it is important to retain a proportion of innovative young people, for example in innovative projects, which allow local development. On the other hand, environmental rehabilitation and the creation of jobs is most important for immigration areas. The large and well-based *West African Long-Term Prospective Study* (WALTPS) – which involved many researchers from both the North and the South – demonstrates in an impressive way the necessity to view and attack the Sahel problems in a wider territory, the West African area (Cour & Snrech eds. 1998). The following measures are proposed: the opening up of frontiers inside West Africa, the reinforcement of economic and political regional integration, and the promotion of smaller and middle-sized cities in coastal zones to allow greater emigration out of the Sahel region.

For the second aspect, the fundamental fight against desertification, widely supported proposals also exist. Among them are the *United Nations Convention to Combat Desertification (UN-CCD)*, which was strongly influenced by West African experiences (UNESCO 1994), but also by existing regional, sub-regional and national programs to fight against desertification (CILSS & CEDEAO 1999). In these approaches, the fight against desertification and migration is interpreted as a transversal task including different policy domains.

For both approaches, regional and sub-regional organisations are very important (Alao 2000), as desertification and environmental migration are not national, but regional phenomena. With supra-national collaboration, co-ordination and solidarity, it is easier to implement conflict management and create special funds (Hammer 2000b). With prevention and instruments for rapid intervention, the vulnerability of the societies can be reduced, more stability created, and environmental migration and its negative impacts possibly prevented.

For a specific political ecology to fight against environmental migration in the Sahel region, the following *main principles* can be established (Hammer 1999b and 2001):

- the extensive promotion of rural areas and their integration with the urban areas by equal relationships of exchange, in order to create the socio-economic basis for the adaptation of methods and techniques, and to allow more attention to the protection of resources,
- the abandonment of development according to the destructive model of European modernisation, which is neither possible nor ecologically (and socio-culturally) sensible, and
- the development of approaches adapted to specific local conditions, and which take into consideration the different spheres of influence in policy, economy, society, and environment.

The intention behind these principles is to assure a favourable political climate in favour of a fundamental combat against desertification. Based on the previous experience, the following *transversal principles* are important (Hammer 1998):

- the development and application of specific, locally adapted techniques, methods, approaches, and forms of land use (postulate of the development of local know-how instead of technical transfer),
- the strengthening of local communities, the promotion of training and continued education, support for self-organising and direct responsibility (postulate of capacity building and promotion of local human resources), and
- the decentralisation of public services, the de-concentration of administration functions, the appropriation of self-administrated municipalities, and support for the bottom-up creation of municipalities, in order to encourage partnerships between public and private organisations in rural areas and local projects (postulate of populism and

of the hierarchic delegation of public tasks according to the principle of subsidisation).

These three principles should enable and qualify local actors and decision makers to take responsibility, to launch their own initiatives, and to develop and implement locally adapted solutions (ENDA GRAF Sahel 1993). For the different fields of policy, the following challenges must be met:

- *agrarian policy:* producer-oriented instead of consumer-oriented policy; regulation of imports relating to endogenous potential of production; promotion and improvement of local and national subsistence production; promotion of decentralised manufacture of agrarian raw material;
- *economic policy:* upgrading of the agrarian policy; home market-oriented support of small and medium enterprises; formalisation of informal activities to allow more investment and innovations; intensified promotion of the relatively advanced sub-regional (Sahel states) and regional (West Africa) economic and political integration (Hammer 2000b);
- *policy of land and resource utilisation:* promotion of resource-oriented, local land utilisation planning which includes all user groups and user claims; transfer of responsibility for use of land and resources to the local communities and political authorities; improvement of the regulations, rights and agreements concerning use of land and resources; promotion of discursive approaches in the regulation of user conflicts in the sense of the "palaver"; increased inclusion and strengthening of local strategies in conflict regulation (Hammer 1997b, Rünger 1999, Stamm 1998);
- *health, population, and social policies:* specific promotion of human resources in rural areas; upgrading and strengthening of communities and several-generation families as a guarantee for social and economic security in rural areas; expansion of sensitisation and educational work in health, population and family questions; combination of health, population and social policy, for example in family planning and gender equality (Benedick 2000);
- *environmental policy:* specific promotion of the cultivation of the resource "wood" as the most important energy source in rural areas, and often in urban areas, too; promotion of the substitution of wood by other (renewable) energies in urban areas (which is responsible for much of the deforestation in rural areas); extension of the environmental information and sensitisation programmes; transfer of responsibility concerning use of resources to local communities;

– *international policy and co-operation:* increased support for the *International Convention to Combat Desertification* (UNESCO 1994); harmonising of the co-operation and the international relationships on the basis of the convention; creation of a coherent international policy towards the Sahel region; direct support of local actor groups in the fight against desertification and co-operation in general; supplementary reduction of debt for the Sahel states.

Coherent and complementary measures and approaches are necessary (Tersiguel & Becker 1997). Even so, there is no guarantee that the Sahel region will be a "saving shore" for later generations, as the etymological derivation of the notion *Sahel* means (arab. *as-sahil* = shore, coast). The aridification of the Sahel region, the potential climate change, the decreasing (fossil) groundwater reserves, the high population growth and the marginal position in the economic and political world system and new world order are indications for an extremely uncertain future (Cour & Snrech 1998, Giri 1989).

## 5. CONCLUSION

Important factors and interactions of environmental migration can be identified, and solutions have been proposed, as for example with the *Convention to Combat Desertification*, ratified by more than 175 states, and the regional, sub-regional and national programmes for desertification control. Such propositions promise a holistic and integrated approach. But they are limited to general intentions and do not penetrate into the various relevant political fields. There are many indications of the need for a coherent policy on the different political levels, between the levels, and between the different relevant political fields, in order to attack the problem of environmentally motivated migration and desertification. The combat of migration and desertification has become a transversal challenge for the relevant political fields and levels, covering agrarian policy, land and resource use policy, environmental policy and international co-operation. A coherent multi-layered policy from the local to the national, international and global levels is essential. Structural changes and investments in human resources are the main priority. Technology and humanitarian aid alone are insufficient to solve the problem in the long term. What the Sahel region needs is support and financial resources, so that adapted concepts and strategies can be developed and implemented by the actors concerned.

# REFERENCES


Afrique Contemporaine. (ed.) 1992. *L'environnement en Afrique* 161. Paris.

Aloa, A. 2000. The role of African regional and sub-regional organisations in conflict prevention and resolution. *New Issues in Refugee Research*, Working Paper 23, UNHCR. Geneva.

Aménagement et Nature. (ed.) 1998. *La désertification*, no. 129. Paris.

Anhuf, D. 1991. "Umweltflüchtlinge in der Sahelzone. Mythos und Opfer einer ökologischen Krise?" Pp. 99-113 in *WüstenErde. Der Kampf gegen Durst, Dürre und Desertifikation*, editors P.E. Stüben and V. Thurn. Giessen.

Aumüller, J. 1993. *Umweltbedingte Flucht und Migration in Südasien – eine Folge sozialer und ethnischer Marginalisierung*. Berlin.

Benedick, R.E. 2000. "Uncharted Territory. Environment and Population Beyond Six Billion." Pp. 41-60 in *Global Change – Causal Structures and Indicative Solutions*, editors R. Kreibich and E. Simonis. Berlin.

Black, R. 2001. Environmental refugees – myth or reality? *New Issues in Refugee Research*, Working Paper 34, UNHCR. Geneva.

Blaikie, P. 1999. "A Review of Political Ecology." *Zeitschrift für Wirtschaftsgeographie* 43(3-4):131-147.

Boubacar, Y. 1995. "Essai d'interprétation de l'échec des programmes environnementaux au Niger." *Revue de Géographie Alpine*, no. hors-série: Au contact Sahara-Sahel 2:65-82.

Bryant, R.L. 1999. "A Political Ecology for Developing Countries?" *Zeitschrift für Wirtschaftsgeographie* 43(3-4):148-157.

Bryant, R.L. and S. Bailey. 1997. *Third World Political Ecology*. London.

CILSS (Comité Inter-Etats de Lutte contre la Sécheresse au Sahel) and CEDEAO (Communauté Economique des Etats de l'Afrique de l'Ouest). 1999. *Programme d'action sous-régional de lutte contre la désertification en Afrique de l'Ouest et au Tchad.* Lomé.

Club du Sahel. 1995. *West Africa Long Term Perspectives. Regional Opportunities and Policy Issues. Perspectives à long terme en Afrique de l'Ouest. Une réflexion régionale sur les politiques de développement*. Paris.

Cour, J.-M. and S. Snrech. (eds.) 1998. *Preparing for the Future. A Vision of West Africa in the Year 2020.* West Africa Long-Term Perspective Study. Paris.

Crisp, J. 2000. Africa's refugees – patterns, problems and policy challenges. *New Issues in Refugee Research*, Working Paper 28, UNHCR. Geneva.

Crousse, B. et al. (eds.) 1991. *La vallée du fleuve Sénégal. Evaluations et perspectives d'une décennie d'aménagements*. Paris.

Dumont, R. 1986. *Pour l'Afrique, j'accuse. Le journal d'un agronome au Sahel en voie de destruction*. Paris.

ENDA GRAF Sahel. 1993. *La ressource humaine, avenir des terroirs*. Paris.

Geist, H. 1999. "Political Ecology and International Research on Global Environmental Change." *Zeitschrift für Wirtschaftsgeographie* 43(3-4):158-168.

Giri, J. 1989. *Le Sahel au XXIe siècle*. Paris.

Giri, J. 1993. "Rompre avec un demi-siècle d'anti-développement." *Politique Africaine*, 49:93-105.

Hammer, T. 1992. *Angepasste Strategien zur Entwicklung des ländlichen Raums. Das Beispiel Burkina Faso*. Bern.

Hammer, T. 1997a. *Aufbruch im Sahel. Fallstudien zur nachhaltigen ländlichen Entwicklung.* Hamburg.

Hammer, T. 1997b. "Raumplanung im Sahel. Landnutzungsplanung als Ausweg aus der sozio-ökologischen Krise?" Pp. 273-283 in *Werkschau Afrikastudien*, editors B. Sottas, T. Hammer et al. Hamburg.

Hammer, T. 1998. "Erfolgsfaktoren ländlicher Entwicklungsstrategien im westafrikanischen Sahel." *Zeitschrift für Wirtschaftsgeographie* 42(1):11-21.

Hammer, T. 1999a. *Nachhaltige Entwicklung im Lebensraum Sahel. Ein Beitrag zur Strategietheorie nachhaltiger ländlicher Entwicklung*. Münster, Hamburg, London.

Hammer, T. 1999b. "Zukunftsfähiger Sahel? Bausteine einer Strategietheorie nachhaltiger ländlicher Entwicklung." *Die Erde* 130(1):47-65.

Hammer, T. 1999c. "Desertifikation und Überlebensökonomie im Sahel. Das 3-Sektoren-Modell als theoretischer Erklärungsansatz." Pp. 87-99 in *Brücken und Grenzen*, editors L. Roost Vischer. Münster.

Hammer, T. 2000a. "Desertifikation im Sahel. Lösungskonzepte der dritten Generation." *Geographische Rundschau* 52(11):4-10.

Hammer, T. 2000b. "Strategie nachhaltiger Entwicklung der Sahel-Staatengemeinschaft CILSS." *Geographische Rundschau* 52(11):18-26.

Hammer, T. 2001. "Politische Ökologie der Desertifikation. Ein Beitrag zum Erklärungs- und Lösungskomplex im Sahelraum." *Geo-Öko* 22(3):79-90.

Hsü, K.J. 2000. *Klima macht Geschichte. Menschheitsgeschichte als Abbild der Klimageschichte*. Zürich.

IPCC, Intergovernmental Panel on Climate Change, 2001. *Special Report on The Regional Impacts of Climate Change. An Assessment of Vulnerability*. Chapter 2, African Climate Trends and Projections. UNEP and WMO.

Klaus, D. 1986. "Desertifikation im Sahel. Ökologische und sozialökonomische Konsequenzen." *Geographische Rundschau* 38(11):577-583.

Kohout, F. 1999. "Politische Ökologie und internationale Politik." Pp. 109-138 in *Politische Ökologie. Eine Einführung*, editor Mayer-Tasch. Opladen.

Kriebich, R. and U.E. Simonis. (eds.) 2000. *Global Change – Causal Structures and Indicative Solutions*. Berlin.

Krings, T. 1994. "Theoretische Ansätze zur Erklärung der ökologischen Krise in der Sahelzone Afrikas." *Zeitschrift für Wirtschaftsgeographie* 38(1-2):1-10.

Krings, T. 2000. "Das politisch-ökologische Analysekonzept in der Umweltforschung. Beispiel der städtischen Brennstoffversorgung in Dakar (Senegal)." *Geographische Rundschau* 52(11):56-59.

Mayer-Tasch, P.C. (ed.) 1999. *Politische Ökologie. Eine Einführung*. Opladen.

Nuscheler, F. 1988. *Nirgendwo zu Hause – Menschen auf der Flucht*. München.

OCDE, Club du Sahel and CILSS. 1988. *Le Sahel face aux futurs. Dépendance croissante ou transformation structurelle*. Etude prospective des pays sahéliens 1985-2010. Paris.

Pourtier, R. 1992. "Migrations et dynamique de l'environnement en Afrique." *Afrique Contemporaine* 161:167-177.

Reichelt, R., H. Faure and J. Maley. 1992. "Die Entwicklung des Klimas im randtropischen Sahara-Sahelbereich während des Jungquartärs." *Petermanns Geographische Mitteilungen* 136(2-3):69-79.

Richter, R.E. 1998. "Umweltflüchtlinge in Afrika. Naturkatastrophen und anthropogene Einflüsse als Ursachen." Pp. 42-73 in *Kampf um die Natur. Umweltzerstörung und die Lösung ökologischer Konflikte*, editors J. Scheffran and W.R. Vogt. Darmstadt.

Richter, R.E. 2000. "Umweltflüchtlinge in Westafrika. Ursachen, Ausmass und Perspektiven." *Geographische Rundschau* 52(11):12-17.

Roithner, T. 1998. *Eine Welt in Bewegung. Ursachen und Folgen umweltbedingter Migration*. Linz.

Rünger, M. 1999. *Traditionelle Bodenrechte und Entwicklungsplanung im ländlichen Sahel.* Eschborn.

Stamm, V. 1998. *Structures et politiques foncières en Afrique de l'Ouest.* Paris.

Tersiguel, P. and C. Becker. (eds.) 1997. *Développement durable au Sahel.* Paris.

UNEP (United Nations Environment Programme). 1991. *Status of Desertification and Implementation of the United Nations Plan of Action to Combat Desertification.* Nairobi.

UNEP. 1997. *World Atlas of Desertification.* London.

UNESCO (United Nations Educational, Scientific and Cultural Organisation). 1994. *La Convention des Nations Unies sur la lutte contre la désertification dans les pays gravement touchés par la sécheresse et/ou la désertification, en particulier en Afrique.* Paris.

Vernet, J. (ed.) 1994. Pays du Sahel. Du Tchad au Sénégal, du Mali au Niger. *Autrement,* série monde, no. 72. Paris.

# VALUES, MIGRATION, AND ENVIRONMENT

## *An Essay on Driving Forces behind Human Decisions and their Consequences*

Walter Leimgruber.
*Department of Geosciences, University of Fribourg, Switzerland*

## 1. INTRODUCTION

Ahead of the Bonn summit, held in July 2001 and intending to save the Kyoto Protocol on climate change, expectations had been high that the problem of increasing global warming could be solved by a general decrease of CO2 and other emissions of greenhouse gases to a level below that of 1990, and maybe even further. However, the Bonn summit succeeded only to a minor extent. Once more, little more than lip service has been paid to the warnings of the scientific community. The compromise which has been reached after tough and tiring negotiations is the absolute minimum of what has to be undertaken to prevent global warming from going on as before. Even so, there is no guarantee that the ecosystem will not capsize and put an end to humanity. Arrogance and narrow-mindedness continue to dominate human decisions.

The outcomes of the Bonn summit and of the subsequent conference at Marrakech in November 2001 illustrate the everlasting political dilemma: to find the right balance between what is urgently needed and what politicians are prepared to concede. It also demonstrates that the ecosystem is still perceived as a bottomless reservoir from which we can scoop an endless quantity of resources, and an equally bottomless sink into which we can dip all our waste – naturally everything at zero cost.

This observation marks the point of departure of this essay: the driving forces behind environmental change and environmentally induced mass migration lie with human decisions and actions. Based on the subjective perceptions of the reality, they reflect the worldview and the underlying value systems of a given society. Solutions towards the mitigation of human-induced global warming effects and of environmentally induced migration will therefore have to root in a change in attitude of the human race. Everybody is concerned by such a change; politicians alone cannot be made responsible for the near-failure in Bonn and Marrakech. As representatives of their countries and populations, however, they bear an enhanced share of



*J.D. Unruh, M.S. Krol and N. Kliot (eds.),*
*Environmental Change and Its Implications for Population Migration,* 247–266.

responsibility: they ought to take the lead in all questions and show new ways ahead. Gouverner c'est prévoir.

As a consequence of the Bonn agreement, the industrialised countries will have to come to grips with the problem of greenhouse gas emissions, both technically and morally: "The Protocol will enter into force and become legally binding after it has been ratified by at least 55 parties to the Convention, including industrialised countries representing at least 55% of the total 1990 carbon dioxide emissions from this group. So far, 36 countries have ratified, including one industrialised country (Romania)." (Press release 2001). If this condition can be fulfilled, we may speak of a relative success. The chance for the survival of humanity continues to exist, and maybe large-scale migration for the reason of global warming need not happen.

This paper is characterised by a moral undertone and is very critical towards our current Western way of thinking. It is therefore not a scientific paper in the narrow sense of the word. In particular, it is not concerned with a discussion of the predictability of environmental migrations or of potential technical measures. Döös (1997) has discussed the former and shown its limits – migrations are to a considerable extent the result of personal decisions, even if external causes must not be ignored. As to the latter, humanity has equipped itself with a formidable set of technological instruments to solve almost every problem, but we have learnt in the course of the 20th century that most technical solutions carry within themselves new predicaments. We cannot foresee the consequences of the application of a particular technology, and we usually tend to narrow down the choice of possible solutions to a problem to a single one that is considered the only realistic one. We think in a 'mono-cultural' way and leave no room for alternatives – the Indian ecologist Vandana Shiva (1993, p. 5) calls this the "TINA syndrome" ('There is no alternative'). A change in attitude is more pressing than new techniques, and it is precisely at this point where the value system has to be discussed.

## 2. VALUES, WORLD VIEWS AND THINKING

All processes influenced by Man, diverse as they may be, are ultimately driven by a value system. The term of values has itself been subject to widespread discussions, since it is used in a variety of disciplines. Habitually, the word first evokes its economic context of use-value, a perspective that has influenced our thinking very deeply. The ethical and socio-cultural aspects (which can be called 'non-use values'; cf. Edwards &

Abivardi 1998) come second only, but it is from this perspective that the present discussion will be conducted.

Values can be called general principles legitimating the rules of behaviour set by norms (Chazel 1988, p. 125). They lie behind our varying interests and guide our knowledge systems, social behaviour, consumption, etc. The bearers of values are the individuals, but every individual is part of a society to which it transfers his/her values, which, in turn, will become a property of the respective society. It is important, however, to stress their subjective character, as certain individuals with their personality are able to transmit or even impose their particular value system on a group or even an entire society. In such cases, the desire of an individual actor is very important relative to the value he/she defends (cf. Livet 2001).

Values exist as a duality or couple, i.e. they are characterised by fixed terms. From their contents, we can imagine them as two extremes lying at either end of a continuum. Becker (1959, quoted in Hillmann 1989, p. 141 ff.) uses the two terms of 'sacred' and 'secular' to describe comprehensive value systems (Fig. 1). These two terms must not be interpreted as religious notions in the sense of a particular institutionalised religion, but rather as general spiritual terms, referring to specific modes of acting. An individual or a society dominated by sacred values will experience problems when dealing with innovations, as it tends towards the conservation of inherited ways of life, whereas a secular society embraces changes and considers them indispensable elements of social progress.

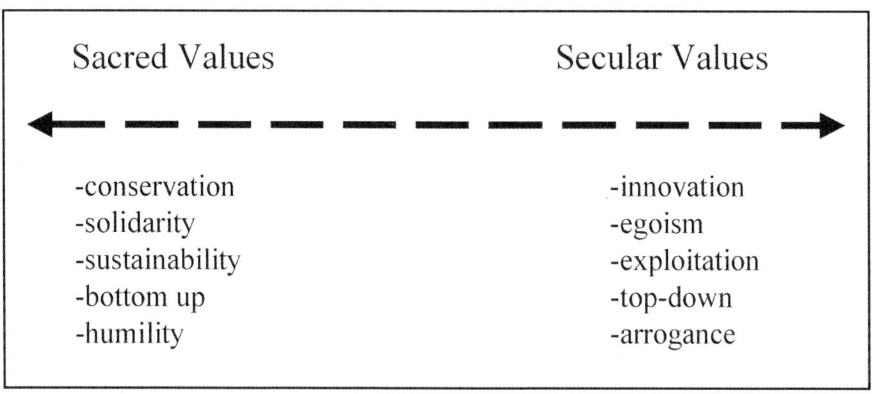


*Figure 1: The value continuum.*

For the same phenomenon, other terms have been used. Cunha (1988), for example, prefers to speak of the existentialist and the productivist paradigm, whereas Fernandes & Carvalho Tomás (2001) use the couple 'ecocentrism' and 'technocentrism'. From another philosophical perspective, secular values correspond to the male or yang view, sacred values, for their part, to the female or yin perspective. Whatever terminology we apply, they two extremes are complementary and represent extreme positions that are never fully attained; yet they have to be taken into consideration. Neither is good or bad, but imbalance in either direction can be detrimental: too much reliance on sacred values leads to stiffness, too much confidence in secular values results in restlessness and uncertainty.

Throughout history, from the pre-Neolithic predator to the Neolithic cultivator and past the middle Ages, the predominant values have always oscillated between the two extremes (Fig. 2). Every innovation (the plough, bow and arrow, the wheel etc.) meant a shift towards secular values, but the impact of such early innovations on the ecosystem and on societies was negligible, given the limited number of people who applied them in daily life. In the Age of Enlightenment (since the 17th century), the philosophical basis for the natural sciences were laid, and Man began to dominate nature, a process that conducted him towards the Industrial Revolution with increasing negative impacts on the environment. Domination was reinforced by the technological breakthrough of the 20th century and led to reckless exploitation. The pendulum swung right towards the secular values that came to govern human actions. The ecosystem was thus viewed as a domain at the exclusive service of mankind, as possessing a simple use value: ecocentrism was replaced by technocentrism (Fernandes & Carvalho Tomás, 2001).

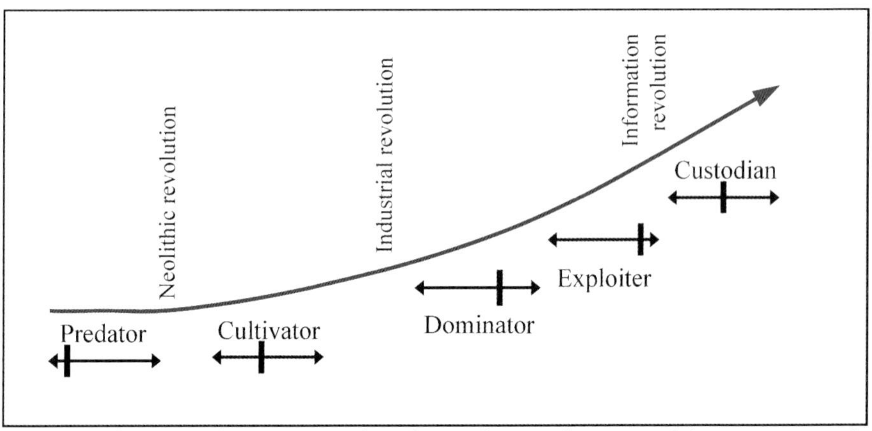


*Figure 2: Changing value systems in the relations between humans and the ecosystem in the course of time. A vertical line marks the dominating values: sacred values to the left, secular values to the right*

While it is no use criticizing the Enlightenment revolution ex post, it must nevertheless be pointed out that its way of thinking has radically changed human attitudes. Appropriation and exploitation became the prevailing traits in the European worldview. Individual actions were put over collective ones; reason triumphed over traditional authority, and through "the disengagement of nature from a supernatural worldview" (Shea & Huff 1995, quoted in Withers & Livingstone 1999, p. 5), the position of Man within the Creation was reversed. The biblical saying "replenish the earth and subdue it" (Genesis 1/28) was taken too literally: to subdue was equalled to exploitation that usually results in destruction. While John Locke advocated that subduing nature is a means to avoid scarcity (Van Dieren 1995, p. 19), he did not deliver a passport to destruction: to subdue does not exclude management.

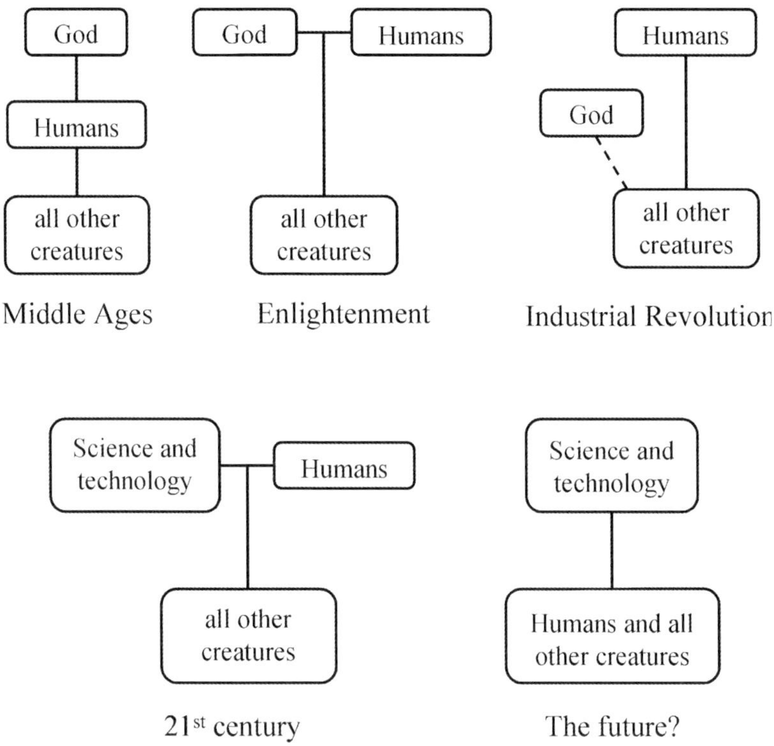


*Figure 3: Man – God relationship (after Leimgruber 2000, p. 12, modified)*

Humans replaced supernatural elements in their worldview by rational ones. This process has been described by Lea (1994) who pointed out that since the Middle Ages, God was moved from His position in the Creation and gradually replaced by Man and, ultimately, by the new gods Science and Technology, who may eventually take control over the human race (Fig. 3). Gould (1999, p. 406) confirms this point when he says "We are in a

European world where the God giving meaning is being displaced from the centre and being replaced by the human as the measure of all things, positing increasingly the way the world will be, positing even the god being displaced." (Emphasis in the original). In other words, humanity plays god, and the 'old God' is given His (or Her) place just somewhere (or even nowhere).

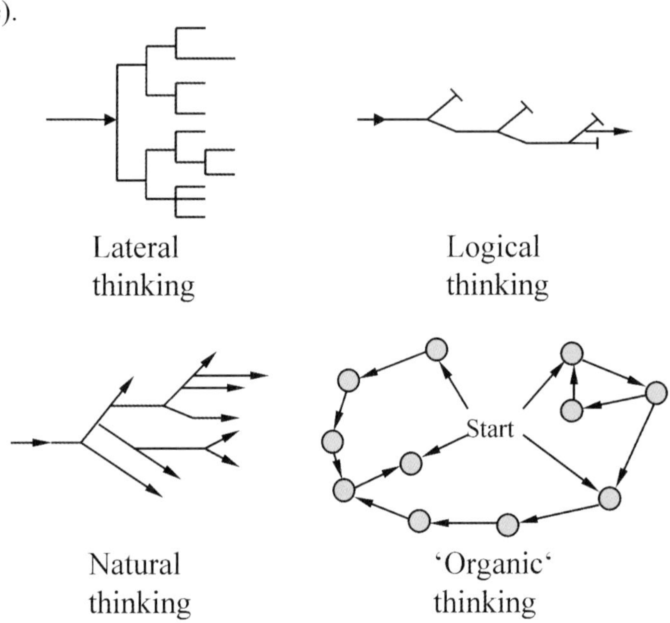


*Figure 4: Modes of thinking (after Wood 1987, p. 291).*

Values influence our thinking, an activity that usually precedes decisions and actions. In the second half of the 20th century, we have acquired the tool of general systems theory that enables us to understand and foresee complicated processes, to avoid a dead end, and to beware of negative outcomes. Modern simulation techniques would be unthinkable without this instrument, which owes its potential to the human brain, by far the most elaborate of all living creatures. Its complexity with six layers (ten if we count the subdivision of layers 4 to 6) and two halves distinguishes us from the animal realm (Aubier 1992, p. 200) and is the basis for complicated reflections. Bourguignon & Koupernik (2001) emphasise this point: "Le cerveau humain, du fait de son extraordinaire richesse, est le seul cerveau de la série animale capable de concilier la répétition et le changement." (Thanks to its extraordinary richness, the human brain is the only one in the animal realm, which is capable of combining repetition and change). This particularity enables us to think in different ways (Fig. 4): laterally by successively opening up new lines of thought, logically by proceeding along a strict 'yes – no' line, or naturally by allowing our thoughts to flow freely (Wood 1987, p. 290 ff.) However, it is important to think 'organically', i.e. see the integration or mutual dependencies of individual lines of thought to understand systemic processes.

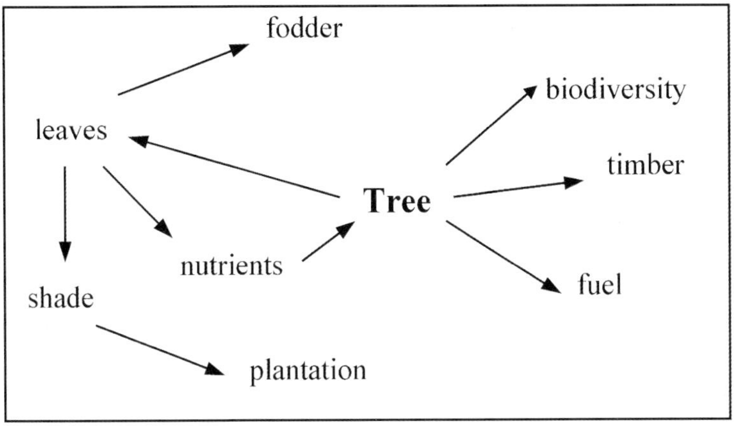


*Figure 5: Diversity in thinking.*

It is a commonplace that we do not know how and what other people think. We should, however, respect the different ways of thinking. All our actions originate in thoughts, but the way we think is to a substantial amount influenced by the society and the culture we live in and the common value system. Every culture faces specific problems of survival, and the response to a particular natural event may vary from one group to another – even from one individual to another. This diversity of thinking is mirrored by the diversity of knowledge systems that are adapted to given social circumstances, and which can be applied to local and regional food support systems as well as to the management of local and regional ecosystems. This diversity, however, has been replaced by 'mono-cultural' thinking (Shiva 1993) which has spread over the globe as a consequence of European colonisation. This point must be stressed because during the era of colonization, European countries not only conquered territories overseas with populations to dominate and resources to exploit, they also diffused their own value systems and considered other peoples' cultures as inferior. This is what Gould means when he speaks of the 'European world' (1999, p. 406). One of the ideas spread was that of use value attributed to a particular resource, a unilateral use from the perspective of European industry. This monoculture of the mind is exemplified by the conviction that one given plant serves one single purpose only (a tree for timber or paper wood; Shiva 1993, p. 13) or by the erroneous assumption that Eucalyptus "is faster growing than all indigenous alternatives" (ibid., p. 31). Mono-cultural thinking wants to make a complicated world simple, tailored exclusively to specific ('western') human needs. A tree, however, not only consists of wood but has also leaves which can serve for fodder and give shade to a plantation, and it holds an important position in a local ecosystem, adding to its biodiversity (Fig. 5). It combines use values and non-use values.

## 3. THE DYNAMIC NATURE OF THE ECOSYSTEM

Environmental change is a normal phenomenon because life is by its very nature dynamic. A static situation is the exception, and its duration is limited. When we study structures, we observe a momentary state in a steady process. As humans, however, we tend to feel uneasy about constant changes. We prefer a stable situation within which we can organise our lives peacefully without having to adapt to new circumstances all the time. Every law and regulation testifies to this desire for stability: we want to be sure what happens in a given moment under certain circumstances, we want to control the situation at any moment. Unexpected events, however, do occur and force us to modify our rules according to the new conditions

The natural environment is an ever-changing system, following its own rule of self-sustainability – this is a commonplace. Humans (who have been transforming their living space in an intentional way for more than ten thousand years) have always tried to construct stable social conditions for themselves – this is another commonplace. But since the 17th century, we have gradually forgotten the first commonplace and have attempted to adapt the environment to the second: mankind longed for a stable nature to permit a stable (i.e. safe) way of human life. Nature had to be controlled; it had to obey human wishes and needs. To do so, technical means were developed to influence, check or repair certain types of changes (e.g. flood control, slopes stabilisation, afforestation, earthquake proof constructions etc.) in order to protect settlements and transportation routes. Inconspicuous at first, they increasingly resulted in artificial landscapes and disturbed the functioning of the ecosystem. Currently we assist expensive processes of landscape rehabilitation: watercourses are given natural banks, and hedges are replanted. Large-scale afforestation programs in many countries put the trees back that have been felled but do not recreate the original forest ecosystems with their dynamic equilibria. "The result is often a simpler ecosystem in which genetic resources are only partially preserved." (Attfield 1999, p. 84). The technical measures primarily fight the symptoms without going to the roots of the problem.

The processes changing our environment broadly fall into three categories:

1. Natural processes beyond human control (earthquakes, volcanic eruptions, long-term climate change, hurricanes),
2. Natural processes as unintended consequences of human actions (soil erosion following forest clearing, desertification due to overgrazing, climate change due to the emission of greenhouse gases),
3. Intentional transformations (flood control, dams, afforestation).

We accept and promote man-made transformations of the environment (the third category) if they bring or enhance security, comfort, and welfare. Any failure can be repaired by technical improvements. We have to live with events of the first category because there is no way of avoiding them. It is possible to predict certain events to some degree – the locations of the seismic zones and of volcanoes on earth are known, and sophisticated models permit us to forecast the weather, but despite the advanced state of science, we shall never arrive at certainty. The true violence of the hurricane Lothar, for example, which struck parts of Europe on Boxing Day 1999, could not be foreseen in time.

The real difficulty lies with the second category because it is precisely here that humans can intervene and change themselves, and it is from this category that emerge the multiple threats to the environment, which may induce people to leave their homes. Resource use and research in the natural sciences look first and foremost towards the wishes and (real or imaginary) needs of society. Satisfying demand is a practical task, based on research and resource use. While scientific results, application, and commercial success may turn up in a relatively short time, the unintended consequences of technical measures will appear much later. DDT and the FCHC, for example, have been applied successfully for many decades until their negative impacts became visible.

The uncertainty about the dynamics of the ecosystem calls for the application of the precautionary principle, particularly in decisions and actions on technology influencing directly or indirectly the natural environment. This includes, e.g., the careful selection of settlement sites or transportation routes, taking into account not only present-day risk evaluations but also past events and local knowledge. Information is an important instrument to publicise precautionary measures, but we have to bear in mind that the public has not the same perception of a potential event as scientists. Our western society has gradually become accustomed to the possibilities, quality and reliability of modern technology; we expect things to function in a certain way and can hardly imagine another scenario than the one we have in our minds. Research on the perception of flood hazard in the

US as early as the 1950ies have shown that "increased spending on flood prevention … was actually paralleled by greater losses from flood damage, possibly because people modify their attitude to flood plain occupancy in the light of changing beliefs as to the power of available technology." (Walmsley & Lewis 1984, p. 109). Too much confidence in technology may be as wrong as excessive scepticism about the efficiency of technical flood prevention methods. The inhabitants of the Beijiang region in rural northern Guangdong province in China believe that floods are unavoidable and prefer to adapt themselves to the risk (Wongh & Zhao 2001). Only one in five rates flood prevention structures useful (ibid. p.197). In many instances, people flee from their home region struck by a natural catastrophe, but want to return home as quickly as possible. The attachment to their homes as well as to symbolic places (including cemeteries and tombs of ancestors) prevails over the fear of further events. This can be observed when earthquakes strike densely populated regions. In the Friuli region (northeast Italy), the population started to rebuild their houses after a first quake in May 1976; they were evacuated after the second earthquake in September of the same year, but the reconstruction was continued (Geipel 1989). In the case of the Kobe earthquake in 1995, it was obviously deemed impossible to relocate an entire city; the authorities promoted improved building technology in order to ensure a safe living space. – People also tend to forget past events, in particular if they lie several generations back. A severe flood hit the small town of Brig in Switzerland in 1993, but nobody had remembered that similar events had occurred in the centuries before, the last in 1912.

Applying a precautionary approach is a necessity for a society that has lost the intimate knowledge about the relations between human actions end the scope allowed by the ecosystem. All activities have significant impacts, and they should not be considered negligible unless proved to be so. The complex and highly dynamic ecosystem will never be perfectly understood. As a consequence, scientific advice to management is always affected by uncertainty. Management decision processes in turn have their own uncertainties, and in the end the impacts on the system are difficult to predict accurately. The consequences of management errors thus may take a long time to be corrected (FAO 2000, chapter 'Indicators of sustainable development and the precautionary approach in marine capture fisheries', section 'The precautionary approach'). What about political decisions?

Politicians find themselves inside a complex network of obligations and interests, and they do not necessarily practise precaution. They tend to think in terms of electoral periods, want to stay in power, and try to avoid unpopular measures even if they were necessary. They may accept that nature has its own ways but want to apply the principle of market liberalism to the relationship Man – ecosystem. The reasoning is simple: if the cost for a specific good exceeds the price customers are prepared to pay, this good will no longer be demanded. This, however, would mean internalising environmental cost, something, which so far has hardly happened. In this way, nature could regulate itself according to the free-market rules. This, at least, is the simplistic view of the ardent defenders of unlimited human freedom. The idea of common responsibility remains outside their thinking.

## 4. CHANGING HUMAN PERSPECTIVES

The human pressure on the ecosystem can be viewed from two angles, both based on the notion of 'needs' (Anderson 1996, p. 85 ff.), but following two diverging paths. First, the overall population growth calls for more food, water, and other goods to satisfy the basic needs of an increasing number of people. Countries with a heavy demographic pressure (countries in the 'South') already now have severe problems of supply, and the future looks bleak for many. Increased demand puts additional stress on the ecosystem and favours its degradation. This fact alone may suffice to propel mass migration. Critical meteorological conditions such as droughts (Richter 2000) will aggravate the problem. Secondly, the increase of affluence in the 'North' continues to provoke new desires for consumer goods. The economy is invited to provide a steady flow of (seemingly) new and better products. People in the North demand unlimited fulfilment of their wishes; otherwise they will run into what I should call the problem of 'social scarcity'. Human creativity for material goods is under a constant challenge, new resources have to be opened up, and the 'one-way-society' generates superfluous waste.

It is a commonplace that this kind of exploitation will eventually overstretch the potential of the ecosystem, both as a provider of resources and as a sink for waste, and it will inevitably collapse at a certain moment in time. Its carrying capacity in the strict sense can only be tapped indefinitely if its limits are respected and its inertia or elasticity used sparingly (Fig. 6). The consumption of the ecosystem by human demand can be seen as passing through three stages, of which the third need not arrive, if adequate measures are being taken at the moment demand passes from stage I into stage II or at the latest when it is being satisfied by the elasticity of the system. The three stages are:

Stage I Demand varies within the carrying capacity of the natural system

Stage II Demand and/or way of satisfying it pass beyond the threshold of sustainability. The inertia of the system guarantees satisfaction for a limited period

Stage III Satisfaction of demand is reduced due to declining performance or collapse of the natural system.

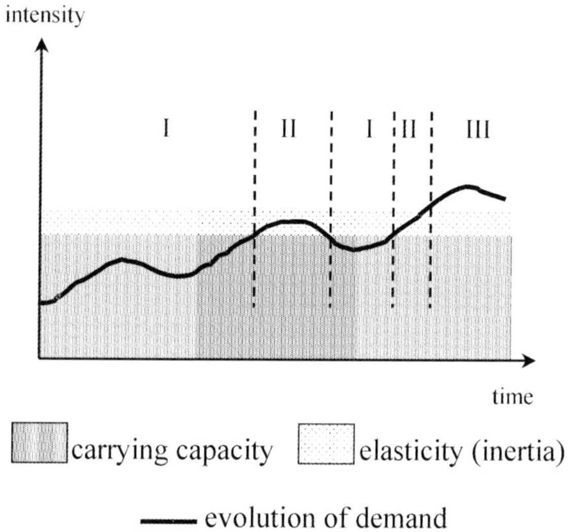


*Figure 6: Demand and the ecosystem (source: Hauser 1990, vol. I, p. 43).*

The model is not deterministic as the drawing might suggest. If the growth of demand can be kept inside the carrying capacity, there will be no permanent degradation, and sustainability can be guaranteed (Attfield 1999, p. 104). Under the prevailing secular value system, this is not so easy, and Hauser reminds that "long-term survival of mankind is only possible within a dynamic equilibrium, and in a dynamic equilibrium, growth can only be tolerated during a short period." (1990, II, p. 584). His remark calls for zero growth – the nightmare of every shareholder and company executive who view growth from a purely quantitative perspective.

Unfortunately for humans, the ecosystem is very complex, and our needs vary considerably over space and time. We do not know where precisely its threshold lies, and we tend to go beyond the limits – a point made more than a generation ago by Meadows et al. (1972) and confirmed twenty years later (Meadows et al. 1992). His ignorance in this matter prevents Man from applying the precautionary principle consistently and considering restrictions, for example. Enhanced resource efficiency could help to solve many problems. What has been possible to some extent with energy after the oil shock of the 1970s should be carried further into other fields.

The basic problem has been recognised, and numerous counter movements have sprung up long before the Rio summit sounded the alarm bell. Top-down initiatives combining interventions on ecological and development questions (by the Club of Rome, the Club of Lisbon, the South Commission etc.) go back into the 1960ies, but in the eyes of many, they seem to be of little effect as they work too slowly. As a consequence, numerous bottom-up movements have emerged in many countries (Ekins 1992), initiating local actions in order to offer immediate relief and to draw the attention of the general public to social and ecological distress. Bottom-up movements demand solidarity and hard work, often without remuneration. Their existence demonstrates that a new way of thinking is making its way. It has been understood that "clean water and air, unpolluted soils, are not available 'freely' in nature once human beings have had a hand in economic development." (Redclift 1995, p. 12), but that a personal sacrifice is required. Such measures go beyond technology and are often taken without political support. They may receive help from NGOs within the scope given them by the political class.

## 5. ENVIRONMENTALLY INDUCED MIGRATION

At this point it does make sense to briefly discuss the problem of 'environmental migration'. This term itself is far from explicit and calls for clarification. Baechler noted a few years ago that the distinction between people who flee from natural disasters and those who migrate because of environmental degradation is fluctuating (1995, p. 74). It remains an open question if they prefer to migrate for good or if they see their movement as temporary, until a return is possible. Temporary migrations have been the rule rather than the exception in semi-arid regions where rainfall determined the agricultural potential, and in shifting cultivation regular displacements were part of the system. The duration of migration depends on the type of environmental change which drives people away: the permanent flooding of a valley due to the construction of a dam is different from a flood due to a specific meteorological situation and from climatic variations whose temporal aspects are difficult to asses.

It has been said above that environmental change is essentially the result of human decisions and actions. Northern (European) thinking has contributed to the deterioration of living conditions in the ecological, economic, and social domain all over the globe, and the industrialised countries have to assume their share of responsibility (Scheffran 1994, p. 23). These three domains are intimately connected: wars are fought over natural resources or over the dominance of an ideology, and they result in the destruction of the land and in poverty among the refugees from the war zones (e.g. Sudan, Vietnam; we could also recall the risks which persist to agriculture and forestry in Northern France due to unexploded shells from the heavy battles in 1916 during World War I; see Amat 1987, 1988); unsafe industrial plants threaten the populations living in their surrounding (Seveso, Bhopal), overexploitation leads to desertification (Goudie 1990, p. 48 f.; Döös 1997; Richter 2000, p. 13), mono-cultural farming results in soil degradation and soil erosion, unconsidered deforestation disrupts regional ecosystems and subsequently promotes soil erosion – the list could be continued. In all such cases, the degradation of the environment acts as a push-factor for the populations concerned. Emigration may be the only possible way for survival.

It is true that mass migrations have been caused by natural events beyond human control. Soil and climate variability – hence economic instability – were partly responsible for frequent nomadic incursions into China (Brauer, [p. 24] speaks of an "ecological discontinuity") which sometimes resulted in lengthy occupation periods, and in the 20th century, repeated migration waves occurred in the Sahel region in Western Africa (Richter 2000), although migration in that region has a long tradition (Hammer 1999, pp. 256 ff.). Following an earthquake, emigration may be temporary only (such as in the case of the Friul; Geipel 1989), but also floods or droughts do not necessarily drive people away for good. Even if emigration for environmental reasons is a decision based on an external factor, it will often be looked at as a constraint that will not last forever. The wish to return home after the end of the threat may be stronger than the perceived risk to survival.

The present-day long-range human modification of world climate, which can now be measured with some accuracy, however, may result in future migration flows that are likely to be irreversible. Coastal lands submerged by the rising sea level cannot be recovered unless one is prepared to construct coastal defences to protect the lands. The Dutch could afford to finance such massive constructions, Bangladesh, however, cannot. Once the ecological basis for survival is destroyed (i.e. the land is drowned or swept away, the soil is degraded or eroded), there is no way to return.

The relations between environmental and human factors have been described by Döös (1997, p.43). He puts great emphasis on the human factor, represented by population, society (with economics and politics), and minority conflicts. Behind them, however, lies the value system according to which a population evolves, organises its economy and political system, and views minority groups. Civil war and international conflicts are also directly related to the three factors and the value system (Fig. 7).

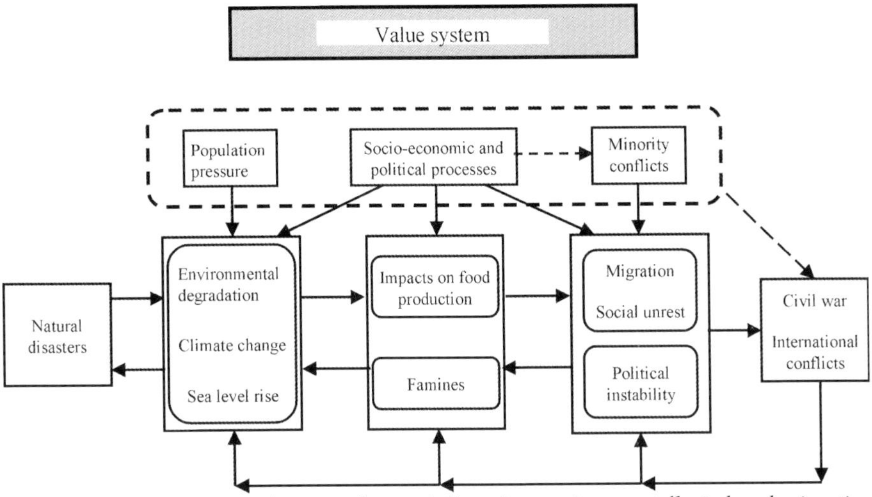


*Figure 7: Interconnections between factors influencing environmentally induced migration (modified after Döös 1997,p. 43).*

The emphasis on values is justified by the role they play in our life. Our actions vary according to the values that guide us: exploitation or conservation, individual profit or common welfare, i.e. secular or sacred values. From the colonial epoch onwards, rationalist European thinking and secular values have been imposed on peoples of the South who, however, were not prepared to this and lacked the intellectual and material means to cope with the problems stemming from this new value system. The North has exploited the South (e.g. by promoting mono-cultural cash crops for the World market and repatriating the profits obtained); it is therefore at least partly responsible for the degradation of the environment and the mass movements occurring from time to time. Since the industrialised world has become aware of the ecological trap into which the World is about to fall, it wants to dictate the South certain ways of behaviour to save our planet – a very cynical attitude which has little to do with a true concern about the future of the Earth but is rather a way to maintain its power over the South without manifesting the intention to do the same! Basically, however, it is not our task to tell other people what to do; we have contributed to the degradation of their environment without trying to understand their world view, now we shall give them a chance to solve the problems not according to our northern ideas but in combination with theirs: the diversity of thinking is offering alternatives (Shiva 1993; Zweifel 1995), not the technocratic northern dictates.

Environmentally induced migration may have several origins. It may root directly in natural changes in the environment, or it may be economically motivated based on unintended consequences of human interventions on the ecosystem. Depending on the intensity of impact and on the perceived duration of absence, a migration movement may be looked upon as a temporary or the final solution to the human distress, and the distance of migration will be related to this perception.

## 6. POLICY IMPLICATIONS

International mass migration has many political consequences, and the North is gradually beginning to understand them. At present, no solutions to the expected flows have been found, and the laws have not yet been adapted to this new 'threat'. So far, migration has been dealt with from the perspective of the labour markets; laws and regulations sufficed to control the flows of immigrants.

Environmentally induced migration, however, is a new phenomenon, and we are ill prepared for it. The category of 'ecological refugee' does not exist in the political and legal vocabulary, nor does that of the 'economic refugee'. Refugees are defined according to the UN Convention as people persecuted for ethnic, ideological or political reasons. A new reality has been emerging since the 1980ies, and we shall have to live with it despite all efforts of the countries of the North to prevent these two new groups from being included in the refugee definition. This means rejecting our responsibility for the ecological and economic causes of such mass movements.

Solutions have to be sought. Barring the borders of industrial countries is not the answer to the growing pressure on our boundaries. It makes the North even more attractive and results in clandestine immigration and trafficking with immeasurable consequences. Human tragedies are happening every day on the outer borders of the European Union, in particular in the Mediterranean, but also in other parts of the World. This mass movement, whether ecologically, politically or economically motivated, is the reaction on the exportation of Northern values and the refusal of the North to take the problems of the South seriously.

New approaches have to be evaluated. Preventing emigration instead of fighting off immigration requires rethinking Northern attitudes in general and towards the South in particular. In addition, the cooperation of the elites in the countries of origin must be secured. Investments must be profitable for the population and the country as a whole, not for the industrial countries and the ruling class alone. Large-scale democratisation is needed, i.e. a redistribution of wealth and social equity – a change of attitude. The pendulum has to swing back from an excessive emphasis on secular values towards the sacred side. There are signs that this is happening, but the process has still to increase in force. Consumers of the North develop solidarity with producers of the South, and the opposition to globalization (or, more precisely, global trade liberalism) is increasing (Hoad 2001). Most opponents – those who demonstrate peacefully and those who silently disagree – are motivated by genuine worries about the future of our planet, both in ecological and in social terms. Politics ought to take these protests seriously and, instead of repressing them, look at the positive core and take action accordingly. This, of course, might prejudice the delicate links between political power and economic hegemony and endanger corporate interests. Courage would be required for such a forward-looking policy.

## 7. CONCLUDING REMARKS

What is to be done to remedy such a dramatic situation? I should like to launch two reflections that may point ways we shall have to go in the future. They are to some extent related.

1. In the film 'Après la tempête' (After the storm), the French philosopher Dominique Aubier discusses, among other things, the trees felled by the hurricane 'Lothar' in December 1999 (Van Effenterre 2000). In a rather unconventional comment, she asserts that the storm had simply stopped the growth of the trees, and that after the passage of 'Lothar' new life will grow in the same forest. The trees are, according to her, not dead, they have simply been 'stopped' in full life ("Nous avons un stop, ce n'est pas la mort que nous avons là, c'est de l'arrêt. Ils sont tous stoppés en pleine vie comme s'ils avaient fait un pas de trop."). From this statement we can infer that to halt one particular process can be the source of a new and better process. To prevent inventions from becoming commercialised (among other things, she refers to biotechnology) may be obvious to a philosopher who takes a wide look at the phenomena in life but not 'normal' to the inventor or the economic interests behind. However, certain politicians and company managers should learn from philosophers and widen their outlooks as well. The same applies to science and technology – it may be better to stop before a road becomes a way of no return. This sounds like heresy, but the current success of organic farming illustrates how a new way is overcoming the deadlock of conventional mono-cultural farming.

2. The World is an open ecosystem, functioning according to deterministic rules and receiving solar energy to keep it going. Open systems

stand good chances of survival because of the constant energy input they receive. The global economy, on the other hand, is probabilistic and appears to function as a closed system. Its sole energy input is self-generated, and if it continues in this way, it will inevitably wear out and come to a standstill. Renewal out of itself is impossible, as the collapse of the State-planned Soviet economy demonstrated. The so-called 'free market economy' is theoretically open to external energy inputs, but monetary profit as the sole fuel of an economy is no external energy. The protesters against globalisation (who are themselves globalised and opposed to the neo-liberal paradigm) remind us that a change is imperative. While Seattle 1999 has been "widely perceived as an ideological turning point." (Smith 2000, p. 2), most protesters offer no solution to the problems evoked, but they are an indicator of a widespread concern about the excessive liberalisation of trade and the growing global disparities.

Environmental change and potentially ensuing mass migration cannot be solved by technical measures alone but require a change in thinking and attitudes. Their driving forces lie with us, the human society, in particular with the industrialised world. Looking at the World's current state and in particular at a number of political leaders, I am rather pessimistic for the future, although I am basically an optimist. We are reaping the fruits of worldwide colonisation, of the export of the European ways of thinking, of reckless exploitation of the Earth for the benefit of a few.

## REFERENCES

Amat, J.-P. 1987. "Guerre et milieux naturels: les forêts meurtriés de l'est de la France, 70 ans après Verdun." *L'Espace Géographique* 3:217-233.

Amat, J.-P. 1988. "La forêt et la guerre, un exemple de sylvofaciès sur les champs de bataille de la Grande guerre en Argonne." *Bull. Assoc. Géogr. Franç.* 3:191-201.

Anderson, E.N. 1996. *Ecologies of the heart. Emotion, belief, and the environment.* New York: Oxford UP.

Aubier, D. 1992. *La face cachée du cerveau.* Paris: Editions Dervy.

Baechler, G. 1995. "Umweltflucht – Auflösung von Lebensordnungen und Konfliktpotential." Pp. 74-86 in *Bevölkerungsentwicklung, Lebensstil und Umweltverantwortung. Allgemeine Ökologie zur Diskussion gestellt* 1, editor M. Flury. Bern: IKAÖ.

Bourguignon, A. and C. Koupernik. 2001. "Cerveau humain". In *Encyclopédia Universalis* 7 (DVD-ROM edition). Paris.

Brauer, R. manuscript. *Boundaries and frontiers in three pre-industrial empires.*

Chazel, F. 1988. "Normes et valeurs sociales." In *Encyclopédia Universalis* 13:124-127.

Cunha, A. 1988. "Systèmes et territoire: valeurs, concepts et indicateurs pour un autre développement." *L'Espace Géographique* 3:181-198.

Döös, B.R. 1997. "Can large-scale environmental migrations be predicted?" *Global Environmental Change* 7(1):41-61.

Edwards, P.J. and C. Abivardi. 1998. "The value of biodiversity: where ecology and economy blend." *Biological Conservation* 83(3):239-246.

Ekins, P. 1992. *A new world order. Grassroots movements for global change*. London: Routledge.

FAO. 2000. *The state of world fisheries and aquaculture 2000* (SOFIA). Internet: http://www.fao.org/sof/sofia/index_en.htm (21.07.2001).

Fernandes, J.L. and T. Carvalho. 2001. "Conservation, development and the environment: a conflictual relationship or a different view for new geographies?" Paper, presented to the IGU Commission on Evolving Issues of Geographical Marginality in the Early 21st Century World, Stockholm, 25-29 July.

Geipel, R. 1989. "Friuli: Ten years after the earthquake of 6 May 1976." Pp. 54-70 in *Population and disasterI*, editors J.I. Clarke, P. Curson, S.L. Kayastha and P. Nag. Oxford: Blackwell.

Goudie, A. 1990. *The human impact on the natural environment*. 3rd ed. Oxford: Blackwell.

Gould, P. 1999. "Lisbon 1755: Enlightenment, catastrophe, and communication." In Livingstone and Withers.

Guerrier, Y., N. Alexander, J. Chase and M. O'Brien. (eds.) 1995. *Values and the environment. A social science perspective*. Chichester: John Wiley.

Hammer, T. 1999. *Nachhaltige Entwicklung im Lebensraum Sahel. Ein Beitrag zur Strategietheorie nachhaltiger ländlicher Entwicklung*. Münster: LIT-Verlag.

Hauser, J. 1990. *Bevölkerungs- und Umweltprobleme der Dritten Welt*. 2 vols. Bern: Haupt (UTB).

Hillmann, K.-H. 1989. *Wertwandel. Zur Frage soziokultureller Voraussetzungen alternativer Lebensformen*. Darmstadt: Wiss. Buchgesellschaft.

Hoad, D. 2002. "The World Trade Organisation, corporate interests and global opposition: Seattle and after." *Geography* 87(2):148-154.

Lea, D.R. 1994. "Christianity and Western attitudes towards the natural environment." *History of European Ideas* 18(4):513-524.

Leimgruber, W. 2000. "Land use and abuse: On the ecological and spiritual marginalization of land." Pp. 7-24 in *Environment and marginality in geographical space*, editors R. Majoral, H. Jussila and F. Delgado Cravidão. Aldershot: Ashgate.

Livet, P. 2001. *Valeurs (philosophie). Encyclopédie Universalis*, DVD version 7, Paris.

Livingstone, D.N. and C.W.J. Withers. (eds.) 1999. *Geography and Enlightenment*. Chicago, London: The University of Chicago Press.

Meadows, D.H., D.L. Meadows, J. Randers and W.W. Behrens. 1972. *The limits to growth*. New York: Universe Books.

Meadows, D.H., D.L. Meadows and J. Randers. 1992. *Beyond the limits*. Post Mills.

Press release. 2001. Governments adopt Bonn agreement on Kyoto Protocol rules. Internet http://unfccc.int/press/prel2001/pressrel230701.pdf (consulted July 24, 2001).

Redclift, M.R. 1995. "Values and global environmental change." Pp. 7-17 in Y. Guerrier, N. Alexander, J. Chase and M. O'Brien (eds.).

Richter, R.E. 2000. "Umweltflüchtlinge in Westafrika." *Geographische Rundschau* 52(11):12-17.

Scheffran, J. 1994. "Kriegs- und Umweltflüchtlinge." Pp. 23-38 in *Migration und Ausländerfeindlichkeit*, editors G. Böhme, R. Chakraborty and F. Weiler. Darmstadt: Wissenschaftliche Buchgesellschaft.

Shiva, V. 1993. *Monocultures of the mind. Perspectives on biodiversity and biotechnology*. London, New York: Zed Books.

Smith, N. 2000. "Global Seattle." *Society and Space (Environment and Planning D)* 18(1):1-5.

Van Dieren, W. (ed.) 1995. *Taking nature into account. A report to the Club of Rome*. Springer, New York.

Van Effenterre, J. 2000. *Après la tempête. Portrait d'une femme extraordinaire*. Tohu-Bohu Film.

Von Weizsäcker, E.U., A.B. Lovins and L.H. Lovoins. 1997. *Faktor vier. Doppelter Wohlstand – halbierter Naturverbrauch*. Der neue Bericht an den Club of Rome. Droemer Knaur, München.

Walmsley, D.J. and G.J. Lewis. 1984. *Human geography. Behavioural approaches*. London: Longman.

Wong, K., and Zhao, X., 2001. Living with floods: victim's perceptions in Beijiang, Guangdong, China. *Area* 33(2), 190-201

Wood, T.F., 1987. Thinking in geography. *Geography* 72(4), 289-299

Zweifel, H., 1995. Bevölkerungswachstum, Armut und Umweltzerstörung – Ursachen und Folgen. In Flury, M. (ed.) *Bevölkerungsentwicklung, Lebensstil und Umweltverantwortung. Allgemeine Ökologie zur Diskussion gestellt*, 1, pp. 25-36. Bern: IKAÖ

# ENVIRONMENTAL - ECONOMIC INTERACTION AND FORCES OF MIGRATION

*A Case Study of Three Counties in Northern China*

Hong Yang[1], Xiubin Li[2], Yili Zhang[2] and Alexander J. B. Zehnder[1]
[1] *Swiss Federal Institute for Environmental Science and Technology, Uberlandstrasse 133, 8600 Dübendorf, Switzerland*
[2] *Institute of Geographical Sciences and Natural Resources Research, Chinese Academy of Sciences, 917 Building, Datun Rd, Beijing, 100101, China*

## 1. INTRODUCTION

Migration on a permanent or temporary basis has always been one of the most important survival strategies adopted by people in the human history. However, the literature devoted to the complex relationship involving environmental change as both a cause and a consequence of migration is rather limited (Hugo, 1996). The knowledge of how economic and environmental factors interact and impinge upon economic development, social change, conflict as well as migration decision is also inadequate. With the acceleration of environmental degradation and the aggravation of environmental stress during the past decades, unprecedented numbers of the population have seen migration as an option. Against this backdrop, attention directed at the relationship between environmental change and migration has arisen among both researchers and policymakers (Richmond, 1993; Doos, 1994; Kibreab, 1994; Black, 1994; Hugo, 1996; Swain, 1996; Xu, et al, 1996; Ma, 1997; O'Lear, 1997 and Hempel, 1997).

The migration literature is replete with typologies which differentiate migrants and migrations according to the relative permanency of the move, the distance traversed, the nature of the boundaries crossed, the causes of the move, the characteristics of the movers, etc. (Hugo, 1996). Government policies and direct administrative intervention in the origin and destination of migration can have strong, sometimes even decisive, impacts on the scale of all types of migration.

Based on causes of the move, migration can be classified into three categories: economic, political and environmental (Xu, et al, 1996 and Charnley, 1997). Economic migration has occupied a large space in the migration literature. In the neo-classical economic framework, economic migration is defined as a move that takes advantage of better economic prospects in terms of employment and income. Migration models established



*J.D. Unruh, M.S. Krol and N. Kliot (eds.),*
*Environmental Change and Its Implications for Population Migration*, 267–288.

within this framework conceptualize population movement as a geographical mobility of workers who are responding to imbalances in the spatial distribution of land, labor, capital and natural resources (Wood, 1982). Among various such models, the Harris-Todaro (Todaro, 1976) model has been widely applied. The model postulates that migrants make rational economic choices. They choose to migrate when expected earnings or welfare are higher if they move away from home. In principle, the Harris-Todaro model is capable of including any factor that affects the welfare of those who migrate or of those who stay behind, so long as the definitions of costs and benefits are broad enough and the impact of constraints on the probability of employment is adequately addressed (Rozelle, et al, 1999 and Ingene, 2001).

Politically-induced migration is mainly related to war, imposition of political persecution on individuals or a group of people because of their political views, religious adherence and ethnic background. Migrants driven by these forces are conventionally referred to as political refugees (Akokpari, 1998). As such phenomena are often studied in the political science, they have been ignored by most migration literature (Schmeidl, 1997).

Compared with the relatively clear definitions of economic and political migration, the definition of environmental migration is rather contentious. Swain (1996) defined environmental migrants as those who are forced to move away from their homes as a result of the loss of their livelihood and/or living space because of environmental changes. The underlying basis of this definition is that people generally choose to stay in their native land and struggle to survive the impact of environmental disruptions until their hope of survival fades away. Hugo (1996), however, defined environmental migration as a movement induced by the degradation and deterioration of the environment they rely on. By this definition, environmental factors that induce migration can be natural disasters and man-made hazards. They can, however, as well be related to ecosystem degradation, reduction of resources and environmental pollution. Apparently, Hugo's definition has a much broader scope. We consider that this definition can better reflect the forces of environmental migration in the contemporary world. The following analysis therefore adopts this definition.

In spite of the different definitions of economic and environmental migrations, differentiating the two types of migration for individual migrants is often difficult. The unprecedented scale of migration of people from mountainous and inhospitable hinterland to the eastern coastal areas in China during the period of economic reform, incepted in the late 1970s, is a good example in point. Many people have classified this migration as economically driven (Croll, 1997 and Rozelle, 1999). However, this view can easily be challenged by the fact that most of the migrants are originated from the areas plagued with severe environmental degradation, water scarcity, pollution and intensified natural cataclysms. It is undeniable that the environmental factors have been important forces in pushing people away from their home. This situation at least suggests that there is an overlap between economic and environmental migration.

Migration will has also been widely considered in the migration literature. Based on the degree to which a move is 'forced' (Peterson, 1958), migration is divided into two broad groups: voluntary and forced. Economic migration has generally been considered voluntary and political migration forced. However, assessment of the will in environmentally-induced migration is rather complicated. Natural disasters, such as flood, earthquake, volcano eruption, drought, etc., can cause damages to the means of living either temporarily or permanently, generating effective forces for migration. For the migration induced by these events, the push factor completely overwhelms the pull factor. Migrants have no other choice but to leave their homeland. Clearly, this migration is forced. On the other hand, migration induced by the reduction in resources as a result of population growth, depletion of resources, environmental degradation, climate change, etc. often takes place in a gradual fashion. In this process, people affected can make their decision to move or to stay by judging benefits and costs of migration in a relatively long time frame. For this type of migration, the movement is rather voluntary than forced.

In the contemporary world, population migration in all countries has been influenced, to various degrees, by the governments and their representative agencies through policy measures and direct administrative intervention. This influence can be either facilitative or constrictive to the migration. In China, since the early 1980s the so-called ecological migration has been carried out in some extremely poor and environmentally hostile areas by the government. Residents are moved out and relocated to other places. The relocation is often accompanied with a provision of basic means of living, typically housing, land and in some cases, irrigation facilities (Xu, et al 1996; Ma, 1997; Fu, 2000 and Runnstrom, 2000). For the people relocated, the migration will can range widely from voluntary to forced.

The present study looks into the migration as a result of environmental change and the uneven distribution of resources. The role of governments in the population movement is emphasized. The study aims to contribute to a better understanding of the decision-making process of environmentally-induced migration. Two focal issues to be addressed are: 1) characteristics of the will of environmental migration with the presence of government intervention, and 2) interactions of environmental and economic forces in driving migration. A case study from three counties in northern China is used to elaborate these issues.

Three counties selected for the case study are: Huailai, Zhangbei and Fengning, all in Hebei province in Northern China. They are located within 250 km from Beijing. Natural conditions are generally unfavorable. Environmental degradation, especially desertification, is severe. The study at the county level examines the scale of migration and its relations to environmental-economic conditions. In each county, a village is selected to investigate the migration mobility of individual households in association with site-specific conditions. The selection of these villages, however, is not meant to represent the conditions of their respective countries. Data for the three counties and villages were collected primarily from field interviews with county officials, village cadres and farmers during two trips in May and November 2001. Other data used for the analysis are from the population census and official statistics of the counties and the province concerned.

The study begins with an examination of the environmental factors as driving forces of migration and the causal relations among the factors. This is followed by an overview of characteristics of environmental migration in the context of China. The attention then turns to look at migration as an independent variable in the environmental-economic relationship by conducting a case study for the three counties in northern China. A conceptual model depicting interactions of environmental and economic forces in driving migration is built upon the case study. Some of the policy dimensions of emerging migration-environment trends and processes are addressed.

# 2. CONCEPTUALIZING ENVIRONMENTAL MIGRATION

## 2.1 Environmental driving forces in migration

Depending on the ways the impacts are exerted, environmental factors causing migration may be divided into two broad categories: predisposing/pervasive conditions and precipitating/intensive events (Burton, et al, 1978 and Richmond, 1993). Figure 1 lists the factors involved in the two categories.

Predisposing conditions refer to factors relating to adverse living conditions, environmental degradation and resource reduction. These factors are generally pervasive with high frequency, long duration and low energy. The factors in the three groups are often closely interrelated. For example, poverty can promote overexploitation of natural resources. It encourages eco-simplification through destructive effects on habitat and on vulnerable species as a result of poor people in search of land, fodder, and fuel. Similarly, desertification and deforestation can reduce the productivity of land, leading to decreases in land and fuel resources.

Environmental migration precipitated by particular environment-related events can effectively force people to move. These events are intensive with low frequency, short duration and high energy. Such events can be both naturally induced and man-made. When an event strikes, environmental carrying capacity in the area affected can drop sharply either temporarily or permanently, forcing people away from home.

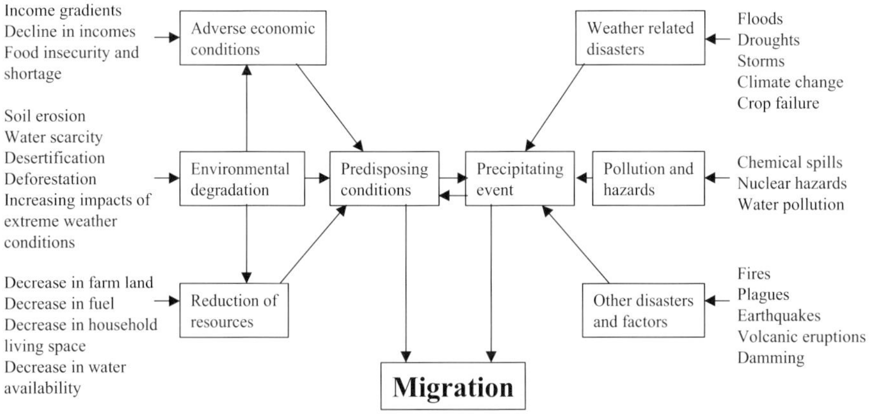


*Figure 1: Environmental Driving Forces in Migration*

Close relationships also exist between predisposing conditions and precipitating events. While a precipitating event can on itself induce migration, the existence of predisposing conditions can reinforce the precipitating event in frequency, intensity and impact on livelihood. For example, deforestation, land reclamation and soil erosion can increase the occurrence of floods on the one hand, and intensify the impact of disasters on the other. The severe flood in the Yangtze River in 1998 has been partly attributable to the encroachment of the lakes along the river in response to the population growth and hence the demand for agricultural land. The activity has substantially reduced the capacity of the river in cushioning peak flows (Wong and Zhao, 2001).

In general, predisposing conditions and precipitating events are more likely to occur in less developed countries than in more developed countries. Likewise, within a country, these factors are likely to impose stronger impacts on less developed regions than on more developed areas (Myers,

1997). Given the close links among the environmental factors specified in Figure 1 and the involvement of economic factors in environmental migration, a holistic perspective that addresses the interactions of these factors is necessary for understanding the process of environmentally-induced migration and dealing with the problems that arise.

## 2.2 Characteristics of Environmental Migration in the Context of China

Environmental migration can range along a continuum from absolutely voluntary to totally forced as shown in Figure 2 (Doos, 1994; Xu, et al, 1996; Hugo, 1996; Swain, 1996 and Hempel, 1997). The former refers to the situation where the choice of the migrants is the overwhelmingly decisive element driving people to move. The latter describes the situation that the migrants are faced with death or severe consequences if they remain in their present place of residence. Most of the migration, however, falls between these two extreme ends (Hugo, 1996). This section presents an overview of the characteristics of environmental migration in China with a specification of the role of government.

*Figure 2. A continuum of voluntary and forced migration. Source: Hugo, 1996.*

### 2.2.1 Spontaneous Choice/Voluntary Migration

Degradation of living environment creates potential migrants. After balancing gains and losses, they make the decision of whether to stay or to migrate. This process can be explained by the classical population migration theory. The income differences between the origin and destination of migration, the employment opportunities and the knowledge of potential migrants of the destination to which they consider to move are important factors in their decision-making. In addition, the distance between the origin and destination and the actual transport conditions also have effects on migration. The massive movement of rural labor from interior areas to the coastal areas in China during the last two decades primarily belongs to this category. As the migration decision is mainly a choice of the people involved, the migration is voluntary. In China, the term 'spontaneous' is often used for this type of migration. The main motivation of the migrants is to seek better living conditions, either environmental or economic, often both. The boundary between environmentally induced migration and economically motivated migration is obscure in this case. The migration mobility of individual people, however, can be strongly influenced by their cultural background, tradition, religion and the level of education. These factors have often impeded the migration of ethnic minorities in China (Xu, et al, 1996).

### 2.2.2 Guided Migration

This type of migration involves a certain level of insurance and compensation to migrants from the government and its representative organizations. For example, the government can encourage people to move by covering partial or entire removel cost, providing resettlement areas, co-ordinating the immigrants and local resident relations, creating employment opportunities and other preferential policies. In some ecologically fragile areas where environmental degradation is severe, restrictions on human activities can be imposed to push residents to move away. These measures are particularly preferred when the investment for rehabilitating the environment is considered to be greater than the cost of relocating the people from the government's point of view. In China, guided migration has taken place mainly in poor areas. The relocation of the people has often been carried out in conjunction with broader social-economic objectives of the government, typically poverty alleviation. As the people involved are allowed to choose not to move to the places designated by the government, the migration can still be classified as voluntary, though the degree of the willingness to move may be lower than the spontaneous migration.

### 2.2.3 Forced Migration with Compensation

This type of migration occurs as a mandatory displacement. In this case, the decision of whether or not to migrate is solely the will of the government. Such a migration is often linked with the large-scale infrastructure construction, environmental rehabilitation and reduction of population pressure in particular areas. The displacement caused by infrastructure construction, particularly the building of dams and reservoirs, has been massive in China. The construction of the Three Gorges Dam has forced over one million people away from their homeland (Xu, 1996). Also, in the effort of rehabilitating heavily degraded ecosystem, mandatory enclosure has been imposed in some areas (Ma, 1997). Residents in the areas closed are forced to move out irrespective of their will, though the resettlement assistance and a certain amount of cash compensation are usually provided. The migration will in this case, is located in the right-hand side of the continuum in Figure 2.

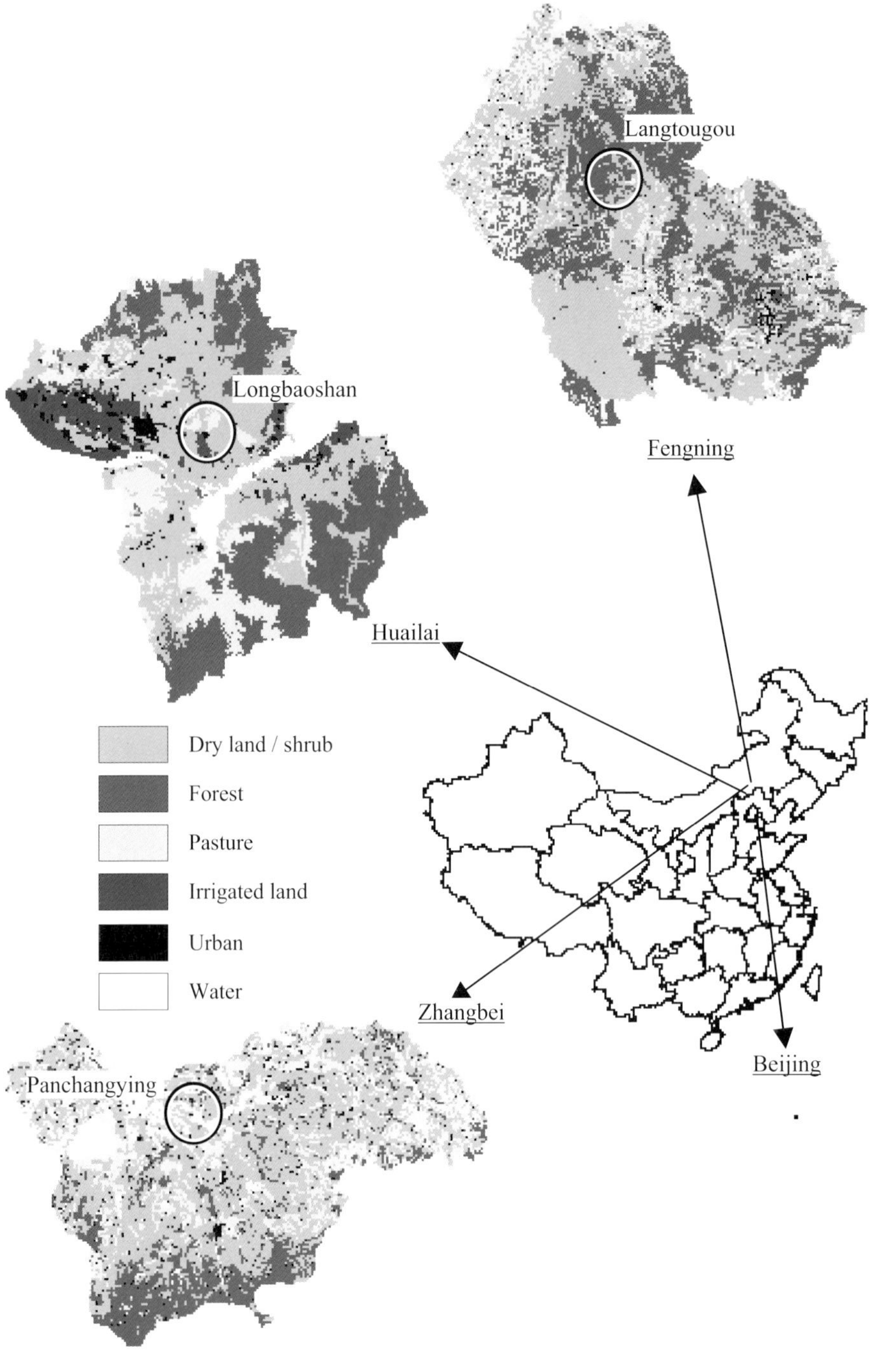


*Figure 3: Spatial location of the three counties in relation to Beijing.*

#### 2.2.4 Migration Forced by Environmental Disasters and Loss of Livelihood

Extreme disastrous events can cause entire loss of means of livelihood. People affected are forced to move, often to the nearest possible place in search of survival. In this case, the migration will is located at the right-end of the continuum. In China, the scale of this type of migration is rather small and the distance of the travel is also short. This is partly because of the limited spaces that allow exoduses to relocate without the permission of the government. When a catastrophic event strikes, the response of the government is typically to provide disaster relief, including food, clothing and temporary shelter to those affected and later to help them rebuild homes on the original sites or in the areas nearby. This has prevented the eruption of large-scale disaster-forced migration. In some remote areas, however, the government relief can be inadequate and ineffective. A case study by one of the authors of this paper found that in Yushu and Guoluo prefectures of Qinghai province in western China, overgrazing, together with droughts during the last decade, has led to collapse of the pastoral ecosystem. Local herdsmen are forced to leave their home without a clear destination in mind. Lack of government assistance and the remoteness of the region have made many of them environmental refugees (Zhang, 2001).

Overall, environmentally-induced migration in China mostly takes place in a gradual fashion. Government policies and direct administrative intervention have played an important role in this process. The balance of gains and losses assessed by individual migrants bound with specific social background, however, determines the degree of the willingness to migrate. This makes it necessary to look at the environmental migration in a broad social-economic context. In the succeeding two sections, we elaborate the interactions between the environmental and economic forces in mobilizing migration using a case study for the three counties in northern China.

## 3. ENVIRONMENTAL-ECONOMIC FORCES AND THE SCALE OF MIGRATION - AN OBSERVATION ON THE THREE COUNTIES

### 3.1 Social-Economic and Environmental Conditions of the Case Study Counties

The three counties: Huailai, Zhangbei and Fengning (Figure 3), are located in the temperate continental monsoon climate zone. The cold and windy winters and the warm and dry summers are the general characteristics of the climate. The annual average rainfall ranges between 300 mm and 600 mm. Most of the rain falls between May and August. Lack of water is a common predicament in the three counties. Agriculture is mainly rainfed and irrigation is only available in limited areas and for commercial crops, such as grapes and vegetables. Droughts, especially during the spring, often cause crop failures. Strong wind in winter and spring is also a common hazard. It damages seedlings on the one hand and increases evapotranspiration on the other. Meanwhile, the strong wind blows away the topsoil, reducing the land productivity. Overgrazing, reclaiming grassland for farmland and deforestation have exacerbated the problems.

*Table 1. Natural environmental conditions of the three counties*

| | Zhangbei | Huailai | Fengning |
|---|---|---|---|
| January average temperature (°C) | -15.6 | 8.9 | -15.8 |
| July average temperature (°C) | 18.3 | 24.1 | 19.8 |
| Annual average precipitation (mm) | 401 | 431 | 425 |
| Percentage of rainfall during Jun.-Aug. | 70% | 70-85% | 70-80% |
| Annual frost-free days | 106 | 149 | 117 |
| Arable land per capita (ha) | 0.38 | 0.10 | 0.19 |
| Percentage of desertified area in total area | 50.4% | 18.5% | 33.3% |
| Percentage of disaster-free cropland in total cropland | 5.45% | 42.28% | 8.20% |
| Major weather hazards | drought, wind, frost, hail | drought, wind, hail, frost | drought, hail, flood, frost, wind |

Source: Ministry of Civil Affairs and Ministry of Construction, 1992.

Table 1 gives some indicators of natural environmental conditions of the three counties. While the difference in annual rainfall is relatively small, the average temperature varies significantly among them. The higher temperature in Huailai provides a much larger number of annual frost-free days than the other two counties. Also, although the major weather hazards in the three counties are similar, the intensity and frequency of adverse weather conditions in Huailai are lower. This is reflected by the relatively large percentage of disaster-free cropland in total cropland.

The climate record shows that in the region where the three counties are located, the weather has become warmer and drier during the last 50 years. While the rainfall did not change significantly, the aridity index increased due to the higher temperature and thus greater evapotranspiration (Figure 4). The global climate change is thought to be responsible for the increasing aridity index (Ministry of Science and Technology, 2000).

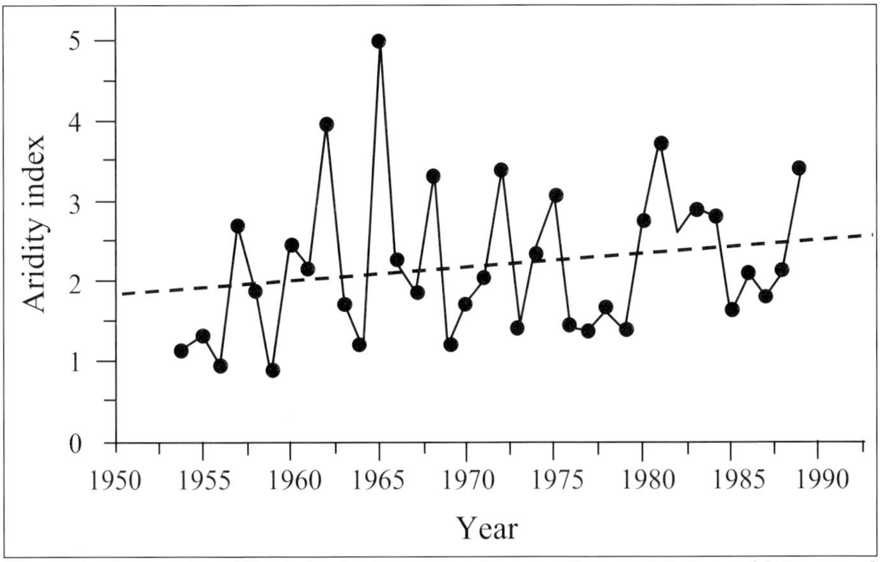


*Figure 4. Changes in aridity index in the case study areas. Source: Ministry of Science and Technology, 2000.*

Over the years, the three counties have experienced continuous environmental deterioration, typified by desert advancement and desertification. Currently, over half of the territory of Zhangbei is affected by desertification. In Huailai and Fengning, the areas affected are also substantial, 18 percent and 33 percent, respectively (Huailai Government, 2001; Fengning Government, 2001 and Zhangbei Government, 2001).

Situated to the northwest of Beijing and Tianjin, the geographical links between the three counties and the two economic centers are intrinsic (Figure 3). Fengning is the origin of the two major rivers, Chaohe and Baihe, that supply drinking water to Beijing and Tianjin, respectively. Huailai holds one of the two reservoirs for Beijing's water supply. Environmental degradation in these counties has imposed increasing threat to the water sources of the two cities. The situation has drawn much concern from the central government and the governments in the two municipalities.

Table 2 presents some indictors of social-economic development of the three counties. In spite of their proximity to Beijing, the economic development is generally retarded. Agriculture and animal husbandry, sectors relying largely on natural resources, remain the major sectors of the economy. Township and village industries, which have flourished in southeast coastal regions, are virtually absent in these three counties. Against the generally low levels of economic development in the three counties, however, there are differences among them. The income level in Huailai ranked at the middle of the province and the other two counties at the bottom. In fact, Fengning and Zhangbei are among the poorest counties in the nation (State Statistical Bureau, 2000). While agriculture is important in the economy, the level of its development is low. In Zhangbei, the application of chemical fertilizer is only 34 kg/ha. The share of irrigated areas in total crop areas is also small. The gradient of economic development among the three counties appears to have a high correlation to their natural-environmental conditions.

*Table 2. Social and economic indicators of the three counties.*

| | Zhangbei | Huailai | Fengning |
|---|---|---|---|
| GDP per capita (yuan) | 2086 | 5824 | 4093 |
| Rural GDP per capita, 1999 (yuan) | 1324 | 2549 | 1338 |
| Total population, 1999 (1000 persons) | 371 | 328 | 373 |
| Rural population, 1999 (1000 persons) | 331 | 271 | 331 |
| Population density (persons/km2) | 88.54 | 182.12 | 42.56 |
| Total GDP (million yuan) | 773.52 | 1901.08 | 1523.59 |
| Share of primary industry in GDP (%) | 40.46 | 16.88 | 28.94 |
| Chemical fertilizer application (kg/ha) | 34.18 | 246.30 | 114.26 |
| Irrigated area in total crop area (%) | 13.36 | 52.44 | 18.64 |
| Grain yield (kg/ha) | 1414.50 | 3521.08 | 2465.02 |
| Grain per rural capita, 1999 (kg/capita) | 335.84 | 300.66 | 394.43 |

Source: SSB, 2000.

## 3.2 Population Changes and the Scale of Migration

Studies of China's population movement cannot be proceeded without understanding its population registration system. The system was first enacted in the 1950s. It had effectively restricted the population movement until the 1980s. Under this system, the population movement was strictly controlled and the move from one place to another was prohibited without the permission of the government (Selden, 1993 and Yang, 2000). The economic reform during the past 20 years has brought about a gradual relaxation over the control of population movement. The entry threshold, however, remains high for rural residents to move to cities. Due to the constraints on the mobility, an increase in population is seen almost everywhere in the country over the years. This is also the case in the three counties studied. One way to observe the population movement, therefore, is to count for the relative increases in their total population.

Figure 5 shows that Zhangbei has the smallest growth of total population among the three counties and Huailai has the largest. A similar situation is also exhibited among the rural population. Given China's strict one child policy and the similar population sizes in the three counties (Table 2), the difference in the population growth reflects primarily their different scales of migration. Down this line, the small growth of the total population in Zhangbei indicates a large number of emigrants in comparison to the other two counties.

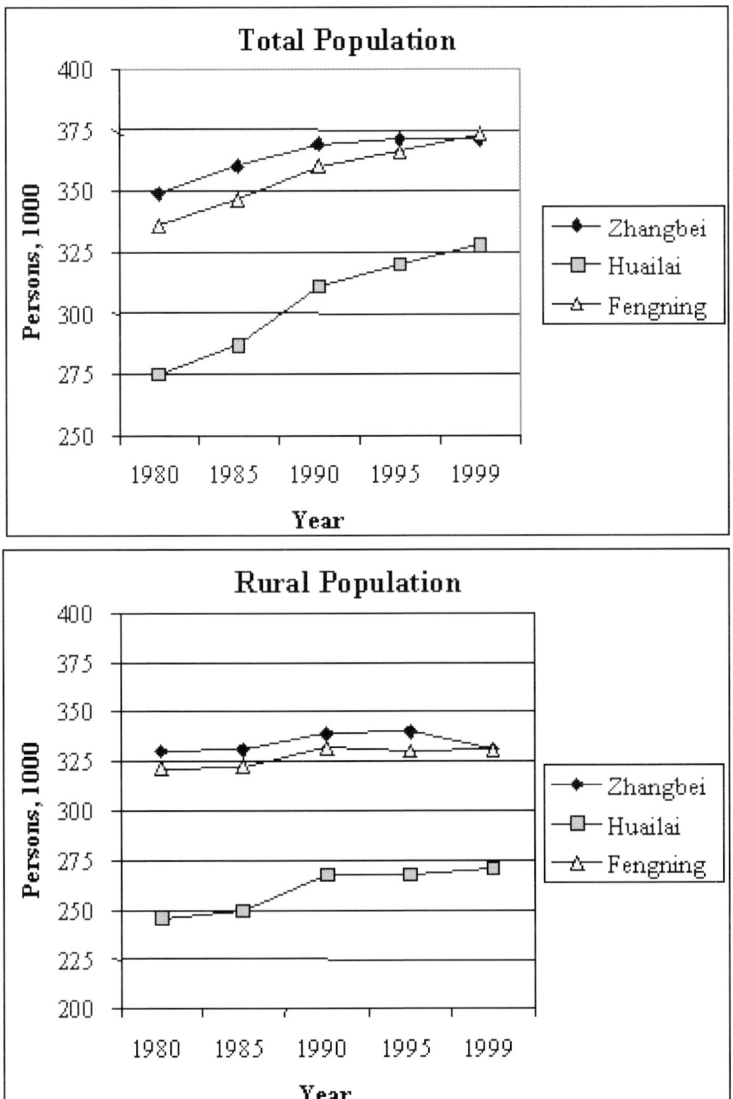


*Figure 5. Changes in Population, 1980-1999. Source: SSB 1985-1990.*

Table 3 shows the actual numbers of migrants in the three counties between 1985 and 1990 derived directly from 1990 population census. Zhangbei has the highest net emigrant figure. In Huailai, the net emigrant figure is much lower. This is in agreement with the trend shown in Figure 5. The inferior environmental conditions and the poor economic development in Zhangbei in comparison to Huailai would mean a stronger push of migration in the former than in the latter. This result seems to support the conventional notion that poorer people have a higher propensity to migrate (Rozelle, et al, 1999). However, it is also noticed that the lowest figure for net emigrants appeared in Fengning, despite its inferior natural and economic conditions in comparison to Huailai. The situation implies impacts of site specific conditions on the migration mobility. In Fengning, half of the inhabitants are the Man minorities. The associated culture and tradition have, to some extent, impeded their mobility.

*Table 3. Number of migrants, 1985-1990.*

| | Zhangbei | Huailai | Fengning |
|---|---|---|---|
| Emigrants | 20230 | 10934 | 6388 |
| Immigrants | 2170 | 4610 | 1042 |
| Net emigrants | 18060 | 6324 | 5346 |

The numbers in the table are the migrants who have stayed at the current destination for more than a year. The migration is more or less permanent. Source: Office of Population Census of Hebei, 1993.

# 4. PERCEPTIONS OF ENVIRONMENTAL CHANGES AND PATTERNS OF MIGRATION - AN OBSERVATION AT THE VILLAGE LEVEL

## 4.1 General Characteristics of the Three Villages

To gain insights into the factors influencing migration mobility, three villages in the respective counties are selected for a household survey. They are Panchangying (Zhangbei county), Longbaoshan (Huailai county), and Langtougou (Fengning county). The three villages are known for their inhospitable natural conditions and severe environmental degradation, particularly desertification. Longbaoshan is a village moved to the current location in the early 1990s when the original village was buried in sand. Villagers were assigned housing and land in the resettlement area by the government. At the time when the village was first relocated, the nearest sand dune was about 1 km away from the village. 10 years later, today the sand dune has intruded the village and the backyards of many households have been half buried. The village is again faced with the ordeal of being forced to move. In the other two villages, the livelihood is also seriously affected by the intensification of desertification.

All the three villages possess neither industry nor irrigation facilities. Rain-fed agriculture and animal husbandry are both important in Panchangying and Langtougou. In Longbaoshan, agriculture is the sole activity. Rapid expansion of agricultural land in the three villages occurred in the 1960s and 1970s. Since the 1990s, the reclamation has dampened. Due to the population growth, the agricultural land on per capita basis has been shrinking.

*Table 4. General information of the three villages.*

| | Panchangying | Longbaoshan | Langtougou |
|---|---|---|---|
| Number of households | 308 | 187 | 70 |
| Number of registered residents | 948 | around 600 | 218 |
| Number of residents currently in village | around 400 | - | - |
| Average cropland per capita (ha) | 0.47 | 0.33 | 0.40 |
| Estimated average annual income (yuan/person) | 1600 | 1000 | 700 |
| Level of income with reference to country average | average | low | Low |

Source: own data from the household survey.

Table 4 shows some demographic and socio-economic features in the three villages. It is noticeable that the ratio of total number of residents to the number of households is low in all the villages, indicating a small average size of families. Most of the households surveyed have only one child. This situation is no doubt attributable to the one child policy enforced since the early 1980s. The small average family size in Longbaoshan and Langtougou is, however, also related to their impoverished conditions. There are quite a few 'single man families' in the villages because they are too poor to find wives. Most of households surveyed have at least one family member working outside of the village, mostly in Beijing and surrounding town centers. Crop farming provides mainly staple food for household self-consumption. Cash income is meager and primarily from animal husbandry and/or non-agricultural activities.

The average size of cropland per capita in the three villages, though varying, is significantly larger than the national average of less than 0.08 hectare/capita. However, the land figures here are highly inflated due to the fact that spring droughts often allow no chance for farmers to sow the seeds, not to mention the failures caused by disasters on crops during the growing period. In Longbaoshan, for example, a severe spring drought in 2001 left most of the cropland unplanted. Similar situation was also seen in the other two villages. Agricultural production is essentially an activity that relies merely on prevailing climatic conditions.

Under China's household registration system, a person's resident status is defined by the location of his/her registration, instead of the actual location of the person. Hence, the total number of residents in the villages surveyed refers to those who have the local resident registration. Village cadres can easily tell the population figures in their villages under this definition (shown in Table 4). The actual number of people residing in the villages at any given time, however, is much smaller than the number registered. This is because many young and educated labors leave to nearby cities and towns to undertake non-agricultural jobs. In Panchangying, for example, about half of the registered population was absent for this purpose at the time of the survey.

In all the three villages, a significant portion of the migrants is seasonal and the length of each spell is relatively short. Most of them come back to the village of origin after several months. Working on farms during the peak season and going out to seek off-farm jobs during the slack season is a general pattern of migration in the three villages. This kind of migration is somewhat different from the conventional concept of labor movement in the migration literature, in which the migration is more permanently based. Due to the high mobility of the village members, village cadres all had difficulties in providing a good estimate of the population in their villages at the survey time.

While all the villages surveyed are poor, the degree of the impoverishment varies. Judging by the income level, the size of the average cropland per capita, access to transport facilities, together with the impression of the on-site observation, Panchangying has better natural environmental and economic conditions than the other two villages. By the same criteria, Langtougou has the most inferior conditions among the three.

## 4.2 Environmental Changes and Migration Mobility

In the household survey, farmers were asked questions pertaining to the environmental change and the migration mobility. As the survey was conducted in the origin of migration, the response of interviewees represents those who had chosen to stay at the time of the survey. Table 5 shows the result of the household survey in the three villages.

The household survey shows clearly that the intensification of desertification has been well perceived by villagers. An overwhelming majority of the interviewees considered the increasing aridity of the climate an important reason for the expanding desertification. This is consistent with the climate records in the region, though it remains a question as to what degree the two factors have reinforced each other (Glantz, 1994). Other factors considered to be important include population growth, land reclamation, overgrazing and desert advance.

*Table 5. Major Survey Questions and Responses.*

| Questions | Panchangying (8 households) | Longbaoshan (11 households) | Langtougou (20 households) |
|---|---|---|---|
| Do you think that desertification has intensified over the years? | Yes 100 %<br>No 0 % | Yes 100 %<br>No 0 % | Yes 100 %<br>No 0 % |
| What is the main reason(s) for the intensification of desertification* | Increasing aridity of climate 90 %<br>Land reclamation 0 %<br>Overgrazing 20 %<br>Population growth 10 %<br>Desert advance 0 % | Increasing aridity of climate 90 %<br>Land reclamation 20 %<br>Overgrazing 0 %<br>Population growth 0 %<br>Desert advance 20 % | Increasing aridity of climate 60 %<br>Land reclamation 60 %<br>Overgrazing 55 %<br>Population growth 35 %<br>Desert advance 15 % |
| Do you think that desertification has affected your income? | Yes 90 %<br>No 10 % | Yes 87 %<br>No 13 % | Yes 90 %<br>No 10 % |
| Would you consider migration if the desertification further intensifies? | Not sure 0 %<br>No 70 %<br>- no place to go 50 %<br>- other reasons 20 %<br>Yes 30 % | Not sure 20 %<br>No 60 %<br>- no place to go 60 %<br>- other reasons 0 %<br>Yes 20 % | Not sure 0 %<br>No 90 %<br>- no place to go 85 %<br>- other reasons 5 %<br>Yes 10 % |
| Do the village cadres and the county government encourage labors to seek jobs outside? | Encourage 70 %<br>No body cares 30 %<br>No 0 % | Encourage 30 %<br>No body cares 70 %<br>No 0 % | Encourage 10 %<br>No body cares 90 %<br>No 0 % |
| Is there any member(s) in your family working outside the village? | Yes 70 %<br>No 30 % | Yes 70 %<br>No 30 % | Yes 40 %<br>No 60 % |

*. The sum of the percentages in each village is over 100 percent. This is because some interviewees pointed out more than one reason for the intensification of desertification. Source: Own data from the household survey.

Except for the increasing aridity, factors identified to be responsible for desertification vary among the three villages. In Langtougou, a large percentage of the interviewees pointed out the population growth as an important factor. In contrast, none or few of the interviewees in the other villages considered so. The difference may be partly related to the different scales of migration in these villages. In Longbaoshan and Panchangying, over 70 percent of the surveyed households have at least one member working outside the villages. During the survey in Panchangying, we only saw few people in the streets (this is one of the reasons for the small number of households in the survey sample). Some families hired labors from nearby villages to help with the farm work. Compared with Panchangying and Longbaishan, Langtougou has much fewer people working outside. Only 40 percent of the households have at least one member working outside the village. This is despite the fact that the latter has a lowest average income and the natural environmental conditions are the most inferior. When asked the reason for not leaving, a common answer was that they could not earn much more money outside. During the survey, we heard very often that someone was not paid for his labor, or someone was cheated by his/her employer. When asked the interviewees whether or not they themselves have been cheated or not paid, most of them answered no. This unverified broadcasting of the employment situation outside has, to some extent, deterred the migration of the villagers.

Land reclamation is considered an important factor for the intensified desertification in Longbaoshan and Langtougou, but not a concern in Panchangying. This may be partly because in Panchangying, no further reclamation has been conducted since the 1990s. In the other two villages, reclamation has continued during this period though on a relatively small scale.

Overgrazing is considered to cause desertification in Langtougou and Panchangying. This is in close relation to the fact that raising goats and cattle are an important source of cash income in these villages. The market price for each goat can be as high as 200 yuan. The return is considerable given the average income of 700 yuan/capita and 1600 yuan/capita in the respective villages. It is interesting to note that although the links between overgrazing and desertification are well recognized by the farmers, most of them still prefer to increase the number of their goats simply because they have no other alternative source of income.

Almost all the interviewed farmers considered that the intensification of desertification has affected their income. However, when being asked whether or not they will be forced to migrate if the situation deteriorates, an overwhelming majority replied that they will not. It is interesting to see that the main reason for not moving away is because of the perception that they have 'no place to go', instead of they don't want to move. Here, 'no place to go' is a metaphor of no chance to find a job outside. There is no significant difference in the answer to this question among the three villages. The result suggests that the employment opportunities and attainment of the expected incomes at the destination are more decisive than the 'push' forces (deteriorating environment and poverty) in driving migration in the villages surveyed.

The encouragement of the officials seems to have a positive impact on migration. In Panchangying, where the migration ratio is high, most of the interviewees said that they are encouraged by village cadres to seek employment outside. In contrast, a majority of the interviewees in Langtougou said that nobody takes care of the matter. Seeking jobs outside relies primarily on the helps from their relatives and friends.

## 5. ENVIRONMENT-ECONOMY NEXUS AND MIGRATION DECISION

In spite of the typologies that differentiate environmentally-induced migration from the economically motivated migration, our case study has shown that there are intrinsic links between the two. In addition, the discrepancies in the mobility and scale of the migration are also influenced by site-specific conditions. Figure 6 depicts the interactions of environmental-economic forces in driving migration based on the case study of the three counties and villages.

Population growth can have a significant impact on predisposing conditions and precipitating events. Overpopulation can also aggravate poverty. Predisposing conditions, precipitating events and poverty together generate push forces of migration in the origin. On the destination side, the existence of employment opportunities and the chance to earn the expected incomes create pull forces of migration. The push and pull forces are the foundation underlying the population migration.

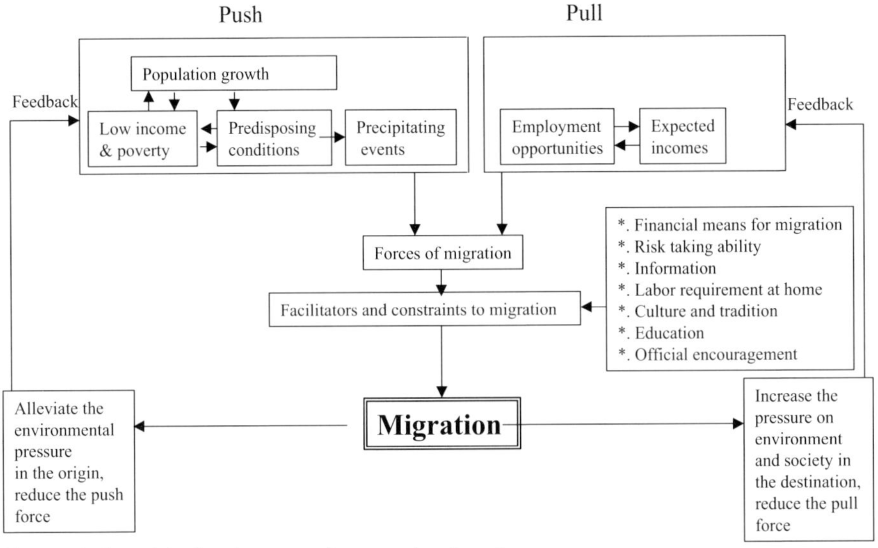


*Figure 6. A model of environmental-economic migration.*

However, the migration decision is a multi-objective process and can be influenced by many site-specific conditions beyond the push and pull forces specified above. These conditions can act as facilitator or constraints to the migration. In our case study, Beijing and surrounding town centers are the common destinations of migration for the three villages, making the pull forces more or less identical for the three villages. On the origin side, the lower income and inferior environmental conditions in Langtougou are meant to create a stronger push to migration than in Panchangying. However, the survey shows that the migration mobility is lower in the former than the latter. The employment information and official encouragement have facilitated the migration in Panchangying. Conversely, the ethnicity, the lack of government assistance and the unfavorable broadcasting of employment information have deterred the migration in Langtougou.

The income level can also impose impacts on the migration decision. A study by Hare (2001) found that a 100 yuan increase in the per capita value of the household's production assets increases spell duration by about two percent. This positive correlation is a result of the substitution of machinery for labor among households with greater accumulation of production equipment. Although our case study does not have sufficient information to verify this relationship, it is conceivable that the extremely low income in Longtougou may have a negative impact on the population mobility. Low income, thus low mechanization level, requires more labor on per unit of land. Meanwhile, low income also reduces the risk taking ability and financial means necessary for making a move.

## 6. CONCLUDING REMARKS

Using the case study for the three counties and villages in northern China, we explored in this paper the multi-objective characteristics of environmental migration and the interactions of environmental and social economic forces in driving the population movement.

Precipitating events and predisposing conditions can induce environmental migration. In China, also likely in most countries in the world, the scale of environmental migration caused directly by precipitating events is relatively small. A majority of migrants are pushed away by predisposing factors. In both cases, the government has played an important role through policy mechanisms and direct administrative intervention. In our case study, environmental degradation, climate change and the decline in resources in association with population growth have been the forces pushing people away from their hometowns and villages. On the other hand, the employment opportunities and the attainment of expected income in Beijing and surrounding town centers have been pull factors attracting migrants to move to. The interaction of these push- and pull-forces leads to a high degree of overlap between the environmental migration and economic migration. Separating the two types of migration is difficult, if not impossible.

For the three counties in the case study, the highest ratio of migrants in the poorest county Zhangbei supports the conventional notion that poorer people have higher propensity to migrate. However, the lowest migration ratio in Fengning suggests that site-specific conditions are also important for the migration mobility. The village level survey identified that the income level, the information network, the cultural background as well as the attitude of the officials towards migration are all influential factors facilitating or impeding the migration.

Environmental migration is essentially a response to the intensification of the conflicts among economic aspiration, population growth and environment degradation/resources reduction. Such a migration on its own, however, cannot solve the environmental problems as shown in our case study. On the contrary, many studies have demonstrated that the migrant influx can have negative environmental impacts on the destinations (Swain, 1996 and Zhao and Tong, 2000). It is, therefore, necessary for policy makers to incorporate environmental migration with objectives of improving the environment and a sustainable use of natural resources in both the origin and the destination of migration.

## REFERENCES

Black, R. 1994. "Forced migration and environmental change: the impact of refugees on host environments." *Journal of Environmental Management* 42:261-277.

Burton, I., K.W. Kates and G.F. White. 1978. *The Environment as Hazard*. New York: Oxford University Press.

Croll, E. and H. Ping. 1997. "Migration for and against agriculture in eight Chinese villages." *The China Quarterly* 152:128-146.

Doos, B.R. 1994. "Environmental degradation, global food production, and risk for large-scale migrations." *Ambio* 23(2):124-130.

Fengning Government. 2001. *Report on the Implementation of the Project on Controlling Sandstorm Sources of Beijing-Tianjin*. Leaflet of the Fengning government.

Fu, C. 2000. "Combating desertification: action program in China." *Ambio* 29(6):359-361.

Glantz M.H. (ed.) 1994. *Drought Follows the Plough: Cultivating Marginal Areas*. Cambridge: Cambridge University Press.

Hare, D. 2001. "Push versus pull factors in migration outflows and returns." *The Journal of Development Studies* 35(1):45-72.

Hempel, L. 1997. "Population in context: a typology of environmental diving forces." *Population and Environment* 18(5):439-461.
Huailai Government. 2001. *Introduction of Ecosystem Comprehensive Protection in Huailai County*. Leaflet of the Huailai government.
Hugo, G. 1996. "Environmental concerns and international migration." *International Migration Review* 3000(1):105-131.
Ingene, C.A. 2001. "The state of the art in modeling migration in LDCs: a comment." *Journal of Regional Science* 41(3):529-543.
Kibreab, G. 1994. "Migration, environment and refugeehood." Pp. 115-130 in *Environmental and Population Change*, editors B. Zaba and J. Clarke. Belgium: IUSSP.
Ma F.X. 1997. „Village relocation: finding a land for prosperity." *Ethnic Unity* 2:27.
Ministry of Civil Affairs and Ministry of Construction. 1992. *Encyclopaedia of Chinese Counties*. Vol. North China, China Social Publishing House, Beijing.
Ministry of Science and Technology (Expert Group of Desertification Protection Plan). 2000. *Analyses of Causes of the Intensification of Wind and Sandstorm and Focusing Areas in Desertification Protection*. Ministry of Science and Technology, Beijing.
Myers, N. 1997. "Environmental refugees." *Population and Environment* 19(2):167-182.
Office of Population Census of Hebei. 1993. Hebei Population Census, 1990. China Statistics Press, Beijing.
O'Lear, S. 1997. "Migration and the environment: a review of recent literature." *Social Science Quarterly* 178(2):606-618.
Richmond, A. 1993. "The environment and refugees: theoretical and policy issues." A paper presented at the meetings of the International Union for the Scientific Study of Population, Montreal, August.
Rozelle, S., L. Guo, M. Shen, A. Hughart and J. Giles. 1999. "Leaving China's farms: survey results of new paths and remaining hurdles to rural migration." *The China Quarterly* 158:367-393.
Schmeidl, S. 1997. "Exploring the causes of forced migration: a pooled time-series analysis, 1971-1990." *Social Science Quarterly* 78(2).
Selden, M. 1993. *The Political Economy of Chinese Development*. Armont, New York: E.M. Sharpe.
State Statistical Bureau (SSB). 1985, 1986, 1999, 2000. *Hebei Economic Statistical Yearbook*, State Statistical Publishing House, Beijing.
Swain, A. 1996. "Environmental migration and conflict dynamics: focus on developing regions." *Third World Quarterly* 17(5):959-973.
Todaro, M. 1976. *Internal Migration in Developing Countries: A Review of Theory, Evidence, Methodology and Research Priorities*. International Labor Organization, Geneva.
Wong K.K. and X.B. Zhao. 2001. "Living with flood: victim's perceptions in Beijiang, Guangdong, China." *Area* 33(3):190-201.
Wood, C.H. 1982. "Equilibrium and historical-structural perspectives on migration." *International Migration Review* 16(2):298-319.
Xu, J., Z. Ouyang, H. Cheng and Q. Lin. 1996. "On environmental migration." *Environmental Science* 17(3):81-96.
Yang, H. 2000. "A comparative analysis of China's permanent and temporary migration during the reform period." *International Journal of Social Economics* 27(3):173-193.
Zhangbei Government. 2001. *The Situation of Combating Desertification in Zhangbei*. Leaflet of the Zhangbei government.
Zhang Y. 2001. *Notes of Field Survey*. Institute of Geographical Science and Natural Resources Research, Beijing.
Zhao, X.B. and S.P. Tong. (2000) "Unequal economic development in China: spatial disparity and regional policy: 1985-95." *Regional Studies* 34(6):549-561.

# ASPECTS OF URBAN/RURAL POPULATION MIGRATION IN THE CARPATHIAN BASIN USING SATELLITE IMAGERY

Judit Bartholy, Rita Pongracz, Zoltan Barcza and Zsuzsanna Dezso
*Department of Meteorology, Eotvos Lorand University, P.O.Box 32. Budapest, H-1518, Hungary. E-mail: bari@ludens.elte.hu, prita@caesar.elte.hu, bzoli@caesar.elte.hu, tante@ludens.elte.hu*

## 1. MAIN EXTERNAL AND INTERNAL MIGRATION IN HUNGARY

According to different estimations (IOM, 2000) 150 million people are moving and settling down to other places each year. Although Central and Eastern Europe is not the most affected area, several social and ethnic groups are involved in migration (in or out of the region). Before the political changes in 1989 only forced or illegal migration was possible in this region. Forced migration occurred because of political reasons after World War II, and during the Stalin era: (i) members of the formal aristocracy or so-called "political enemies" were forced to leave their homes and all properties and move out from the capital and other large cities to rural farms and distant villages; (ii) Slovakian, German, etc. ethnic groups had to move out from the country. Since the collapse of former political regime migration is legally allowed, and usually driven by economical reasons. The most significant external and internal migration factors are presented on Figures 1 and 4, respectively.

After World War I the historical border of Hungary changed significantly (comparison of the present and the former borders can be seen on Figure 1). Mainly from Romania and Slovakia transborder Hungarians immigrated back to Hungary during the 20$^{th}$ century (1), after 1989 immigration increased (2) from other neighbour countries (former-Yugoslavia, Ukraine), as well. Large number of refugees immigrated to Hungary (3) during the 1990s because of ethnical conflicts and war in the Balkan region. Pensions in the US, Canada, and Western European countries allow higher living standards in Hungary, therefore people with Hungarian roots often move back, buy houses, lands and settle down in Hungary (4/a). Foreign agricultural investors and producers immigrate to Hungary mainly from Austria, they buy and cultivate the cheap lands, and employs agricultural workers for low salary (4/b). Among the external components a general westward direction of moving process is also recognizable, which means that



*J.D. Unruh, M.S. Krol and N. Kliot (eds.),*
*Environmental Change and Its Implications for Population Migration*, 289–313.

from Central and Eastern Europe people migrate to Western Europe. In case of highly educated people (5/a) this process can be called "drain brain", but also, people with low education level emigrate for physical work better paid in Western Europe than in Hungary (5/b). Furthermore, Eastern Europeans tend to move westwards to Central Europe (including the Carpathian Basin). In case of Hungary some of the in-migrates are transborder Hungarians living as a minority in other countries, and the rest are non-Hungarians choosing Hungary for economic reasons – very often they are working illegally in the country (without proper legal immigration and work permissions). Refugees from less developed countries (mostly illegal traffickers from Asia and Africa) immigrate or only stay temporarily and emigrate further to the West (6).

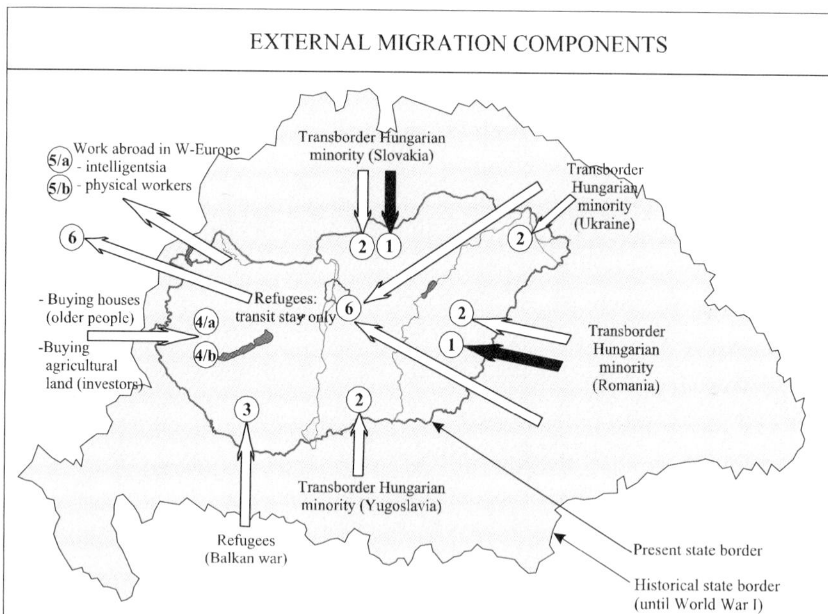


*Figure 1: The main external migration factors in the Carpathian Basin.(Large part of the migration occurs inside the historical state border since the change of the state border resulted in several Hungarian minorities outside the country.) (1): Transborder Hungarians immigrated to Hungary in the 20th century (from Romania, Slovakia); (2): Present tendency of transborder Hungarians migration (from Romania, Yugoslavia, Ukraine, Slovakia); (3): Because of the Balkan war refugees are coming to Hungary from the former Yugoslavian countries; (4/a): After retiring foreign citizens (with Hungarian roots) immigrate to Hungary, buy houses and lands; (4/b): Foreign agricultural investors immigrate to Hungary, they buy and cultivate lands because of cheap land price and low salaries; (5/a): Brain-drain, high educated intelligentsia emigrates to the Western European countries because of better work possibilities and higher salaries; (5/b): Physical workers emigrate to the Western European countries because of much higher salaries; (6): Refugees arrive from less-developed Asian and African countries (mostly illegal traffickers) mainly with transit stay only.*

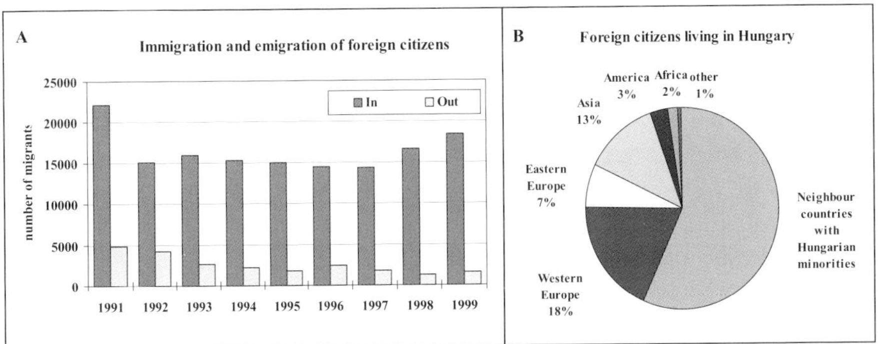


*Figure 2: Recent trends and present status of external in and out migration. Immigrants distribution by region. (After Hungarian Central Statistical Office, 2000.)*

After the recent political changes transborder settling has been liberalized. According to Figure 2/A the number of foreign immigrants increased dramatically from 1989 compared to the foreign emigrants. A large part of these foreign citizens is from the Hungarian minorities of the neighbour countries as Figure 2/B presents (based on Hungarian Central Statistical Office, 2000).

During the last century, migration ratio of Hungarian citizens was usually less than 1%. Exceptions include the large emigration periods after the World War II (about 5% of the inhabitants left the country including prisoners of war to Siberian lagers), after the failed revolution in 1956, and around the beginning of the political changes. These periods can be noticed on Figure 3 showing the statistical data (Hablicsek, 1999) about the total number of Hungarian citizens migrating in and out of the country during the past 110 years.

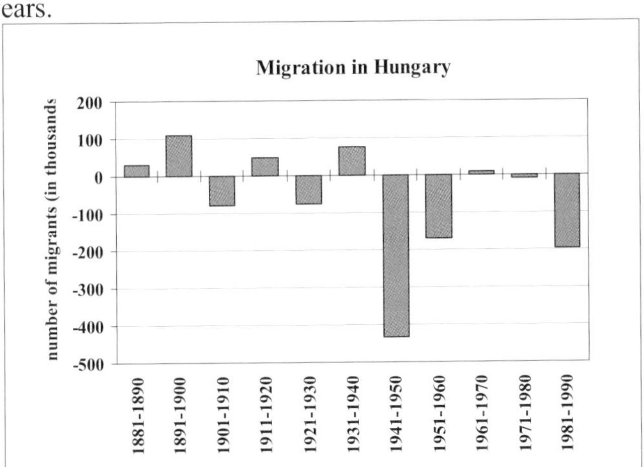


*Figure 3: Number of Hungarian citizens migrating in and out of the country during the last 110 years (after Hablicsek, 1999).*

INTERNAL MIGRATION COMPONENTS

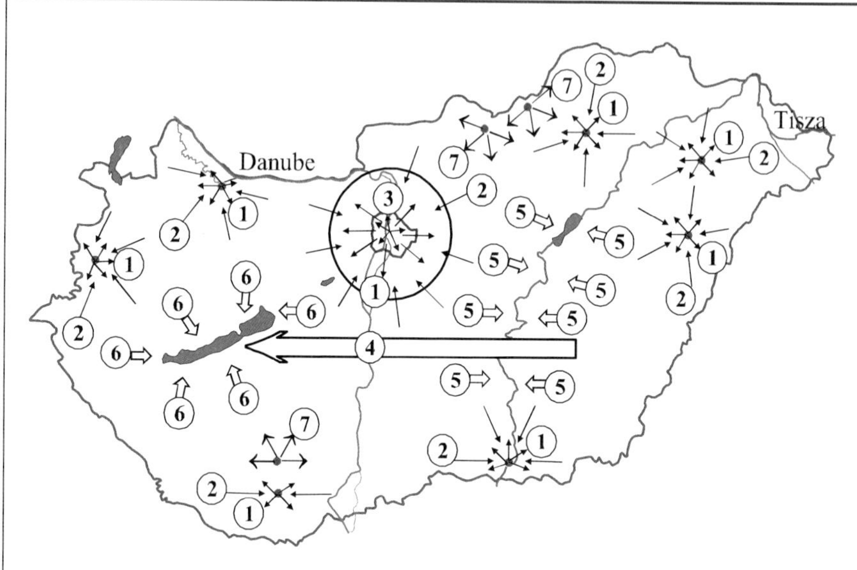


(1) People move out from cities to rural villages
(2) Migration to the city agglomeration
(3) In the capital people move out from the downtown to residential area
(4) Migration from the eastern part of the country to the western part
(5) People move to floodplains for living and agricultural production
(6) People move to the Lake Balaton
(7) Breakdown of mines and large industrial companies results in depopulation of some cities

*Figure 4: Selected important motivations for internal migration in Hungary.*

Nowadays, several migration processes are present simultaneously inside the country. Their structure is mapped on Figure 4 where the most important components are listed. Recently, motivations of migration have changed from being mostly politically driven and often forced to economically driven and entirely based on personal decisions. In general, in case of the large cities people tend to move out to the rural villages located within the agglomeration area (1) because of cleaner air, green and nicer environment. Also, migration from farther rural areas to the city agglomeration can be noticed (2), its main reason being better job opportunities. Since the capital is especially large compared to other Hungarian cities, inside Budapest, people are continuously moving out from the downtown to the outer, greener, safer, and friendlier residential districts (3). Besides moving to the centrally located capital agglomeration a general westward migration can also be noticed in the country (4). One of the popular areas in the western part of Hungary is the region of the lake Balaton where people move because of the rapidly increasing economy and touristical activity (6). Due to precipitation shortage in the last 3-4 decades people settled down in the previously unpopulated floodplains (5), they bought the land, built their houses, and started agricultural production. After the political changes several non-profitable and environment-polluted industrial sectors collapsed resulting in depopulation of regions based solely on these industrial production (7).

Internal migration components related to the capital agglomeration have changed around the early 1990s. Migrating to Budapest was the main tendency before the last decade, but nowadays more and more people are moving out from Budapest mostly to the nearby villages of the agglomeration (county Pest) as shown on Figure 5.

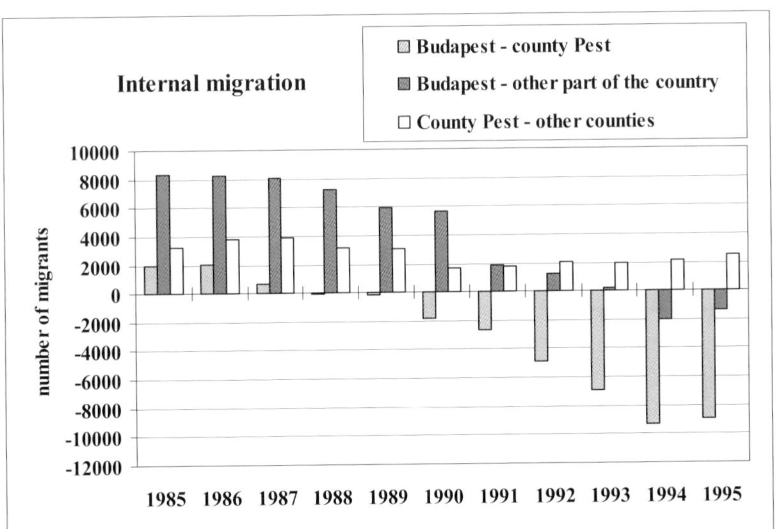


*Figure 5: Turning point between centralized migrations to and around the capital (after Daroczi, 1998).*

Only a small part of migration is related to regional climate change (e.g., factor 5 on Figure 4). Lately, in the Carpathian Basin hydrological extreme events tend to occur more frequently than before. During the last 30-40 years long-lasting precipitation shortage dominated this region resulting in aridification of large floodplains and marshland areas (Bartholy and Pongracz, 1998). Therefore many people decided to move to these places, which temporarily became habitable without any severe natural hazards (i.e., floods). However, in the last 4-5 years the drought spell ended. Flood events have returned (Szalai, 2000) causing large property damage. Several processes can be accounted for, only a few examples are pointed out here: (i) extensive forest clearcuts in the Carpathes – mostly in Ukraine and Romania, (ii) water pollution episodes – cianid poisoning in the river Tisza in 2000 (Index, 2000). These water pollution events hide industrial migration, namely, companies from industrial countries (e.g., Australia, Western European countries) tend to make their production in less developed places because of the lack of strict environment regulations and the high corruption that lessen cheap industrial production and increase their profits.

Important factors related to global change include the regional/local aridification, deforestation, soil degradation processes, and the climatological effects of increasing industrial activity and growing urban agglomeration areas. Therefore, these issues are discussed concerning the Eastern/Central European region in details in the following chapters.

## 2. COMPARATIVE ANALYSIS OF SATELLITE-BASED VEGETATION INDEX VALUES FOR THE ATLANTIC-EUROPEAN REGION AND THE CARPATHIAN BASIN

Several so-called "pull" effects occurred in Eastern/Central Europe recently, which urged to start a land cover/urban climate monitoring project in the Eotvos University. Some of them are listed below:

- with the political and economical changes happened 10 years ago in Hungary the planned governmental control on land use collapsed,
- the lands and most of the forest got back to private owners through casual and chaotic privatization process,
- for the short-term profit large portion of the national forests were cut out,
- the previously more or less balanced agricultural production was disturbed because of the collapsed agricultural cooperative system and the new land owners lacking of experience,
- droughts, floods, and river poisoning disasters occurred in the last few years several times (with significantly higher frequency than it happened before).

All these elements of the above list on agricultural areas are accounted for the need of continuous monitoring of the country. Satellite imagery using NDVI time series provide an excellent tool for this purpose.

Rapid development of satellite use in different topics of research has provided excellent tool to build agricultural monitoring systems and to improve our understanding of the complex links between air, water and land, including vegetation (Lillesand and Kiefer, 1987). The american National Oceanographic and Atmospheric Administration (NOAA) have launched several satellites to polar orbit. One of the sensors aboard these satellites is the Advanced Very High Resolution Radiometer (AVHRR) that was designed to acquire multispectral data for meteorological, hydrologic, and oceanographic purposes. It detects reflected and emitted radiation from the Earth in five bands of the visible and thermal infared intervals of the electromagnetic spectrum (Shimabukuro et al., 1997). Some of these bands can be used for detailed analysis of vegetation and land surface characteristics, while others provide information for fire detection and surface temperature studies monitoring global climate change (Ehrlich et al., 1994).

We have investigated the surface characteristics of the Atlantic-European region and the Carpathian Basin with special emphasis on the vegetation. In order to analyse these selected areas we have used one of the most common vegetation indices derived from remotely sensed data, the Normalized Difference Vegetation Index (NDVI). Spatial and temporal statistical analysis are presented in this paper.

## 2.1 Data

NDVI is one of the most common vegetation indices derived from satellite observations (Tucker, 1979). Values of this index are calculated from the reflected solar radiation in the near-infrared (NIR) and red (R) wavelength bands, 580-680 nm, and 730-1100 nm, respectively. NDVI can be determined using the following formula:

$$NDVI = (NIR - R) / (NIR + R)$$

Thus, NDVI is a nonlinear function which varies between -1 and +1 but it is undefined when the reflectances or sum of R and NIR are zero. Atmospheric corrections and quality control are always needed (Putsay et al., 1998; 2000). NDVI values vary with absorption of red light by plant chlorophyll and the reflection of infrared radiation by water-filled leaf cells (Justice et al., 1985). It is correlated with Intercepted Photo-synthetically Active Radiation (IPAR). In most cases IPAR and NDVI is strongly correlated with photosynthesis. Because photosysthesis occurs in the green parts of plant material the NDVI is usually used to estimate green vegetation (Townsend et al., 1991). Economical use of the derived NDVI includes agricultural applications, namely, yield estimation and forecast (e.g., Hamar et al., 1996; Rasmussen, 1997).

Free access datasets observed by satellites of NOAA, namely, spacecrafts 7, 9, 11 and 14, are available for the period of 1981-2001. Quality controlled (Rao, 1993) observations can be downloaded with coarse spatial resolution of 1°x1° or fine spatial resolution of 8 km x 8 km, in this paper we have used this latter dataset. Composite datasets are created from daily measurements. We investigated both monthly and 10-day NDVI composites for the Atlantic-European region and the Carpathian Basin.

## 2.2 Results and Discussion

In our investigations 10-day and monthly NDVI values for the Atlantic-European region and the Carpathian Basin have been statistically analysed and compared for the last two decades. Variation analysis focus on month-by-month fluctuations, interannual variations, and possible tendencies.

First, annual maximum and minimum NDVI fields have been determined for both areas in the period 1981-2001. Furthermore, annual variation can be characterized by the flush field as being the difference between the largest and the smallest NDVI values of the given year. In Figures 6 and 7 extreme low and high annual maximum, minimum, and flush NDVI fields are presented for the Atlantic-European region and the Carpathian Basin. In order to provide better visualization and improve the evaluation of these fields different grey scales have been applied on the maps.

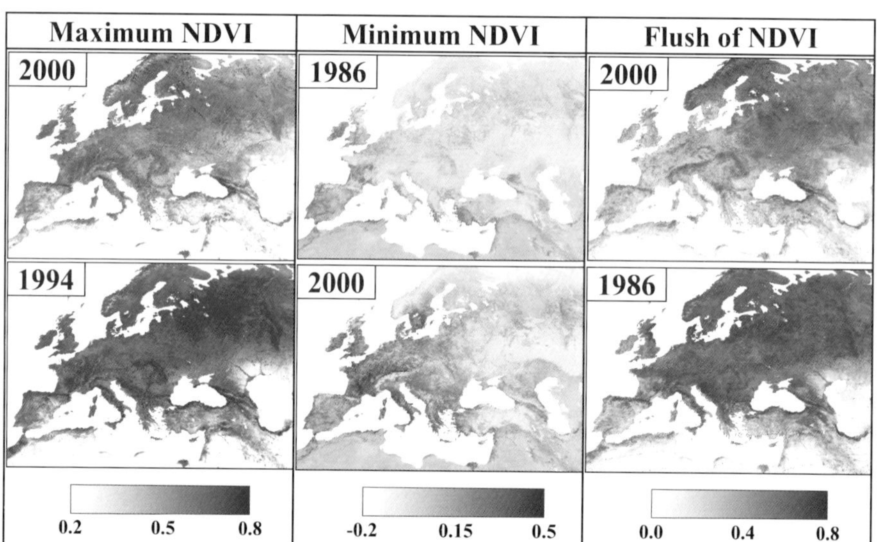


*Figure 6: Range of annual maximum, minimum, and flush of 10-day NDVI fields for the Atlantic-European region (1981-2001).*

Greenness maps of Figures 6 and 7 demonstrate the high annual fluctuations of the NDVI fields for Europe and the Carpathian Basin. Based on the annual maximum, minimum, and flush values for each pixels individual 10-day periods, months, and years can be classified. Further simplified classifications of geographical regions are necessary. Some conclusions can be drawn on the base of the maps.

- The highest greenness of the last two decades occurred in 1994.
- Year 2000 was a very special period in many respects. In largest portion of the entire Atlantic-European region the smallest maximum, the largest minimum, and the smallest flush values occurred in this year.
- During the entire two decades the North-African desert area can be recognized easily having the lowest values of the minimum and flush NDVI fields.

Figures 6 and 7 suggest that years 1986, 1994, and 2000 have extreme behaviour both on large-scale (Atlantic-European region) and on regional-scale (Carpathian Basin).

The lowest maximum values and low minimum NDVI (the second lowest values of the 21 years) resulting the highest flush values occurred in 2000, especially on the Hungarian Great Plains (Fig. 7). This could be explained by the April 2000 flood event and the drought of the second part of the year.

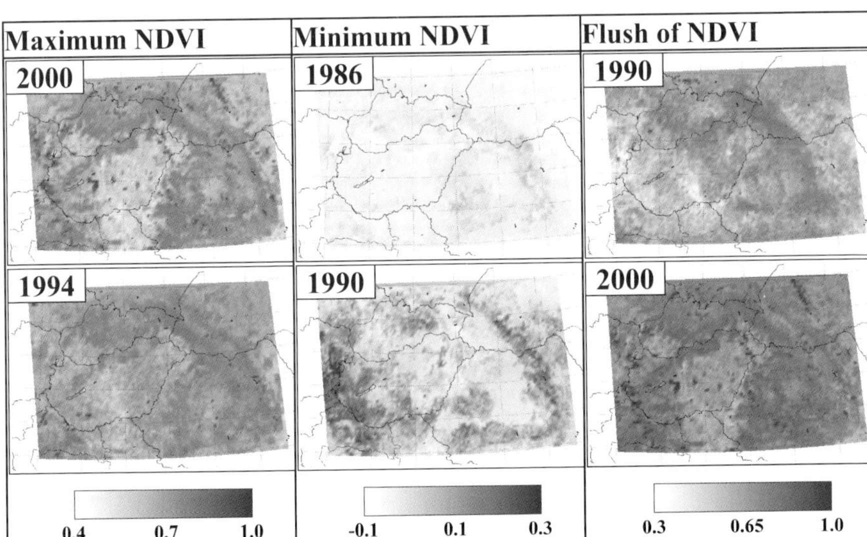


*Figure 7: Range of annual maximum, minimum, and flush of 10-day NDVI fields for the Carpathian Basin (1981-2001).*

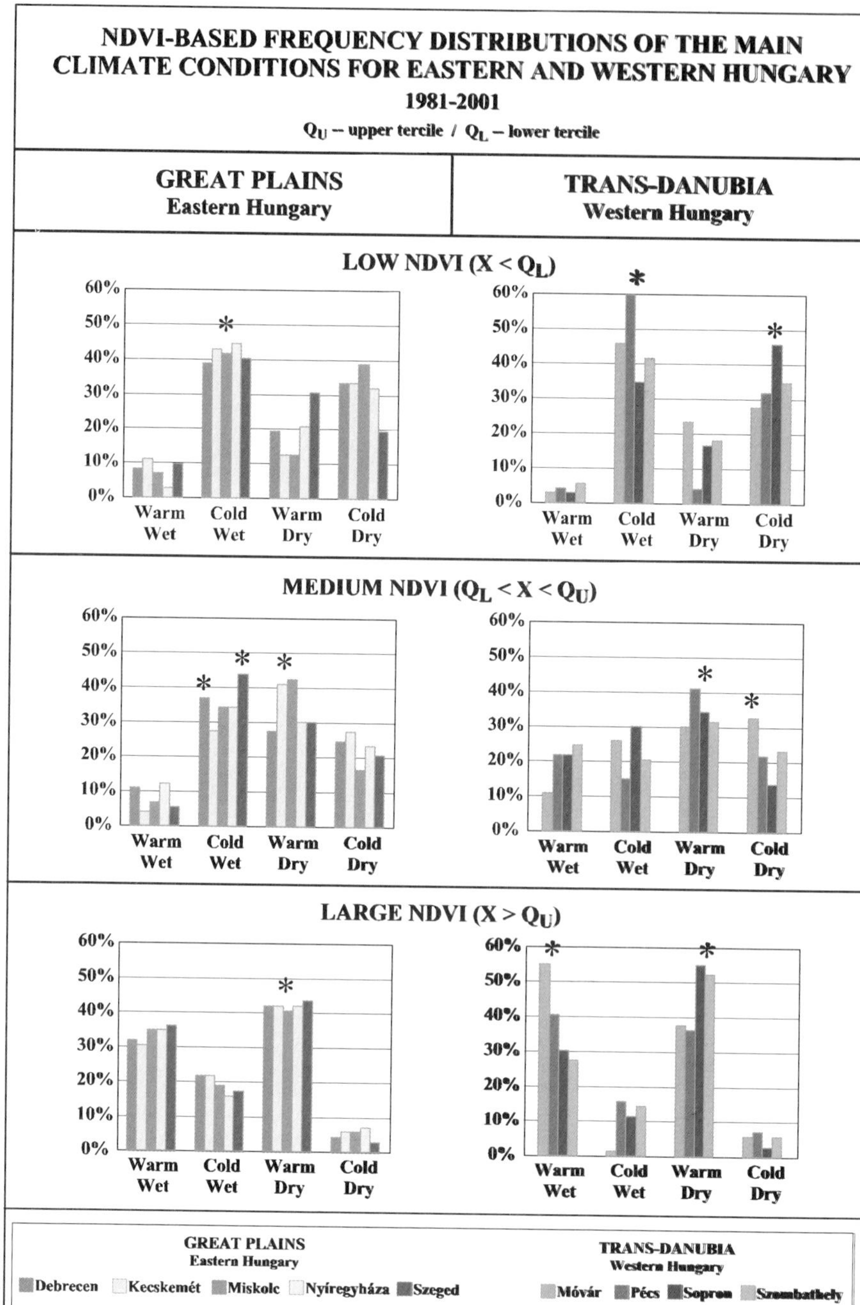


*Figure 8: Distribution of the main climate conditions for Eastern and Western part of Hungary based on NDVI values (1981-2001).*

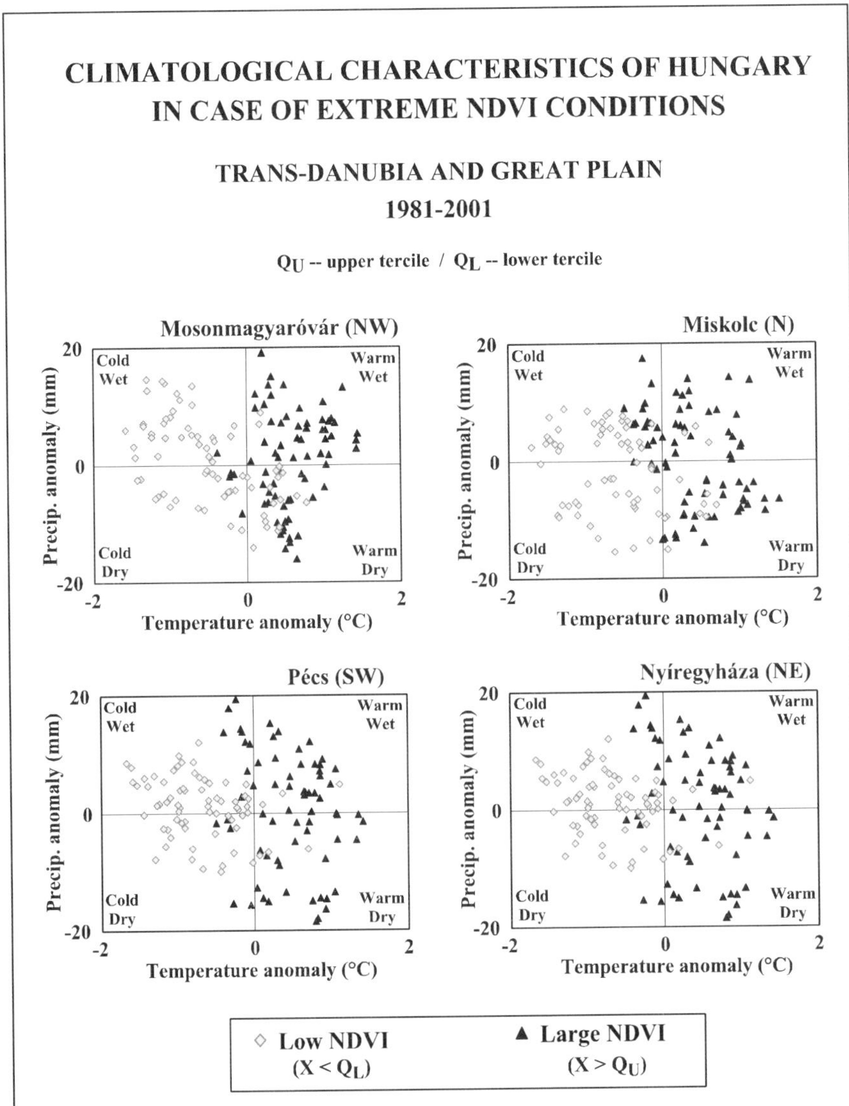


*Figure 9: Climatological anomalies of Hungary related to extreme NDVI conditions (1981-2001).*

The main climate characteristics have been linked to vegetation conditions using upper and lower terciles of the NDVI time series for two Hungarian regions (Great Plain and Trans-Danubia) and nine subregions (5 and 4, respectively). It is demonstrated on Figure 8 that in both regions in case of low NDVI values cold-wet and cold-dry conditions tend to occur, while large NDVI values imply warm-wet and warm-dry climate conditions. Based on the results presented here, subregions of the Great Plains can be characterized by more homogeneous spatial distribution than the Trans-Danubian subregions where relatively large differences may occur in frequency values.

Figure 9 shows the monthly temperature and precipitation anomalies observed in periods of low and large NDVI values (determined by the lower and upper terciles, respectively) for 4 subregions of Hungary. The results suggest that in case of higher than normal temperature large NDVI values are likely to occur, while colder climate conditions provide less favorable environment for the vegetation resulting in low NDVI values.

Next, more than 25 subregions have been selected for detailed temporal analysis both in Europe and the Carpathian Basin. In this paper only a few examples are presented.

The annual variations of 10-day NDVI time series have been determined for each selected subregion. In general, in the vegetation period NDVI show smaller year-to-year variation than in the winter period. Plain and hilly areas of the Carpathian Basin are compared on Figure 10. Small Plains are located at the western part of Hungary, while the Hungarian Great Plains can be found in the middle of the Carpathian Basin. Mountains Mecsek and Bükk are located at southwestern and northern Hungary, respectively.

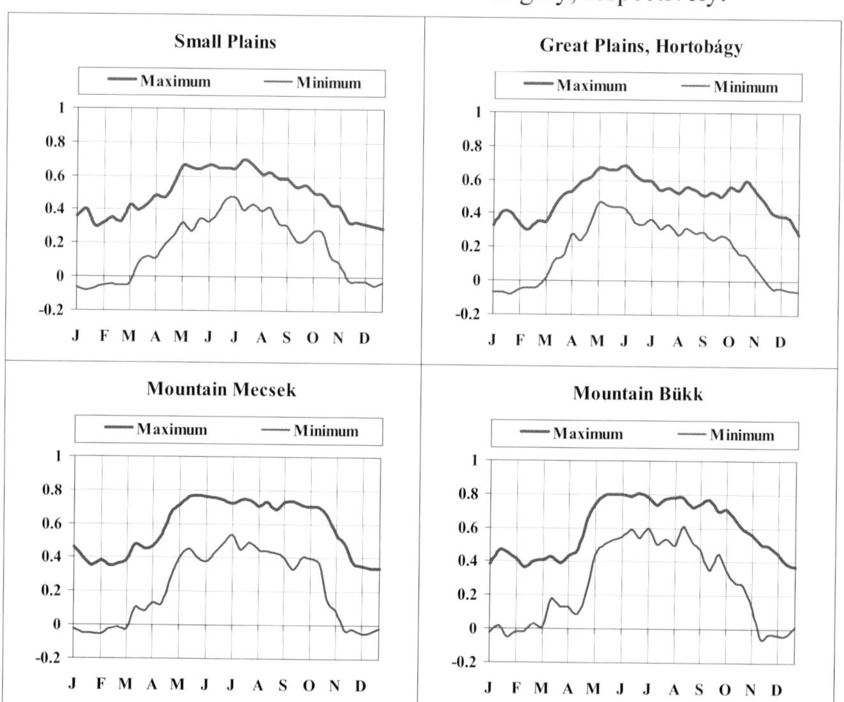


*Figure 10: Annual variation of the range of 10-day NDVI fields in plain regions (upper panels) and hilly regions (lower panels) of the Carpathian Basin.*

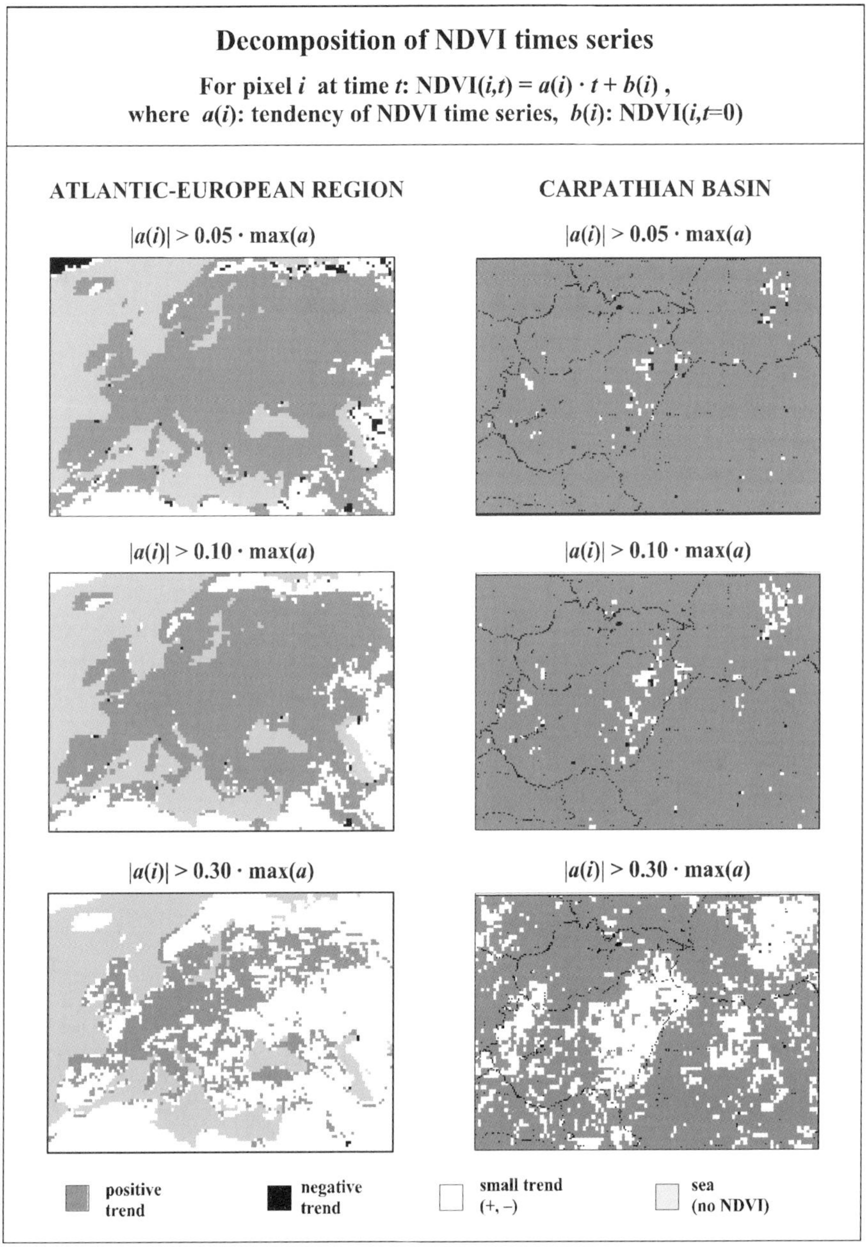


*Figure 11: Tendency analysis of NDVI time series for the Atlantic-European region and the Carpathian Basin (1981-2001).*

Higher NDVI values occur in hilly subregions than on the plains, also the vegetation covering period lasts longer in mountains. The shift from no-vegetation to high-greenness is more abrupt in hills, lasts only 10-15 days instead of 25-30 days that is usual on plains.

Furthermore, detailed tendency analysis of the NDVI fields for the Atlantic-European region and the Carpathian Basin have been carried out. First, the spatial distribution of the pixel trends have been determined, then time series of NDVI anomaly for selected subregions in the Carpathian Basin have been analyzed.

Based on the decomposition of NDVI time series greenness tendencies can be calculated on smaller scale in both regions (Fig. 11). For a given pixel $i$ and time $t$:

$$\mathrm{NDVI}(i,t) = a(i)\cdot t + b(i)\ ,$$

where $a(i)$ is the tendency of NDVI time series, and $b(i)$ is the NDVI value at the beginning of the time series in pixel $i$.

Small trends with almost no change during the 21-year period (1981-2001) are indicated by white area on the maps, while larger negative and positive trends are black and grey, respectively. Figure 11 suggest that significant positive trend of NDVI time series can be detected for both region. Based on the largest tendency value of the entire region three different limit values have been presented. Hence, these maps compare spatial distribution of 21-year tendencies of pixel time series where the 5%, 10% and 30% of the maximum tendency are exceeded.

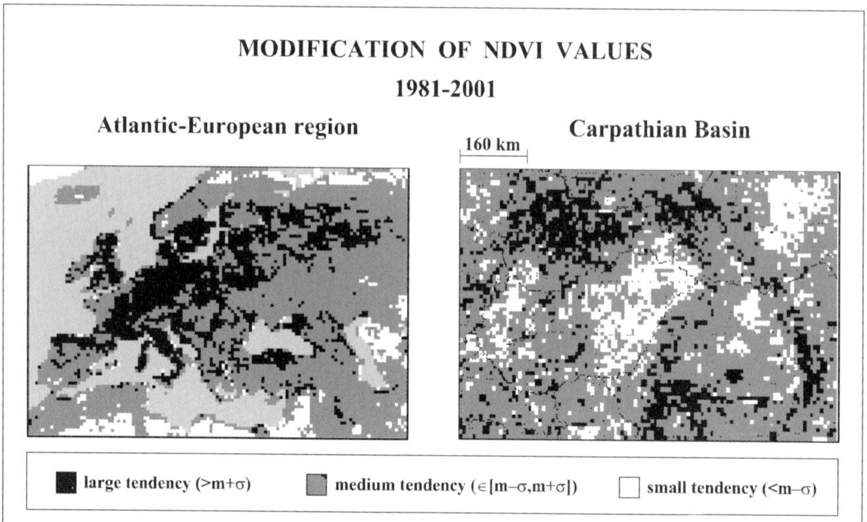


*Figure 12: Tendency of NDVI time series in the Atlantic-European region and the Carpathian Basin (1981-2001).*

Further analysis of positive tendencies can be found on Figure 12 where the mean tendency values and the standard deviation provide finer categories. In case of Europe large tendencies of NDVI time series occur in hilly and low mountain areas, while small tendencies are apparent in the deserts and at higher latitudes. In the Carpathian Basin the forest areas of the Carpathian mountains can be characterized by the largest tendencies, while the smaller trends can be found in flat regions (Great Plains, Small Plains, etc.).

Finally, time series of NDVI anomaly for the selected subregions have been analysed. During the last two decades significant positive tendencies occurred in NDVI anomalies in all subregions. Figure 13 compares two vulnerably sensitive subregions of Hungary, namely Hortobágy at the Great Plains, and Hanság at the Small Plains. In order to smooth the fluctuation of the 10-day anomaly time series 12-month running means are presented.

Furthermore, the interannual fluctuations are filtered providing overall variability of the selected subregions. The 5-year absolute variations are able to demonstrate the tendencies on longer time scales. In case of the Great Plains positive trends can be detected while on the Small Plains variability of the NDVI anomaly decreased in the early 1990s.

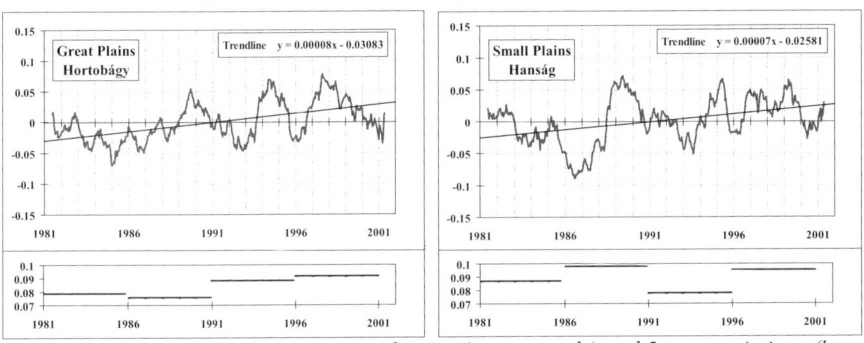


*Figure 13: 12-month running means, tendencies (upper panels) and 5-year variations (lower panels) of 10-day NDVI anomalies in plain regions of the Carpathian Basin (1981-2001).*

## 2.3 Conclusions

Spatial and temporal statistical analyses of NDVI have been presented in this chapter.

Our investigations include calculations of annual maximum, minimum and flush NDVI fields for the Atlantic-European region and the Carpathian Basin. The derived maps can serve as a base for biogeographic regionalization. Furthermore, they suggest that year 2000 was a special period both on large-scale and on regional-scale.

The main climate characteristics have been linked to vegetation conditions using upper and lower terciles of the NDVI time series for the

Hungarian Great Plain and Trans-Danubia. It is demonstrated that in both regions in case of low NDVI values cold-wet and cold-dry conditions tend to occur, while large NDVI values imply warm-wet and warm-dry climate conditions. Based on our results the Great Plains can be characterized by more homogeneous spatial distribution than the Trans-Danubian subregions.

Temporal analyses have been carried out to evaluate the annual variation of greenness in selected subregions of Europe and the Carpathian Basin. Hilly and plain areas have been compared. Then, we have focused on systematic changes of the NDVI time series. Positive tendencies have been found in anomaly as possible local consequences of the global climate warming.

## 3. ANALYSIS OF URBAN HEAT ISLAND EFFECTS IN LARGE CITIES OF HUNGARY

Hungary can be characterized by a unique structure of population density, since around one-fifth of the entire country live in the capital, Budapest (2 millions out of the total 10 millions Hungarian citizens). This centralized structure can be considered a major reasoning why it is important to monitor and describe the urban heat island and other urban climatological modification of the large agglomeration around Budapest.

### 3.1 Data

Part of the Earth Observing System Program of the American National Aeronautics and Space Administration (NASA) satellite TERRA was launched to a polar orbit at 705 km height in December 1999. Main goal of remote measurements of this mission is to improve our understanding of global dynamics and processes occurring on the land, in the oceans, in the cryosphere, and in the lower atmosphere (NASA, 1999). Observations and their use in detailed analysis will play an important role in the development of validated, global, interactive Earth system models being able to predict global climate and environmental change accurately enough to assist policy makers worldwide in making decisions concerning the protection and management of our environment and resources. Planned lifetime of satellite TERRA is about 15 years. The first validated observations started in February 2000 and regular measurements are available from July 2000 via Internet.

One of the five instruments included on TERRA is the sensor MODIS (Moderate Resolution Imaging Spectroradiometer) providing remote observations. It measures biological and physical processes on land and the ocean using a cross-track scanning multi-spectral radiometer with 36 electromagnetic spectral bands from visible to thermal infrared (Barnes et al., 1998). The measurements include surface temperature (both land and ocean), ocean color, global vegetation, cloud characteristics, snow cover, and temperature and moisture profiles. MODIS is capable of viewing the entire globe daily at high resolutions, ranging from 250 m to 1 km per pixel.

Urban heat islands have been investigated using satellite imagery for 25 years, however, the early studies (e.g., Matson et al., 1978; Price, 1979) evaluated coarse resolution satellite data. The use of surface temperature and vegetation index values has been compared in Gallo and Tarpley (1996).

In this paper daytime and nighttime surface temperature time series measured in the Carpathian Basin have been analysed. Surface temperature is strongly related to surface energy budget, latent and sensible heat flux. Calculation of surface temperature data is based on the thermal infrared measurements of MODIS that are quality controlled and calibrated by surface observations (NASA, 1999). The seven spectral bands being used to determine the output are the following: 3660-3840 nm, 3929-3989 nm, 4020-4080 nm, 8400-8700 nm, 10780-11280 nm, 11770-12270 nm, and 13185-13485 nm. Wan and Snyder (1999) developed a model to calculate surface temperature from the spectral observations using atmospheric and surface corrections. This model can be applied in case of clear weather.

## 3.2 Results and Discussion

This paper presents analysis carried out on both daytime and nighttime surface temperature fields of the Carpathian Basin. Spatial structures of the urban heat island of several large cities of Hungary have been determined depending on seasons and different macrocirculation conditions.

First, the ten largest cities of Hungary with more than 80 thousands inhabitants have been selected according to the population data of the Hungarian Ministry of Interior (2000) and their pixel representations (including their rural environment) have been determined from the total 1200x1200 pixels containing the Carpathian Basin at the upper half of the satellite image. Since 20% of the country population live in Budapest (about 2 millions), which means ten times more inhabitants in the capital than in any other large city. Also, the difference between the spatial extensions of Budapest and other cities are similar. Therefore, the agglomeration of the capital is represented by 50x50 pixels from the entire satellite image, while other selected cities are represented by only 30x30 pixels.

Figure 14 provides spatial structures of the surface temperature fields determined for the ten largest cities of Hungary on a clear night (July 30, 2001) when anticyclonic macrosynoptic situation dominated the weather in the Carpathian Basin. The mean temperature increased from west to east. Geographical locations of the cities are shown on the topographical map in the center. The urban heat islands are significant in case of each city, the difference between the mean temperature of urban and rural parts ranges around 1.1-2.7 K. The most intense effect occurred in Budapest on that day. Furthermore, orographical modification of the urban heat islands can be recognised, especially, in case of Pecs.

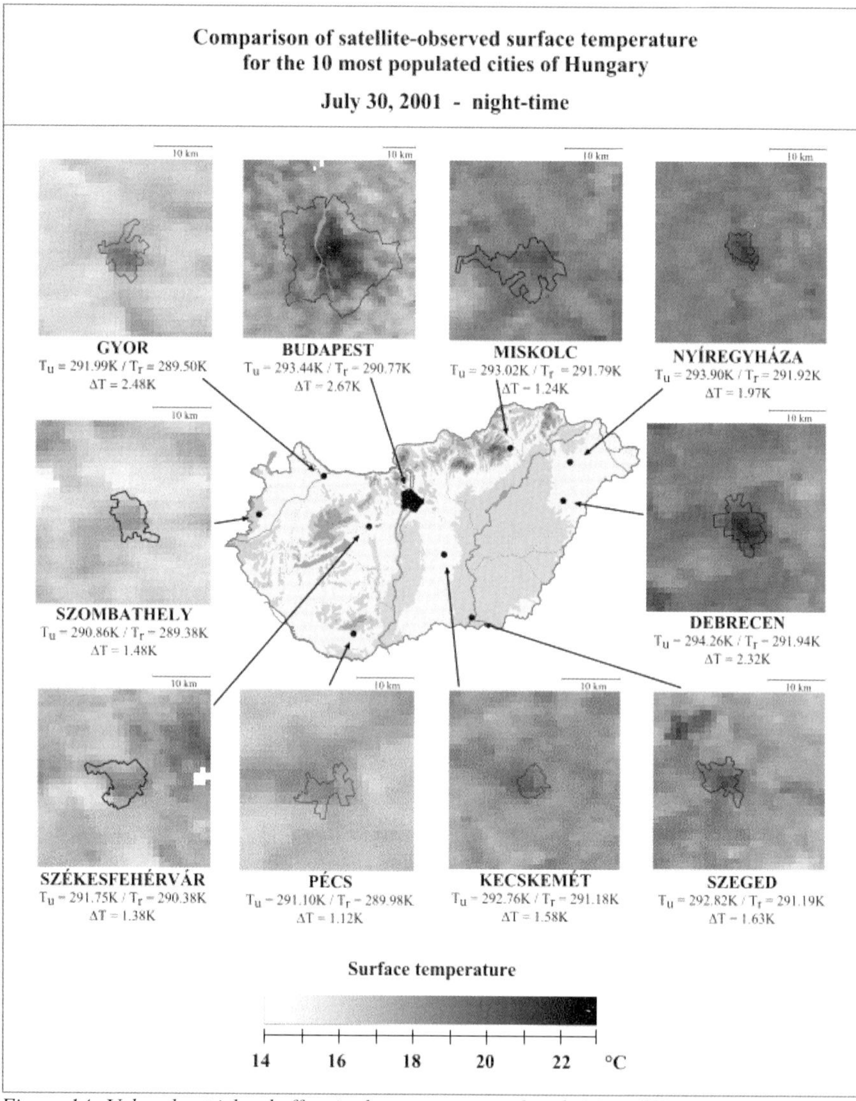


*Figure 14: Urban heat island effect in the ten most populated cities of Hungary.*

Then, the representative areas have been divided into urban and rural pixels in case of each city – hilly regions have been eliminated since they significantly affect the urban heat islands. Comparing spatial averages of observed values converted to surface temperature urban heat island effects have been analysed for the selected large cities located in different regions of Hungary. The results suggest that intensity of the urban heat island detected in the selected Hungarian cities is the highest during the summer season and during the night period (Bartholy et al., 2001). The average values for the selected cities range between 1 K and 3 K.

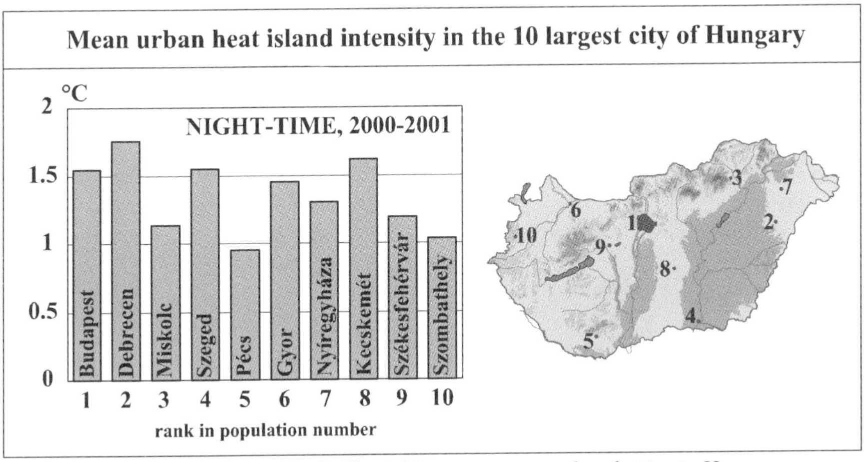


*Figure 15: Mean urban heat island effect of the ten most populated cities in Hungary.*

Mean nighttime urban heat island intensity is shown on Figure 15, where the cities are ranked according to the number of their inhabitants starting with the capital. Basically, more populated cities perform more intense heat island, and only orographical modification disturbs this relationship by decreasing the urban heat island intensity. Since Budapest, Miskolc, and Pecs are located in hilly regions, partly or entirely, their mean urban heat island is less intense than expected. Although city Kecskemet is small, but its central location on the Great Plains may explain the larger mean difference between the temperature of urban and rural areas.

Further analysis have been carried out by identifying special pixels (especially in case of Budapest) and evaluate the time series of surface temperature observed by MODIS. According to the characteristical surface cover several pixels have been selected, namely, residential area with 10-story concrete buildings, airport with large concrete and asphalt cover, cemetery park surrounded by concrete buildings, university campus with concrete buildings and parks, and the Buda Castle that can be characterized by rocks and concrete cover. Their location is presented on Figure 16.

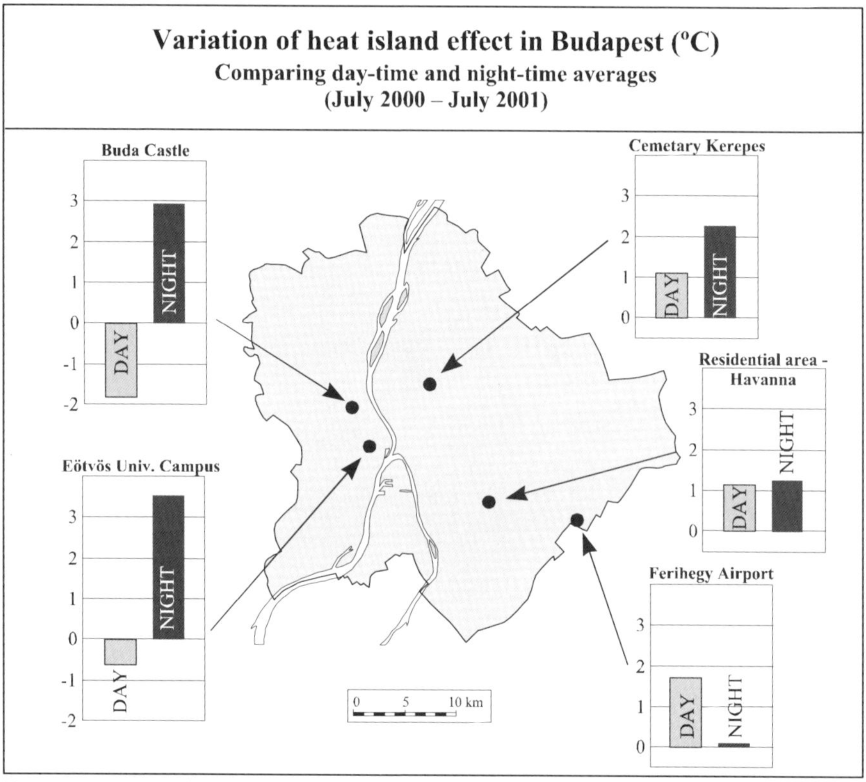


*Figure 16: Mean urban heat island effect during daytime and nighttime in Budapest.*

The airport and residential area are located close to the border of the city, while the other three pixels can be found in the downtown area. Since the Castle and the University campus are near the Danube, the river affect their surface temperature, especially during daytime when rural environment is usually warmer than the urban pixels. During nighttime all urban pixels show higher surface temperature than the rural environment that means urban artificial coverings store significant heat energy.

Finally, satellite and ground based observations have been compared keeping in mind that regularly observed temperature by the meteorological services of the entire world is measured at 2 meter height in a sheltered white box, while satellite data sets are representatives of surface characteristics. According to Roth et al. (1989) values of satellite-based surface temperature are higher with larger standard deviation than ground based temperature observations. Their correlation ranges usually about 0.7-0.8 (Nichol, 1996).

Figure 17 (left panel) represents scatterplot diagram and linear regression analysis of daily minimum temperature based on ground observations in Budapest and surface temperature measured by satellite TERRA. These two variables are very strongly correlated (0.98). Somewhat weaker but still strong linear correlation (0.91) have been found between daily maximum temperature and satellite-based surface temperature. Their scatterplot diagram and linear regression analysis are shown in Figure 17 (right panel).

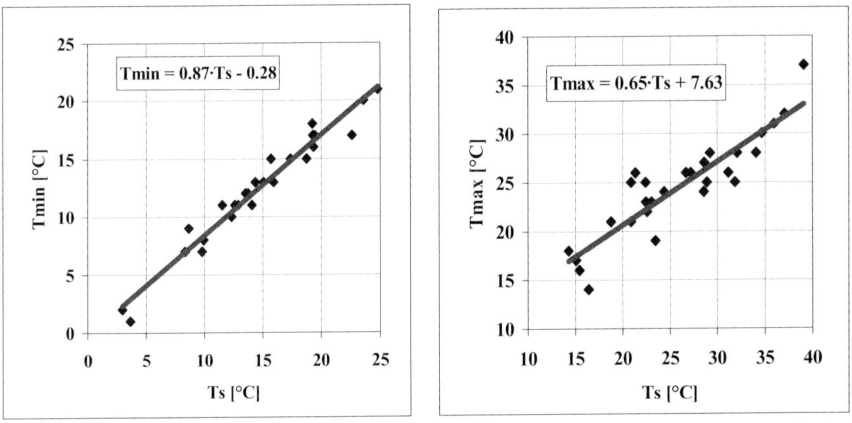


*Figure 17: Linear relationship between the ground based observations of daily minimum/maximum temperature (Tmin/Tmax on the left/right panel) and the surface emperature (Ts) measured by satellite TERRA.*

## 3.3 Conclusion

First, large cities of the Carpathian Basin have been selected and their urban and rural pixel representations have been determined. Daytime and nighttime surface temperature time series measured by the instrument MODIS of satellite TERRA have been analysed. Urban and rural spatial averages have been calculated and compared for each selected city. The analysis suggests that intensity of the urban heat island detected in the selected Hungarian cities perform high variability. Average values of the temperature differences between urban and rural areas range between 1 K and 3 K. Inhabitants number is the main factor of determining the urban heat island intensity that is modified by orography. Furthermore, very strong correlations have been found between ground-based temperature observations (especially in case of daily minimum temperature) and remotely sensed data.

According to the characteristical surface cover (e.g., large concrete buildings, parks, residential areas, airports, etc.) several pixels of the selected cities have been selected and compared to the rural environment. On the entire metropolitan area the night-time heat island effect is stronger than during daytime because of the special radiation properties of densely built-up areas (concrete, asphalt covered surfaces). Furthermore, the river Danube affects significantly the urban heat island of Budapest, especially during daytime.

## 4. FINAL REMARKS

Several parallel on-going migration processes have significant effects on urban heat island effect, environmental change, global warming, and their regional consequences. Although their existence cannot be doubted, it is very difficult to quantify those linkages precisely. The complex phenomena were analysed in this paper using special points of view and different methodological approaches (social science, mathematical statistics, satellite imagery, etc.). It is clear and must be considered that the Carpathian Basin does not belong to the few regions of the world where migration is one of the key environmental factors either due to poverty and epidemic diseases, or due to rapid (and very often not democratic) political changes with radical turns, or due to very extreme meteorological events followed by environmental catastrophes. On the other hand, our small research fragments related to the Carpathian Basin and presented in this paper may provide additional information to understand recent and foreseen environmental changes and processes in the region.

As a final summary of the topics discussed in details in this paper the following conclusions can be drawn.

1. Positive tendency of both monthly and 10-day time series of the Normalized Difference Vegetation Index (NDVI) occurred on the base of the 20-year period analysed – this could be a possible local consequence of the global warming process.
2. Time periods of low NDVI values are accompanied by cold-wet and cold-dry climate conditions in the entire Carpathian Basin, while during the seasons of large NDVI values warm-wet and warm-dry climate conditions are dominant in the region.
3. Analysis of fine resolution satellite imagery (TERRA/MODIS launched in 1999) for the ten largest cities in Hungary suggests that inhabitants number is the key factor determining the urban heat island intensity, which is also modified by orographical conditions.

4. After separating the urban and rural regions of the largest Hungarian agglomeration areas the average heat island intensity is estimated around 1-3 K.
5. Surface temperatures of the urban areas estimated by satellite information are highly correlated to ground-based temperature observations. The strongest correlation has been found in case of daily minimum temperature.

## ACKNOWLEDGEMENTS

The authors wish to thank the Distributed Active Archive Center (Code 902.2) at the Goddard Space Flight Center, Greenbelt, MD, 20771, for producing the NDVI data in their present form and distributing them. The original data products were produced under the NOAA/NASA Pathfinder program, by a processing team headed by Ms. Mary James of the Goddard Global Change Data Center; and the science algorithms were established by the AVHRR Land Science Working Group, chaired by Dr. John Townshend of the University of Maryland. Goddard's contributions to these activities were sponsored by NASA's Mission to Planet Earth program. Also, the authors wish to thank the American NASA for producing the satellite surface temperature data in their present form and the Earth Observing System Data Gateway for distributing the data. Research leading to this paper has been supported by the *Hungarian National Science Research Foundation* under grants T-026629, T-025803, and T-034867, also by the AEROCARB project of *European Union* Nr. 5 program under grant EVK2-CT-1999/0013. Furthermore, supports of the Bolyai Janos Research Fellowship of the Hungarian Academy of Sciences and scholarship of the Hungarian Ministry of the Environment are appreciated.

## REFERENCES

Barnes, W.L., T.S. Pagano and V.V. Salomonson. 1998. "Prelaunch characteristics of the Moderate Resolution Imaging Spectroradiometer (MODIS) on EOS-AM1." *IEEE Transactions on Geoscience and Remote Sensing* 36(4):1088-1100.

Bartholy, J. and R. Pongracz. 1998. "The differing trends of the Hungarian precipitation time series, spatial and decadal changes of extreme precipitation." (In Hungarian). Pp. 62-66 in *Proceedings of the 2nd Conference on Forest and Climate*, editors K. Tar and K. Szilagyi. Debrecen: Kossuth University Press.

Bartholy, J, R. Pongracz and Z. Dezso. 2001. "Evaluation of urban heat island effect for large Hungarian cities using high resolution satellite imagery." In *Proceedings of the Fifth European Conference on Applications of Meteorology ECAM 2001*, editor M. Hunkar. Budapest. CD-ROM.

Daroczi, E. 1998. "Increasing residential mobility in county Pest." (In Hungarian). Pp. 245-256 in*Migration I*, editors S. Illes and P.P. Toth. Budapest: Hungarian Central Statistical Office.

Ehrlich, D., J.E. Estes and A. Singh. 1994. "Application of NOAA-AVHRR 1 km data for environment monitoring." *International Journal of Remote Sensing* 15:145-161.

Gallo, K.P., J.D. Tarpley. 1996. "The comparison of vegetation index and surface temperature composites for urban heat island analysis." *Int. J. Remote Sensing* 17(15):3071-3076.

Hablicsek, L. 1999. "The impact of international migration to the future of population." (In Hungarian.) Pp. 125-143 in *Migration II*, editors P.P. Toth and S. Illes. Budapest: Hungarian Central Statistical Office.

Hamar, D., C. Ferencz, J. Lichtenberger, G. Tarcsai and I. Ferencz-Arkos. 1996. "Yield estimation for corn and wheat in the Hungarian Great Plain using Landsat MSS data." *International Journal of Remote Sensing* 17:1689-1699.

Hungarian Central Statistical Office. 2001. *Hungarian Statistical Data, 2000*. Budapest.

Hungarian Ministry of Interior. 2000. *Local Authorities Database*. http://web.b-m.hu/proba/cimtar.nsf/lakossag.

Index. 2000. *Poison in the Hungarian rivers*. http://index.hu/cikkek/0002/cianx/.

International Organization for Migration. 2000. *World Migration Report 2000*. IOM.

Justice, C.O., J.R.G. Townshend, B.N. Holben and C.J. Tucker. 1985. "Analysis of the phenology of global vegetation using meteorological satellite data." *International Journal of Remote Sensing* 8:1271-1318.

Lillesand, T.M. and R.W. Kiefer. 1987. *Remote Sensing and Image Interpretation*. New York: John Wiley and Sons.

Matson, M., E.P. McClain, D.F. McGinnis and J.A. Pritchard. 1978. "Satellite detection of urban heat islands." *Mon. Wea. Rev.* 106:1725-1734.

National Aeronautics and Space Administration (NASA). 1999. *Science writers' guide to TERRA*. NASA Earth Observing System Project Science Office, Greenbelt, MD.

Nichol, J.E. 1996. "High-resoltuion surface temperature patterns related to urban morphology in a tropical city: a satellite based study." *J. Appl. Met.* 35(1):135-146.

Price, J.C. 1979. "Assessment of the heat island effect through the use of satellite data." *Mon. Wea. Rev.* 107:1554-1557.

Putsay, M., I.Szenyan and J. Kerenyi. 1998. "Atmospheric correction of NOAA/AVHRR visible and near-IR channels in Hungary." Pp. 503-508 in *Proceedings of the 1997 Meteorological Satellite Users' Conference*. Brussels, Belgium.

Putsay, M., A. Merza and J. Kerenyi. 2000. "Atmospherically corrected NDVI maps retrieved from NOAA/AVHRR data and their annual profiles." Pp. 526-533 in *Proceedings of the EUMETSAT Meteorological Satellite Users' Conference*. Bologna, Italy.

Rao, C.R.N. 1993. "Degradation of the visible and near-infrared channels of the Advanced Very High Resolution Radiometer on the NOAAP9 spacecraft: assessment and recommendations for corrections." *NOAA Technical Report NESDIS- 70*. NOAA/NESDIS. Washington DC.

Rasmussen, M.S. 1997. "Operational yield forecast using AVHRR NDVI data: reduction of environmental and inter-annual variability." *International Journal of Remote Sensing* 18:1059-1077.

Roth, M., T.R. Oke and W.J. Emery. 1989. "Satellite-derived urban heat island from three coastal cities and the utilisation of such data in urban climatology." *Int. J. Remote Sensing* 10(11):1699-1720.

Shimabukuro, Y.E., V.C. Carvalho and B.F.T. Rudorff. 1997. "NOAA-AVHRR data processing for the mapping of vegetation cover." *International Journal of Remote Sensing*, 18:671-677.

Szalai S. (ed.) 2000. *Meteorological aspects of floods occurred in Hungary in 1999*. (In Hungarian.) HMS Climatological and Agrometeorological Report III. Hungarian Meteorological Service, Budapest.

Townsend, J., C. Justice, W. Li, C. Gurney and J. McManus. 1991. "Global land cover classification by remote sensing: present capabilities and future possibilities." *Remote Sensing of Environment* 35:243-255.

Tucker, C.J. 1979. "Red and photographic infrared linear combinations for monitoring vegetation." *Remote Sensing of Environment* 8:127-150.

Wan, Z. and W. Snyder. 1999. *MODIS land-surface temperature algorithm theoretical basis document*. Institute for Computational Earth Systems Science, Univ. of California, Santa Barbara.

## Advances in Global Change Research

1. P. Martens and J. Rotmans (eds.): *Climate Change: An Integrated Perspective.* 1999 ISBN 0-7923-5996-8
2. A. Gillespie and W.C.G. Burns (eds.): *Climate Change in the South Pacific: Impacts and Responses in Australia, New Zealand, and Small Island States.* 2000 ISBN 0-7923-6077-X
3. J.L. Innes, M. Beniston and M.M. Verstraete (eds.): *Biomass Burning and Its Inter-Relationships with the Climate Systems.* 2000 ISBN 0-7923-6107-5
4. M.M. Verstraete, M. Menenti and J. Peltoniemi (eds.): *Observing Land from Space: Science, Customers and Technology.* 2000 ISBN 0-7923-6503-8
5. T. Skodvin: *Structure and Agent in the Scientific Diplomacy of Climate Change.* An Empirical Case Study of Science-Policy Interaction in the Intergovernmental Panel on Climate Change. 2000 ISBN 0-7923-6637-9
6. S. McLaren and D. Kniveton: *Linking Climate Change to Land Surface Change.* 2000 ISBN 0-7923-6638-7
7. M. Beniston and M.M. Verstraete (eds.): *Remote Sensing and Climate Modeling: Synergies and Limitations.* 2001 ISBN 0-7923-6801-0
8. E. Jochem, J. Sathaye and D. Bouille (eds.): *Society, Behaviour, and Climate Change Mitigation.* 2000 ISBN 0-7923-6802-9
9. G. Visconti, M. Beniston, E.D. Iannorelli and D. Barba (eds.): *Global Change and Protected Areas.* 2001 ISBN 0-7923-6818-1
10. M. Beniston (ed.): *Climatic Change: Implications for the Hydrological Cycle and for Water Management.* 2002 ISBN 1-4020-0444-3
11. N.H. Ravindranath and J.A. Sathaye: *Climatic Change and Developing Countries.* 2002 ISBN 1-4020-0104-5; Pb 1-4020-0771-X
12. E.O. Odada and D.O. Olaga: *The East African Great Lakes: Limnology, Palaeolimnology and Biodiversity.* 2002 ISBN 1-4020-0772-8
13. F.S. Marzano and G. Visconti: *Remote Sensing of Atmosphere and Ocean from Space: Models, Instruments and Techniques.* 2002 ISBN 1-4020-0943-7
14. F.-K. Holtmeier: *Mountain Timberlines.* Ecology, Patchiness, and Dynamics. 2003 ISBN 1-4020-1356-6
15. H.F. Diaz (ed.): *Climate Variability and Change in High Elevation Regions: Past, Present & Future.* 2003 ISBN 1-4020-1386-8
16. H.F. Diaz and B.J. Morehouse (eds.): *Climate and Water: Transboundary Challenges in the Americas.* 2003 ISBN 1-4020-1529-1
17. A.V. Parisi, J. Sabburg and M.G. Kimlin: *Scattered and Filtered Solar UV Measurements.* 2004 ISBN 1-4020-1819-3
18. C. Granier, P. Artaxo and C.E. Reeves (eds.): *Emissions of Atmospheric Trace Compounds.* 2004 ISBN 1-4020-2166-6
19. M. Beniston: *Climatic Change and its Impacts.* An Overview Focusing on Switzerland. 2004 ISBN 1-4020-2345-6
20. J.D. Unruh, M.S. Krol and N. Kliot (eds.): *Environmental Change and its Implications for Population Migration.* 2004 ISBN 1-4020-2868-7

KLUWER ACADEMIC PUBLISHERS – DORDRECHT / BOSTON / LONDON